# STEM in the Technopolis: The Power of STEM Education in Regional Technology Policy

Cliff Zintgraff • Sang C. Suh • Bruce Kellison
Paul E. Resta
Editors

# STEM in the Technopolis: The Power of STEM Education in Regional Technology Policy

Springer

*Editors*
Cliff Zintgraff
The University of Texas at Austin
IC2 Institute
Austin, TX, USA

Bruce Kellison
The University of Texas at Austin
IC2 Institute
Austin, TX, USA

Sang C. Suh
Texas A&M University-Commerce
Commerce, TX, USA

Paul E. Resta
The University of Texas at Austin
College of Education
Austin, TX, USA

ISBN 978-3-030-39850-7 ISBN 978-3-030-39851-4 (eBook)
https://doi.org/10.1007/978-3-030-39851-4



This Springer imprint is published by the registered company Springer Nature Switzerland AG
The registered company address is: Gewerbestrasse 11, 6330 Cham, Switzerland

*In memory of George Kozmetsky and in honor of David Gibson, for their lifetimes of research into the technopolis and the potential for technology-based economic development and for educating the next generations.*

# Foreword

Education is key to the economic vitality of cities! And it is important to every stakeholder in cities around the world. Education is important to parents and family members of school-age children and to those who teach them. Education is important to college students and their professors and to the leaders of those institutions. Education is important to local employers, to staff of nonprofits who improve quality of life, and to policymakers who make cities worthy of their citizens. Through education, students prepare for the jobs of tomorrow. As educated students become young citizens, they will better understand the challenges faced by their communities, make better decisions about their leaders, and be ready to contribute to the cities they call home.

I've had the privilege to see the importance of education up close in my work at the Annenberg Public Policy Center, as Chairman of Sister Cities International, and as Mayor of San Antonio. San Antonio provides a great example. With a population of 1.5 million people, San Antonio is the second largest city in Texas and the seventh largest in the United States. Between now and 2040, we will add another 1.1 million people to the 1.5 million already living inside our city limits. Another million or more will live in our metropolitan area. This growth is wonderful for our economy. It is less wonderful for the traffic on our roads. It is a challenge to the sustainability and resiliency of our water, land, and other natural resources and to our ability to provide government services to citizens.

San Antonio also exemplifies the importance of education because of culture and demographics. It may be 2019, but San Antonio is already the America of 2050. The city was founded by Spanish missionaries, first populated by Canary Islanders, grown through European immigration and through immigration from Mexico and Latin America. Today, our city's population is over 60% Hispanic. We have a rich Tejano culture, and we are home to the Alamo and all San Antonio Missions, collectively a UNESCO World Heritage site. We are a welcoming city that represents the best of the melting pot that is America.

In the midst of this growth and culture is the industry of San Antonio. We are known for tourism and as Military City USA. We are also known for our industry clusters in health, biomedical research, manufacturing, aerospace, and financial services. We are home to the second highest number of certified information security professionals in the United States, improving the safety of our military, government, and the private sector.

In light of this background, how is our city preparing today's students—*all* of our students—for tomorrow's careers? Of particular importance is STEM education, the integration of science, technology, engineering, and mathematics to prepare citizens for the challenges of the twenty-first century. Also important is STEAM education. STEAM integrates the arts and injects a special degree of creativity. STEM and STEAM create a nexus between scientists, engineers, and the creative people of our communities. STEM and STEAM experiences help students connect their studies to the real world.

In San Antonio, we are aggressively moving on the STEM and STEAM fronts, at all education levels and very much in K–12 (grades kindergarten through 12). In 2010, the city adopted *SA2020*, a comprehensive framework for development of the city. In 2014, leaders of the SA2020 effort founded the San Antonio STEM Council as a gathering place for STEM educators. The council is now the Alamo STEM Ecosystem and a member of STEM Learning Ecosystems, one of over 80 communities in the United States organizing to drive STEM education and prepare citizens for careers.

Before that, in 2008, our robust cybersecurity industry cluster founded San Antonio's version of CyberPatriot, a national cyber defense competition. The strength of the local cluster helps area schools field 300+ teams a year, the highest per capita participation in the nation. San Antonio's flagship university, The University of Texas at San Antonio, leads technical development for the competition. I have the honor of awarding the yearly CyberPatriot Mayor's Cup. CyberPatriot is helping drive a virtuous development cycle, preparing students for cybersecurity education in college. Those students enter local jobs with employers who in turn support K–12 through CyberPatriot. The CyberPatriot program is the subject of this book's San Antonio chapter.

Meanwhile, from 2015 to 2016, I served as Tri-chair of SA Tomorrow, an initiative building San Antonio's long-term development plan. In 2018, with area nonprofit partners, my office started *SA Smart*, a middle and high school research competition focused on the challenges identified in SA Tomorrow. The 2018 topic was transportation, and the 2019 topic was sustainability. It was my honor to award the 2018 and 2019 SA Smart Mayor's Cup to K–12 teams who developed and researched ideas for improving quality of life in San Antonio. It's important to note that these students did more than develop great ideas. They spent months talking about their ideas with their teachers, friends, and families and became ambassadors for important conversations we need to have in our community.

San Antonio provides just one example of how unique circumstances in cities and city-sized regions are great platforms for STEM and STEAM education. On such platforms, partnerships can be built between K–12 schools, local industries, local college programs, city government leaders, and city's nonprofit organizations. These partnerships are key to unlocking the virtuous development cycle illustrated in the volume's first chapter.

In the current volume, *STEM in the Technopolis*, many other examples are shared from cities and regions around the world. The volume also highlights education philosophies and methods, economic development principles, national policies, and equity principles that connect the challenges of cities to great STEM and STEAM experiences for students. In San Antonio, we will be taking these ideas to the next level! We challenge you to do the same.

With best regards,

Ron Nirenberg
Mayor, City of San Antonio, TX, USA

# Preface

If you are looking for a book that crosses disciplines, you have come to the right place. *STEM in the Technopolis* grew from ideas that view wholes as greater than the sum of parts. This is how a global innovation program developer, a computer scientist, an economic development researcher, and a professor of education became coeditors of one volume. We are privileged to share with you the individual stories of this volume, and the story as a whole. Our story is about great STEM education experiences for students that spring from the challenges of cities and regions. We share examples from around the world, common themes, and ways to elicit the same kinds of results we see in the cases from our volume.

Two large themes first drew us together. The first is STEM education. Whether we are building global programs and software, driving economic development, or training the next generation of teachers, the topics of science, technology, engineering, and mathematics are prominent. It is especially important to understand how the topics are co-taught—or, to use the fancy words, how they are taught in an integrated and transdisciplinary manner.

To teach them together, we need context, and few people were better at context than Dr. George Kozmetsky. It is through his influence on Austin, Texas, USA, at The University of Texas at Austin, and through his global network that we, editors, found common ground. His receipt of the US Presidential Medal of Freedom for Technology is one validation of his ideas. He believed strongly in the power of collaboration. He founded the Society for Design and Process Science (SDPS) and the $IC^2$ Institute at UT Austin. These two organizations have led development of the volume, and his DNA is seen throughout our work.

Dr. Kozmetsky would honor our desire to critique that work, build on it, and think of ways to take it to new places and make it bigger and better. In that spirit, we think the *technopolis model* first defined by Kozmetsky and his colleagues needs to grow. For example, imagine that 15 years ago an education advocate (like one of us) walked into a large company and said "You need to educate middle school students so they will come to work for you!" That really happened, and the large company employees considered the idea to be crazy. The exact response referred to the long wait for a return on investment. That company was focused solely on students in college.

The emergence of STEM education has radically changed that perception. STEM is a platform everyone can see themselves in. STEM is a powerful way for diverse stakeholders to think about the root causes for why many students come to the workplace unprepared. There is a big difference between technical knowledge and project know-how. The context of real-world problems motivates teachers and students, provides content, and makes the effort to learn more worthwhile and relevant.

And what better problems are there than the problems of cities and regions? We all know what they are: transportation, water, energy, pollution, health, economic security, physical security, cybersecurity, equity, quality of life for all people on the planet, etc. Every stakeholder group in a technopolis cares about the solutions to these problems.

Our volume illustrates how the challenges of cities and regions form rich context for STEM education, cutting across many disciplines. In this one volume, you will read about STEM, STEAM (STEM+Arts), regional economic development, industry clusters, network building, education philosophy, education theory, educational methods and practice, the structure of schools, the nature of twenty-first-century learning, digital equity, rural development, health, ICT, energy, cybersecurity, the Fourth Industrial Revolution, manufacturing, agriculture, cloud software development, technology commercialization, and innovation—and, in one case, how a city's world-class crisis became their inspiration for great STEM education.

Building effective collaborations among different kinds of people is hard work. It is also one of the most rewarding tasks in life, both professionally and personally. We hope you value the ideas of this volume, but more than anything, we hope the volume inspires you to connect with others as we have during its development. We hope you connect in your communities and that, along the way, you create great educational experiences for students.

Austin, TX, USA Cliff Zintgraff
Commerce, TX, USA Sang C. Suh
Austin, TX, USA Bruce Kellison
Austin, TX, USA Paul E. Resta

# Acknowledgments

We gratefully acknowledge the invitation of Dr. Murat Tanik of the Society for Design and Process Science (SDPS) to contribute to the scholarship on STEM education and economic development in cities and regions and to SDPS for advancing development of the book at their 2017 Annual Conference.

We also thank the IC$^2$ Institute at The University of Texas at Austin (UT Austin) for generous funding to bring together the contributors to this volume for a public conference and discussion on STEM education and economic development policy. In particular, Institute Director Dr. Art Markman and Deputy Executive Director Dr. Gregory Pogue supported the project from its inception and encouraged us to keep the initiative's focus squarely on the characteristics of successful technopolis policy. We thank UT Austin PhD student Songhee Han for her dedication and skillful organization of the conference and for her intellectual contributions based on her experience as an educator. We regularly counted on the superpower administrative skills of Diane Skubal and Inez Traylor to track and complete the myriad tasks that made the conference and this volume possible.

Finally, we recognize the tens of millions of committed STEM educators, leaders, mentors, and volunteers around the world who work every day to transform lives and regions. Some of their stories illuminate this volume but most do their work without the credit they deserve. We honor their commitment to educate students and improve quality of life for their fellow citizens.

# Contents

# List of Contributors

**Ademar Aguiar** University of Porto, Porto, Portugal

**Hemanth Bandi** McKesson Corporation, Richmond, Virginia, USA

**Alejandro Roldán Bernal** Ruta N Innovation Agency, Medellín, Antioquia, Colombia

**Tricia Berry** The University of Texas at Austin, Austin, TX, USA

**Richard V. Butler** Trinity University, San Antonio, TX, USA

**Cheng-Yuan Chen** Taiwan First Girls High School, Taipei, Taiwan

**Ting-An Chou** Taiwan First Girls High School, Taipei, Taiwan

**Paulo Estevão Cruvinel** Embrapa Instrumentation – Laboratory for Precision Agricultural Inputs Application, São Carlos, SP, Brazil

**Silvia Rocha Falvo** São Carlos School, São Carlos, SP, Brazil

**Cristiane Chaves Gattaz** Embrapa Instrumentation – Laboratory for Precision Agricultural Inputs Application, São Carlos, SP, Brazil

**Songhee Han** The University of Texas at Austin, Austin, TX, USA

**Bruce Kellison** The University of Texas at Austin, Austin, TX, USA

**Donna K. Kidwell** Arizona State University, Phoenix, AZ, USA

**Jinoh Kim** Texas A&M University-Commerce, Commerce, TX, USA

**Chung-Hsien Kuo** National Taiwan University of Science and Technology, Taipei, Taiwan

**Sara Pereira** University of Minho, Braga, Portugal

**Anthony J. Petrosino** Southern Methodist University, Dallas, TX, USA

**Gregory P. Pogue** The University of Texas at Austin, Austin, TX, USA

**Guolong Quan** Jiangnan University, Wuxi, China

**Paul E. Resta** College of Education, The University of Texas at Austin, Austin, TX, USA

**Graciela Rojas** Movimiento STEM, Mexico City, Mexico

**Joe Sánchez** CyberTexas Foundation, San Antonio, TX, USA

**Laura Segura** Movimiento STEM, Mexico City, Mexico

**Ravae Villafranca Shaeffer** Education Service Center, San Antonio, TX, USA

**Maximilian K. Sherard** The University of Texas at Austin, Austin, TX, USA

**Sang C. Suh** Texas A&M University-Commerce, Commerce, TX, USA

**U. John Tanik** Texas A&M University-Commerce, Commerce, TX, USA

**Sneha A. Tharayil** The University of Texas at Austin, Austin, TX, USA

**Hsiao-Yen Wei** Taiwan First Girls High School, Taipei, Taiwan

**Chao-Lung Yang** National Taiwan University of Science and Technology, Taipei, Taiwan

**Yun-Chi Yang** Taiwan First Girls High School, Taipei, Taiwan

**Cliff Zintgraff** The University of Texas at Austin, Austin, TX, USA

# Part I
# Foundations

# Chapter 1
# The Virtuous Cycle: Global Cases of K-12 STEM Education in the Technology Policy of Cities

**Cliff Zintgraff**

**Abstract** When leaders in cities and local regions talk about technology policy, they usually talk about government, universities and industry and how the three work together. This relationship is important. But what about the hundreds of millions of young students, taught by tens of millions of teachers, supported by hundreds of thousands of volunteers, who deliver STEM education to primary and secondary students around the world? K-12 STEM education should be a full participant in the building of modern knowledge economies. In this chapter, the opening chapter of *STEM in the Technopolis*, readers are encouraged to think hard about how to make K-12 STEM education a full participant in city/regional technology and development policy. Three global cases exhibiting this integration are introduced. Topics covered include the technopolis model, industry clusters, K-12 STEM pedagogy, social and digital equity, building the brand of cities and regions, the eight indicators of a virtuous cycle, and a model for the virtuous cycle that happens when K-12 STEM education is fully integrated into local policy. A brief introduction is provided to the rest of the volume. This chapter and volume are offered as a platform and encouragement to educators, academics, STEM professionals, policymakers, and all stakeholders. The author encourages the building of effective and inspirational STEM education programs, rooted in local priorities, benefiting students, families, and quality-of-life in cities and regions around the globe.

## 1.1 Introduction

### *1.1.1 The Argument for K-12 Education in the Technopolis*

During the 1980s, Dr. George Kozmetsky and colleagues were completing seminal work on a model for technology-based economic development in regions (Jones 2018). Based on their particular experiences in Austin, Texas, U.S., they would

C. Zintgraff (✉)
The University of Texas at Austin, Austin, TX, USA
e-mail: cliff.zintgraff@utexas.edu



C. Zintgraff et al. (eds.), *STEM in the Technopolis: The Power of STEM Education in Regional Technology Policy*,
https://doi.org/10.1007/978-3-030-39851-4_1

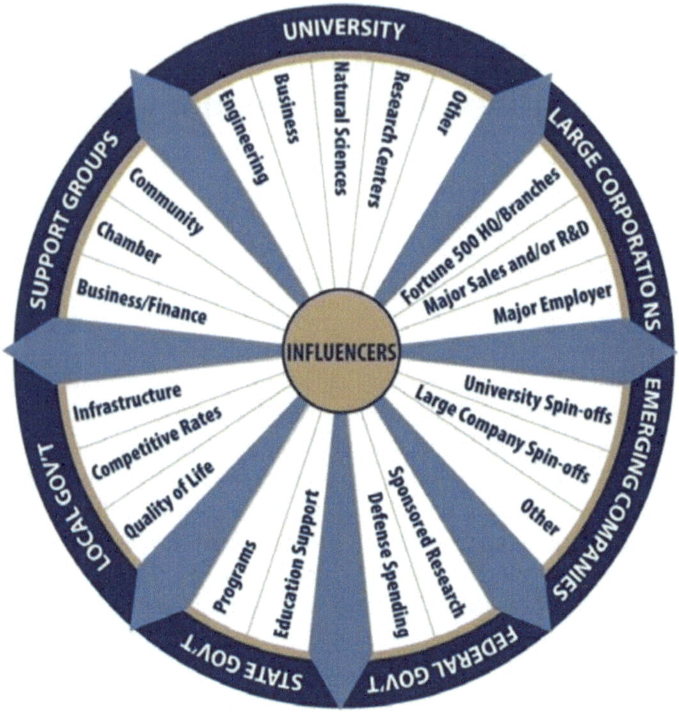


**Fig. 1.1** The technopolis wheel (Smilor et al. 1989)

ultimately publish the *technopolis model*, also known as the *technopolis wheel* (Smilor et al. 1989). The model described a highly collaborative framework for technology-based development that involved all sectors of a city or region. Ultimately, Dr. Kozmetsky would be recognized for his contribution with a Medal of Freedom for Technology from U.S. President Bill Clinton. The transformation of the Austin economy stands as proof of the model's efficacy.

Today, one hears more about the Triple Helix (Etzkowitz and Leydesdorff 2000), but the technopolis model remains broader in its approach. The model incorporates federal, state, and local government; large versus emerging companies; and support groups, also known as NGOs or non-profits; and universities. Figure 1.1 is the original illustration of the model from Smilor et al. (1989).

Smilor et al. (1989) planted a seed that helps start discussion of K-12 in the technopolis. A story was relayed of an early Austin success recruiting the Microelectronics and Computer Technology Corporation (MCC) to Austin. The quality of schools contributed to Austin's quality of life, which contributed to Austin's selection as the site for MCC. The link to technopolis development is indirect, but telling for the future. The children of the people moving to Austin would soon be among the schools' students. The parents would soon be influencers of how their children receive an education.

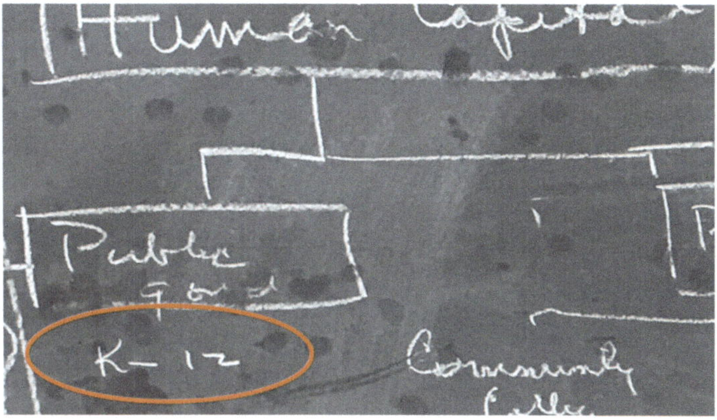


**Fig. 1.2** Dr. George Kozmetsky's diagram and K-12

One finds more than a seed in Jones (2018) biography of Dr. Kozmetsky, which includes early stories of Kozmetsky's economic development laboratory, the $IC^2$ Institute, supporting secondary education until others came to fill the role. $IC^2$ was involved in high school entrepreneurship training programs, and $IC^2$ helped create EnterTech, a technology job training program whose target included high schools. Late in his life, Dr. Kozmetsky helped design CBIRD, the Cross Border Institute for Regional Development, built for Texas' Rio Grande Valley and for northern Mexico. The CBIRD program contained extensive designs for high school activities that would support technopolis development goals (Gibson et al. 2003).

At the $IC^2$ Institute, there is an undeniable indicator of K-12's role. In Dr. David Gibson's former office (of Smilor, Gibson, & Kozmetsky), and Dr. Greg Pogue's current office (co-author of this volume's technopolis chapter), there is a chalkboard with the preserved writing of a Kozmetsky-drawn diagram, an activity he was famous (and sometimes "nicely infamous") for. The legitimate place of K-12 in the technopolis is, in this author's opinion, immortalized in the diagram, as seen in Fig. 1.2.

Is the role of K-12 in the technopolis more widely recognized today? An anecdote from the author's experience can illustrate. In the mid 2000s, there was a meeting at an industry site for discussion of university STEM education, with talent pipelines feeding local industry. The author raised the idea of expanding the program in question into high schools, starting the pipeline earlier, perhaps as early as ninth grade. The idea was not well received. Waiting six years or more for a return on investment was not something industry players in the room could imagine.

This perspective has changed dramatically. With the rise of STEM schools (Scott 2012; Texas Education Agency 2014), there is clear understanding in industry that broad policy intervention needs to happen in middle school (grades 6 through 8) at the latest, before students' perceptions are set, and while a sense of self-efficacy around STEM subjects is easiest to instill. Programs like FIRST Robotics (Melchior et al. 2016) and CyberPatriot (Zintgraff 2016; Sanchez & Zintgraff, this volume) make heavy use of industry support and mentors with programs starting in elementary school.

So when governments think about driving technology-based development, they should think about more than the well-understood and important partnerships between governments, academia and industry that address issues like intellectual property, research funding, and technology transfer (Etzkowitz and Leydesdorff 2000; Smilor et al. 1989). They should also think about the hundreds of millions of students, taught by tens of millions of teachers, supported by hundreds of thousands of volunteers, who deliver STEM education to children around the world.[1]

### 1.1.2 Principles of K-12 in the Technopolis

K-12 in the technopolis is about more than STEM education in schools, and it is about more than development of a talent pipeline for industry, as important as both are. A rational and intentional integration of K-12 impacts the brand of a city or region; creates mutual benefits with priority industry clusters; creates a virtuous cycle of pedagogical reform; and ultimately a virtuous cycle of economic development. Social concerns and equity are addressed whenever education efforts in the technopolis are expanded to include more students from underrepresented populations.

#### 1.1.2.1 The Central Role of Industry Clusters

Michael Porter (1990) elaborated the role of industry clusters in cities and regions. The basic idea is simple, that a cluster of related businesses result in benefits of networking and economy of scale to all cluster participants. As it relates to K-12 in the technopolis, aligning STEM education with industry clusters can be a powerful strategy. For schools, industry clusters can provide content inspiration, vocal support, volunteers, mentors and funding. For industry, programs aligned with their interests provide rich outlets for philanthropic activity, and they feed the talent pipelines that can fuel a virtuous cycle for the cluster.

In fact, using Porter's (1990) ideas, a case can be made that K-12 is just another industry cluster participant, right next to the more traditional actors. Within industry clusters, there is much transfer of knowledge, explicit and tacit. Porter emphasized this cluster attribute, and Gibson and Conceição (2003) also studied knowledge transfer (KT) at length. One only need look at the knowledge that flows from industry to schools, knowledge about real-world applications, about careers, and about the core STEM content itself, to realize that schools act in many respects like just another member of the cluster. Information flows both directions regarding the

[1]The number of students and teachers is extrapolated from the U.S. National Center for Education Statistics (2017). The number of volunteers is almost met by the number of volunteers worldwide in FIRST robotics (FIRST, 2019), with additional support from Zintgraff (2016). The author expects there are actually many more global STEM volunteers.

twenty-first century skills necessary in the workplace and how they can be taught. When programs are intentionally organized around priority industry clusters in a region, that knowledge transfer is enhanced (see Sanchez & Zintgraff; Gattaz, Falvo, & Cruvinel; and Aguiar & Pereira, this volume).

The cyber security cluster in San Antonio, Texas, U.S. presents a strong example. San Antonio has a decades-long history as a community with information security assets. As a result, a high number of professionals in San Antonio are available as mentors; in fact, some retired professionals have second careers as STEM educators in schools (Zintgraff 2016). The cluster's local history, the region's self-image regarding that cluster, and the number of professionals in the region all contribute to the possibility of a virtuous cycle developing between the cluster and STEM programs that serve K-12 students.

#### 1.1.2.2 STEM Pedagogy, a Key Enabler

Compare two scenarios. In one, a teacher follows a traditional lecture-based pedagogy. Students receive content (perhaps very good lecture content), they read, and they pass tests with multiple choice and/or open-ended questions. In the second scenario, a significant portion of the student's time (half or more) is spent on research, inquiries or projects designed to teach content and/or require students to demonstrate understanding, and perhaps do so in collaboration with their peers. Which of these pedagogical strategies is the better platform for collaboration with people from outside the school?

The author argues that inquiry-based or related pedagogy is a richer platform for school-to-technopolis collaborations. While either pedagogy allows for guest speakers and information about careers, deeper understanding and real opportunities for interaction and mentorship come when students deeply engage content hands-on and minds-on. Most of the cases in this volume are good examples of intentional pedagogy enabling interactions between schools and technopolis actors. Petrosino (this volume) provides an in-depth exploration of the educational philosophy, theories and models that enable the type of integration this volume advocates.

Is the definition of STEM education also limited in this manner? The power and beauty of the term *STEM* derives, at least in part, from how diverse stakeholders feel a sense of belonging in the domain—they feel they belong to the STEM club. Does a traditional mathematics class qualify as STEM education? Does a traditional science class? What view is adopted for this chapter and volume? When Judith Ramaley re-coined the term *SMET* to *STEM* while at the U.S. National Science Foundation in the early 2000s, she advanced a narrative of science and mathematics as bookends to engineering and technology (Christenson 2011). Stated differently, STEM education requires integration across the subjects.

In this volume, mathematics education is viewed as important, as is science education. However, when taught with a solo design, such classes are called *STEM-related*. STEM classes happen without qualification when substantial integration (far more than a word problem or five-minute discussion at the end of class) is seen across two or more of science, technology, engineering and mathematics. It is this integration that changes the nature of instruction, and this integration that opens the educational experience up for deep collaborations with partners from across a community.

#### 1.1.2.3 Technology within Pedagogy

STEM fundamentally involves technology, but to consider its role, a more nuanced view is needed. In K-12 STEM education, one sees: (1) how technology is studied as a subject; (2) how technology is used to deliver education; (3) the use of real-world concerns to create context; and (4) how learning theories and methods are co-evolving with technology to make this approach workable at scale.

Take, for example, the Medellín, Colombia STEAM-LABS program. The City of Medellín 2021 Science and Technology Plan identified energy, health, and ICT as the city's high-technology clusters of focus (Echeverri 2014). STEAM-LABS adopted these clusters as a framework for identifying interdisciplinary topics for study.

The Medellín teachers developed curricula around authentic, locally important problems in those clusters. Two examples were: (1) ways to deliver minimum water allocations to poor citizens; and (2) using video games to teach citizens about health (Zintgraff et al. 2015). Students studied how technology can solve these societal problems, which included interacting with the relevant technology.

Technology was used to deliver the Medellín curriculum, from basic tools like cloud storage, presentation software, video delivery, and desktop publishing software, to more advanced educational technology, like learning management systems and simulation environments. The students studied technology of different types (e.g., electromechanical flow systems). They used technology in their projects, including many of those same tools, along with software development platforms.

All instruction was developed consistent with the principles of learner-centered design. In the specific case of STEAM-LABS, instruction was developed based on the Buck Institute of Education's project-based learning model. Instructional design emphasized the Buck Institute's 6 As– authenticity, academic rigor, applied learning, active exploration, adult connections, and assessment practices (Markham et al. 2003).

Likewise, San Antonio's cyber competitions delivered problem-based instruction, with online training content and a live monitoring system on PCs assessing student performance. Technology is both subject and mechanism for delivery. The program addresses a critical societal need, cyber security. Sanchez and Zintgraff (this volume) also describe the pedagogy as mentor-based learning, with mentors serving a central role in the program.

Additional developments in educational technology, instructional theory, and instructional design are moving forward in lockstep. Technology and instructional methods are evolving to support personalized instruction at scale (Powell et al. 2015). Students are being assessed formatively, with assessment emerging from authentic coursework (versus through costly and timely–and often despised–standardized tests). Simulations, virtual worlds, and augmented reality are blurring the gap between artificial classroom exercises and authentic experiences (Wu et al. 2013).

The recent two volumes of Reigeluth's *Green Books* on instructional design (Reigeluth and Carr-Chellman 2009; Reigeluth et al. 2016) focused almost exclusively on learner-centered design, from philosophy to practice.

#### 1.1.2.4 The Technopolis, Social Concerns and Digital Equity

Equal availability of physical resources for education and instructional designs/content is a worldwide challenge (Resta and Laferrière 2015). A functioning technopolis presents possible solutions to the local instances of equity challenges. Successful industry clusters are motivated to build their brand in regions. Their members have the capacity to deliver physical resources and human capital support while meeting philanthropic goals. Resta (this volume) notes the need for culturally responsive content. Content built around the local challenges of cities and regions is fundamentally sensitive to local needs.

A special example of equity concerns in education is seen in Roldán's description (this volume) of STEAM education in Medellín, Colombia. Once the most dangerous city in the world, Medellín has become a worldwide model for innovation, and one distinctive of Medellín is the strategy and organization they use to drive STEAM (STEM+Arts) education reform. They drive STEAM education reform, and run STEAM education projects, from the same agency that drives city space planning, transportation equity, large company expansion, university technology transfer, and entrepreneurship. K-12 education is fundamentally integrated in their innovation agency. Medellín's choice to embrace STEAM education, as opposed to simply STEM, is a direct attempt to deliver education in a manner responsive to culture, economic disparity, and specific local challenges.

#### 1.1.2.5 Building the Local Brand

Throughout this volume's case studies, one sees policymakers working to transform their communities through technology-based development. Wherever the tool of government policymaking is used, policymakers need the political support of citizens to fund initiatives, vocally support initiatives, and see policies implemented to completion. Which approach to building support is most effective: a government official's speech, or the advocacy that flows from schoolchildren learning about the challenges of their community? STEM education's first purpose must always be learning. At the same time, students are powerful communicators to their parents, siblings, extended family and friends.

#### 1.1.2.6 In This Chapter

The remainder of this chapter examines the relationship of K-12 STEM education to the long-term development of industry clusters—the same clusters that have led globally to wealth and improved quality of life. The author argues that a city's or region's programs and policies for technology-based development should incorporate support for robust STEM education experiences for students. The policies should address the knowledge and skills needed to succeed in relevant careers, and they should encourage the framing of STEM education experiences within the context of priority local industry clusters and societal challenges. In doing so, regions can drive a virtuous cycle of education, economic development, and quality of life for citizens.

The author argues for the existence of this virtuous cycle by examining three cases from around the globe. The cases are from Colombia, the U.S., and Taiwan. In each case, the author identifies common indicators seen when a virtuous cycle is in operation. The indicators are: industry cluster inspiration, industry cluster support, individual support from professionals, university support, government support, nonprofit support, societal concerns driving inspiration, and K-12 education feeding the talent pipeline. Based on the analysis, a virtuous cycle model is presented.

Additional discussion covers what is meant by the term *policy* in the current volume. Some examples of policies are shared that reflect the volume's themes. A closing discussion outlines the content of the remainder of the volume.

There is power in the compatibility between the economic, social and political context provided by regions, on the one hand, and increasingly common methods for teaching and learning that are driving educational reform. Their strategic alignment can benefit all stakeholders in a technopolis.

## 1.2 The Eight Common Indicators

When virtuous cycles are nascent or in operation, eight indicators are often found. Recognizing these indicators gives all stakeholders the chance to support them, or at the least, to do no harm to them. Readers will note the similarity of the stakeholders described to those found in the technopolis model/wheel. Table 1.1 lists a summary of the indicators, following the narrative description below.

1. Industry Clusters Inspire K-12 Education

K-12 education is not socially isolated from the communities in which it is delivered. Teachers, administrators, parents, siblings, and the students themselves see the activity that happens where they live. Educators who are inclined toward delivering education in context see these examples and find ways to bring them into classrooms. This is especially true if a cluster aligns well with emerging education trends. It may even be the case that in most communities where an industry cluster is strongly present, there is already a related K-12 education program in operation.

2. Industry Clusters Support K-12 STEM Education

Industry players support K-12 STEM educators in many ways. Support is provided through career speaker programs (Young 2007; Laursen et al. 2007), demonstrations of content (Young 2007; Forssen et al. 2011; Bardeen and Cooke 2011), field trips for students and teachers (Goonatilake and Bachnak 2012; Gamse et al. 2014; Bardeen and Cooke 2011), teacher externships (National Science Foundation 2015), and direct funding (e.g., FIRST in Texas 2016b; CyberPatriot 2013a). Seeing students learn through the mechanisms of one's field is inspiring. Industry support

**Table 1.1** Indicators of a Virtuous Cycle

| Number | Description |
|---|---|
| 1 | Industry clusters inspire K-12 STEM education |
| 2 | Industry clusters support K-12 STEM education |
| 3 | Individual professionals support K-12 STEM education |
| 4 | Universities support K-12 STEM education |
| 5 | Government entities support K-12 STEM education |
| 6 | Non-profits support K-12 STEM education |
| 7 | Societal concerns inspire K-12 STEM education |
| 8 | STEM education feeds universities and industry |

for education also demonstrates corporate social responsibility, and can be an effective marketing and branding strategy for a corporation.

3. Individual Professionals Support K-12 STEM Education

Those industry cluster contributions require the active participation of industry professionals. Through intrinsic motivations, or external ones like self-development or social interaction (Clary and Snyder 1999), professionals get involved as career speakers, field trip hosts, content experts, technical and equipment experts, professional development providers, or as coaches/mentors in project- and competition-based programs (Zintgraff 2016).

4. Universities Support K-12 STEM Education

Not all professionals come from for-profit industry. Volunteers can arrive from universities (Zintgraff 2016)—from university faculty, staff or the student population. Universities are often conducting research as they run programs. University support for STEM programs can help develop the university's reputation locally, providing recruiting benefits.

5. Government Entities Support K-12 Education

Governments are, in the end, responsible for the vast majority of primary and secondary education. Beyond that commitment, government entities with non-education missions often make contributions to education. For example, in the U.S., city governments sometimes provide funding, program and advocacy support to programs, even though separate government entities ("school districts") that are governed by independent boards are responsible for K-12 schools (e.g., CyberPatriot 2013a). Such voluntary behavior suggests the strategic importance governments place on the activities of local schools and their alignment with strategic development priorities.

6. Non-Profits Support K-12 STEM Education

Non-profits organize programs and volunteers that connect to regional STEM education programs. Three examples that operate at large scale are FIRST robotics, VEX Robotics, and CyberPatriot (FIRST in Texas 2016a; VEX Robotics 2013; CyberPatriot 2013a). The society-driven missions of non-profits often overlap with the topic areas that fuel K-12 STEM education programs; for some non-profits, education in these areas is their primary mission.

7. Societal Concerns Inspire K-12 STEM Education

Simultaneously at work is the influence of societal concerns on actors in the system. Cities and regions face numerous challenges: economic development, equity, transportation, public safety, climate, and many others (NIST 2017; IEEE 2017). K-12 STEM education programs use these challenges to inspire student learning (e.g., Markham et al. 2003). The actors involved in those programs frequently overlap those mentioned above.

8. K-12 STEM Education Feeds Universities and Industry

College students and professionals were once K-12 students. Some followed traditional paths to the workplace. Others benefited from programs tightly tied to local workforce needs. The professional economic development community understands well that the presence of a talent pipeline is key to attracting new businesses to a city or region (Bhatnagar 2008). One must consider the talent pipeline from both individual and group perspectives. These pipelines prepare individual people for work in the field, and they create a group perception in the community that drives interest in programs and careers. Stated another way, the virtuous cycle is practical, but it also psychological. Industry clusters create shared social meaning in the minds of citizens. This socially constructed knowledge (Vygotsky 1978) helps drive the virtuous cycle. Table 1.1 lists the eight indicators of the cycle.

## 1.3 Cases

Three geographically diverse cases are presented to support the argument made above, and to illustrate the virtuous cycle in action. The first case comes from Medellín, Colombia. Their STEAM-LABS/STEAMaker and Horizon programs are framed by the city's science, technology and innovation plan (Zintgraff et al. 2015; Roldán, this volume). They use the city's development challenges as subject matter for project-based instruction. The second case comes from San Antonio, Texas, U.S., location of a Center of Academic Excellence for CyberPatriot, a cyber defense competition for secondary students (CyberPatriot 2013a). Their effort is supported by an industry cluster that includes the second-highest number of certified information security professionals in the U.S. (Zintgraff 2016). The third case is from Taipei, Taiwan (Yang, Yang, Chou, Wei, Chen, & Kuo, this volume). Framed by Industrial Revolution 4.0, Taiwanese high schools and colleges have embraced robotics and project-based instruction, with support from the local and national government and other actors in their region.

### *1.3.1 Medellín: STEAM-LABS, STEAMakers and Horizon Programs*

#### 1.3.1.1 Overview

In Medellín, Colombia, a series of STEAM (STEM+Arts) education programs have framed educational experiences within the development goals of the city. Two prior programs, STEAM-LABS and STEAMakers, were completed and helped establish a foundation. The Horizons program is ongoing, run by the city's innovation agency and tightly aligned with city goals. All of these programs demonstrate the principles highlighted in the current volume. For more details about the Horizons program, see Roldán in this volume.

STEAM-LABS (Zintgraff et al. 2015) began as a spin-off of a technology commercialization training program in Medellín, Antioquia Department (State), Colombia, delivered by the $IC^2$ Institute along with the College of Education at The University of Texas at Austin (UT Austin). Framed by the Medellín 2021 science, technology and innovation plan, STEAM-LABS was designed to introduce STEAM-aligned Project-Based Learning (PBL) methods to Medellín educators, with a dissemination plan to grow adoption.

Preliminary planning and vetting of ideas involved meetings with the education ministry; with Proantioquia, the region's premiere industry foundation; with Parque Explora, the local science park and museum; and with other industry representatives and educators. Seven education, government and community leaders visited Austin to see similar programs in action. The STEAM-LABS program trained 23 secondary educators and 8 college educators, organized into 11 teams. The program also conducted activities for school rectors and for industry supporters.

In STEAM-LABS, educator teams developed PBL curricula. The curricula was then delivered to students. Each curriculum unit focused on a problem aligned with one of Medellín 2021's high technology clusters. The UT Austin team delivered two one-week instructional sessions in Medellín, using the same project-based approach being trained; they supported development teams remotely; and they entered each of the 11 schools to visit with teachers, administrators, and/or students receiving instruction. The overall process began with an opening symposium with 130 attendees, mostly educators, but also well represented by university, government, nonprofit and industry leaders.

Figure 1.3 shows a closing activity at the symposium. Consistent with the focus on STEAM, an artist was hired to create a painting reflecting the event. The lady in the painting represents knowledge, and her flowing hair represents the dissemination of knowledge to all citizens of Medellín. The painting was presented to the city's Secretary of Education.

A second symposium was held on program completion, with 100 attendees reviewing progress and making go-forward plans. Going beyond original requirements, the educator participants trained an additional 137 teachers in their own schools, and the program ultimately served 1440 students (Zintgraff et al. 2015).

The STEAMakers program (Parque Exploration 2016) built on the progress of STEAM-LABS. Through a partnership with High Tech High School in San Diego, California, U.S., peer teacher relationships were established. The program added stronger rector training, and introduced maker education methods into instruction. Maker education was seen as adding an important student-as-creator element to PBL, encouraging exploration of solutions to the city's challenges.

**Fig. 1.3** CreAccíon Medellín symposium, December 2013

#### 1.3.1.2 The Eight Indicators

In STEAM-LABS, the indicators described were both (1) partially designed in by the UT Austin team, and therefore self-fulfilling, but also (2) already considered in many respects by Medellín actors. The Medellín 2021 plan was clear about the city's high-technology clusters, and the rationale for using them in pre-college education. STEAM-LABS largely required projects to be aligned to those clusters. STEAM-LABS organized industry supporters, recommended ways to engage those supporters, and organized a training session for them. While outcomes from industry engagement were mixed, some schools had strong outcomes, both through industry provision of resources, and also through the volunteer efforts of professionals. Proantioquia actively supported the program. Empresas Públicas Medellín (EPM), the Medellín-based utility company serving the city and other national clients, continued its long history of education support. University professors were trained and acted as subject-matter experts. The program was funded and managed by the city government's higher education agency and by Parque Explora, which was organized as a non-profit. Sapiencia (the higher education agency) facilitated involvement of university professors, and benefited by positioning themselves as a destination for students interested in the Medellín 2021 clusters. See Table 1.2 for a summary of the eight indicators present in STEAM-LABS.

### *1.3.2 San Antonio: CyberPatriot*

#### 1.3.2.1 Overview

In San Antonio, Texas, U.S., the city's cyber security cluster is the frame for the CyberPatriot program (CyberPatriot 2013a). CyberPatriot is a national competition in which high school teams (grades 9 through 12) and middle school teams (grades 6 through 8) compete in cyber defense challenges. For the city, cyber security is a

**Table 1.2** The eight indicators in STEAM-LABS and STEAMakers, Medellín

| Element | Description |
|---|---|
| Industry clusters inspire K-12 STEM education | Clusters were from Medellín 2021 Science, Technology and Innovation Plan |
| Industry clusters support K-12 STEM education | Industry players provided funding and volunteers to individual schools |
| Individual professionals support K-12 STEM education | Professionals participated, as individuals, and facilitated by employers |
| Universities support K-12 STEM education | Support from higher education agency, and participation of teachers from technical universities |
| Government supports K-12 education | Support and funding from education ministry |
| Non-profits support K-12 STEM education | Medellín program team was from science museum; event space provided by museum |
| Societal concerns inspire K-12 STEM education | All projects aligned to industry clusters and strongly exhibited connection to detailed local challenges |
| STEM education feeds universities and industry | Higher education agency viewed as program to recruit and prepare students |

well-established cluster. Several decades ago, the U.S. Air Force's Air Intelligence Agency was formed and located in San Antonio. The mission grew, and a cluster of information security contractors and startup companies developed, with the city eventually claiming the second highest number of certified information security professionals in the U.S. (U.S. Department of Defense Office of Economic Adjustment 2016). More recently the city competed and won placement of the 24th Air Force and 25th Air Force in San Antonio, both focused on cyber operations (24th Air Force 2012, 25th Air Force 2016). Several institutions of higher education in San Antonio are recognized as Centers of Academic Excellence by the U.S. Department of Homeland Security and the U.S. National Security Agency (Greater San Antonio Chamber of Commerce 2012).

The city's program is recognized as a Center of Excellence for CyberPatriot (CyberPatriot 2013b). The program features problem-based learning, extracurricular instruction, classroom instruction in selected schools, teacher professional development, robust mentoring from industry professionals, and a mature training program for a robust corps of volunteer professionals who support the program. According to program staff, 204 student teams competed in CyberPatriot in 2017. With one exception, the city has placed a team in the national finals each year since 2008. In 2011, a San Antonio team won the national finals. The team was from the Information Technology and Security Academy (ITSA), a program formed in 2002 through a partnership of the city government, city-owned utility company, the Alamo Community College District, the University of Texas at San Antonio, and numerous industry partners.

A special feature of the city's program is the Mayor's Cyber Cup. The cup, awarded to the best performing team in the San Antonio metro area, is part of a larger event that includes a job fair, a college fair for cyber security programs, and an awards ceremony. Figure 1.4 demonstrates the participation of secondary education, higher education, government, industry, and community partners in developing

**Fig. 1.4** San Antonio Mayor's cyber cup awards ceremony

the program. The participation of the Mayor lends a high profile, and it reflects the city's de facto policy of supporting and developing the cyber security cluster.

#### 1.3.2.2 The Eight Indicators

The start of cyber security secondary education in San Antonio was arguably ITSA. This high school academy was created in direct support of the local cyber security (then called information security) cluster. In organizing, members of the cluster created a support system for delivering education related to cyber security. Between the first and second year (11th and 12th grade years) of the program, all students receive internships, which puts them in direct contact with adult professionals. Later, as CyberPatriot became a prominent program, a corps of professionals, some volunteering on their own, and others organized by industry, served as mentors to teams.

Alamo Colleges, the University of Texas at San Antonio, and other institutions supported these and other programs through funding, in-kind resources, and staff support. The CyberTexas Foundation is a non-profit formed in part to support CyberPatriot. The city government provides funding, and through its actions, promotes the importance of the cluster and its educational programs. All these efforts advance the security of computers, networks, infrastructure and privacy; cyber security is a global top-of-mind concern of society. In San Antonio, almost all students in these programs move on to college education, and some enroll in cyber security programs, graduate, and find employment in the cluster. Although a small number overall, selected students receive part-time or full-time jobs immediately on completion of their high school cyber security programs. These indicators are listed in Table 1.3.

**Table 1.3** The eight indicators in CyberPatriot, San Antonio

| Element | Description |
|---|---|
| Industry clusters inspire K-12 STEM education | The cyber security cluster was the direct inspiration for the Information Technology and Security Academy, and for CyberPatriot in San Antonio. |
| Industry clusters support K-12 STEM education | The cluster organized to support both these and other cyber security education programs. |
| Individual professionals support K-12 STEM education | Hundreds of industry professionals have volunteered in the CyberPatriot program. |
| Universities support K-12 STEM education | The University of Texas at San Antonio leads technical development for the national competition content. Multiple local universities provide mentors, space, resources. |
| Government supports K-12 education | City and county government support the program via funding, resources, and promotional support. |
| Non-profits support K-12 STEM education | A non-profit was formed in part to support CyberPatriot. |
| Societal concerns inspire K-12 STEM education | The global concern about cyber security fuels interest in the program. |
| STEM education feeds universities and industry | Some students completing these programs enter cyber security degree programs; selected students are hired straight from high school programs; multiple institutions are U.S. Department of Homeland Security and National Security Agency Centers of academic excellence. |

## 1.3.3 Taipei: Robotics and Industrial Revolution 4.0

### 1.3.3.1 Overview

In Taipei, Taiwan, student teams participate in robotics-based education, extracurricular programs, and competitions. Two programs found in local schools are FIRST robotics (For Inspiration and Recognition of Science and Technology) and VEX Robotics (Team 6191 TFG x FRC 2017; Taipei American School 2017; RobotEvents.com 2016). Both of these programs are global, very large in scale, and enable students to collaborate with peers from around the world. However, before this global nature can emerge, teams require local support. As teams participate, opportunities arise to create local impact.

Taipei First Girls High School (Taipei TFG) is the home of FIRST Team 6191 (Yang, Yang, Chou, Wei, Chen, & Kuo, this volume). This all-girls team is using robotics education for STEM learning, while they also collaborate with the Taiwanese government to address *Industrial Revolution 4.0*. This phrase refers to the increased automation of manufacturing through custom, flexible mass production, and represents a major national trend. Team 6191 has benefited from government support. The school received 300,000 Taiwan dollars (approximately

US$10,000) from the national and Taipei city governments to support their program. The program is viewed as adding important real-world context to the students' educational experience, addressing critiques from companies about entry-level employees' ability to apply their knowledge in real-world scenarios.

The team is also helping build robotics education in the region, connecting to national technology development priorities. The national Minister of Science and Technology learned of the school's efforts and success in competitions. The Minister proposed a plan and included the school as a leader in a project to support makerspace education in Taiwan. The project included support from technology firms, other schools, research programs, and manufacturers. The school helped operate summer camps in the north and south of the country, working with Central Taiwan Science Park. Plans were underway to develop a FIRST regional (national or multi-country) competition in Taiwan.

Team 6191 has other wide-ranging supporters and associated mentors, who assist them technically and disseminate knowledge of their work. They include National Taiwan University of Science and Technology, Omron Corporation, Quanta Computer, Golf Gifts & Gallery, and FabLab. Local manufacturers, blacksmiths and car repair shops help the team build their robots. Figure 1.5 shows a picture of Team 6191 Taiwan First Girls High School.

#### 1.3.3.2 The Eight Indicators

While it is not necessarily clear how these robotics programs began, national technology development priorities are an impetus for continuation and growth. Team 6191 benefits from corporate support, including from manufacturers. They benefit

**Fig. 1.5** Team 6191 Taiwan First Girls High School

**Table 1.4** The eight indicators in Taipei

| Element | Description |
|---|---|
| Industry clusters inspire K-12 STEM education | The national emphasis on Industrial Revolution 4.0 is an impetus for growing programs. |
| Industry clusters support K-12 STEM education | Team 6191 receives support from the local industry and manufacturers listed. |
| Individual professionals support K-12 STEM education | The team has mentors who assist them technically and promote their work. |
| Universities support K-12 STEM education | Support from National Taiwan University of science and technology. |
| Government supports K-12 education | Funding and program support from the local and national governments. |
| Non-profits support K-12 STEM education | The opening of a public makerspace in the school by FabLab. |
| Societal concerns inspire K-12 STEM education | Economic development and quality of life derives from success in the manufacturing industry. |
| STEM education feeds universities and industry | Robotics education helps prepare students to apply what they've learned in college and the workplace. |

from small business support. The team benefits from mentors associated with those supporters. They work with universities, the local and national government, and are receiving a public makerspace from the non-profit FabLab. The evolution of manufacturing will demand an evolution in education and training to maintain economic impact and associated quality-of-life benefits as these students pursue higher education and some enter the manufacturing workforce. Table 1.4 contains the eight indicators for Team 6191.

## 1.4 Connecting to Policy

### *1.4.1 Defining Policy for this Volume*

This volume's title, and the chapter's title, suggest K-12 STEM education should be found in a city's or region's technology policy. What is meant by *policy*? What is meant by *regional technology policy*?

Policies and related plans abound regarding technology, economic development and education. They exist at the national, state and local level. This volume's conversation is primarily about city and regional policy (notwithstanding the national context offered in several chapters, and the national focus in Part 3 of this volume). Under consideration are the policies and plans cities and regions make for their own long-term development, focused on their local needs.

Conversations about STEM and policy must inevitably relate to technology. One might conclude that discussion should be exclusively about formal technology policy documents. As the cases in this volume will demonstrate, the reality in the field is far less restricted. One region may capture policies and plans in a central

**Table 1.5** Examples of policy documents relevant to this volume

| Location/Host type | Level | Name/Description |
|---|---|---|
| Medellín, Colombia/ City government | Regional (city) | Medellín 2021 Science, Technology and Innovation Plan |
| San Antonio, U.S./ Non-profit | Regional (city) | SA2020: A community vision for the future of San Antonio |
| Multiple locations/ National non-profit | National org., regional plans | STEM Learning Ecosystems initiative |
| Austin, U.S/Chamber of commerce | Regional (city) | Example of chamber supporting K-12 STEM education |
| Texas/Foundation | State | Educate Texas: Policy and advocacy positions and publications. |
| United States/Federal government | National | Federal Science, Technology, Engineering, and Mathematics (STEM) education, 5-year strategic plan |
| Mexico/Federal government | National | NIÑASTEM PUEDEN, part of Nuevo Modelo Educativo |

document; others may allow leading non-profits, chambers of commerce, and others to plan and coordinate for the community; there are many approaches found.

A broader definition of *regional technology policy* is used in this volume, including: dedicated technology policy documents; economic development documents; policies and plans from governments at all levels; chambers, industry associations, and non-profits; and more. The goal of this volume is to reflect what is happening in practice. A clear but high-level definition best serves the purposes of the volume. Table 1.5 lists several examples of the kinds of policy documents seen in this volume's cases.

### *1.4.2 Examples of Policy Actions*

What are the policy actions that can be taken to encourage the approaches shared in this paper? They are many and varied, but for those looking for a starting point, Wilson (2005) offered a framework for technology-based policy development. The framework defines levels of intervention: removing disincentives, providing incentives, providing support structures, and providing programs. Consistent with this framework, technopolis actors who control or influence policy can work to, for example: (1) remove or adjust barriers around educational standards; (2) arrange for tax breaks or benefits to organizations supporting STEM education; (3) create recognition programs–for example, challenge competitions, Mayor's awards, etc., to encourage participation; or (4) identify critical needs, and allocate resources, and directly fund programs.

Perhaps most important, actors can explicitly recognize in documents, and by public proclamations, the role STEM education plays in their community, and K-12 STEM education's impact on overall development plans. Technopolis actors can develop policies for STEM education, and they can treat those policies as fundamental to local development.

## 1.5 Modeling the Cycle

The grand idea of this volume is as follows: Strongly including K-12 STEM in regional technology policy can help drive a virtuous cycle. Priority industry clusters directly or indirectly inspire educators to build relevant education experiences and programs. Industry, governments, universities and non-profits provide vocal support, in-kind contributions, mentors and funding. Students get engaged, select more STEM coursework, and get into college. Those students complete and get jobs in local STEM-related industry clusters. Now employees, those former students give back to their schools, and so do their companies.

Figure 1.6 models the virtuous cycle. It includes characteristics and influencers that help speed or slow the cycle. The model contains the following ideas:

- Regions should contextualize STEM education within local development challenges.
- STEM education should both study technology, and leverage educational technology development.
- Appropriate education theories and methods (e.g., project-based learning) are required to achieve the desired benefits.
- Governments and other technopolis actors should remove barriers, create awareness and recognition of contributors, engage in partnerships, and when necessary, fund programs.

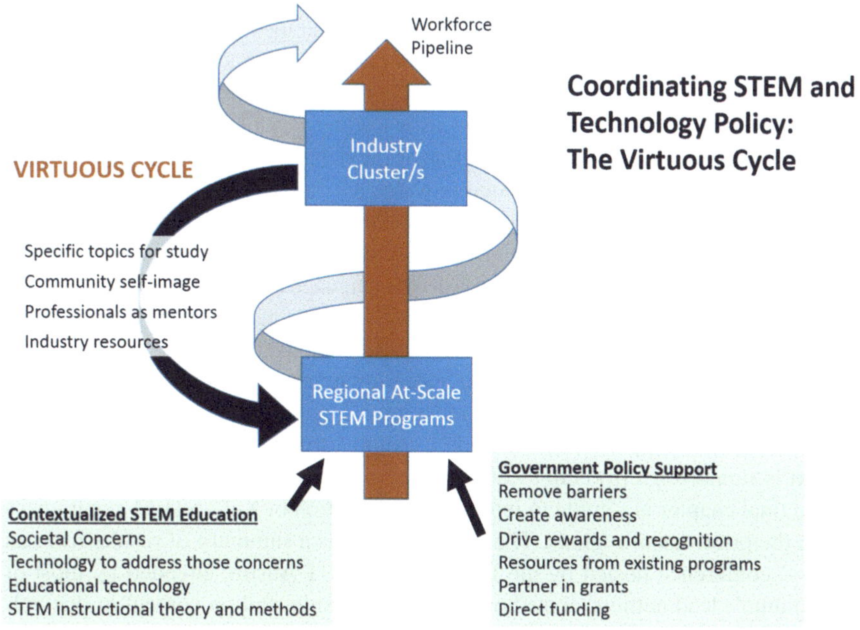


**Fig. 1.6** The STEM education, industry cluster, and society virtuous cycle

- When the virtuous cycle is operating, clusters will contribute topics for study, community self-image, professional mentors, and resources that drive the virtuous cycle.

## 1.6 Chapter Conclusion: About the Rest of the Volume

The subsequent chapters in this volume explore the main themes introduced in this chapter, or they share cases from around the world that demonstrate the *STEM in the Technopolis* virtuous cycle either operational (most cases) or under development. The chapters are organized into three parts: Foundations, Cases at City or Regional Level, and Cases with National Context.

In Part 1: Foundations, the main themes are further explored. Chapters cover the role of regional industry clusters, how emerging and emerged STEM pedagogy enables the virtuous cycle, a revised model of the technopolis, the place of digital equity in the STEM Technopolis, and quantitative perspectives and measures.

In Part 2: Cases at City or Regional Level, diverse cases are explored. The cases are from Medellín, Colombia; San Antonio, U.S.; Taipei, Taiwan; Austin, U.S.; Fundão, Portugal; and Querétaro, Mexico. The cases are diverse in both geography and local context. Medellín's story is about moving from being the most dangerous city in the world to being a world-recognized center of innovation and *la más educada* (the most educated). San Antonio's story is about its world-leading role in cyber security, how that came to be, and its deep impact on educating San Antonio K-12 students. Taipei's narrative surrounds the Fourth Industrial Revolution and the role being played by secondary schools. The Austin case describes not only the local ecosystem, but also the national STEM Learning Ecosystems movement that Austin is implementing, with the added richness of Austin's development as a leading city of innovation. Fundão, a small city in a rural area of Portugal, has transformed its future by embracing ICT business and education. Querétaro, Mexico provides a localized example of Movimiento STEM, a STEM Learning Ecosystems international member and a Mexican national movement supported by Mexican federal government policies driving STEM experiences for students. São Carlos, Brazil has driven language and mathematical literacy in K-12 grades using context and support from its long-time agriculture industry cluster. Like Querétaro, the São Carlos case has significant national and local context.

In Part 3: Cases of National Context, cases are presented with rich national context descriptions. These chapters from India and China offer descriptions of national policy, tools and support. Both include examples that demonstrate how national context is supporting efforts in local regions.

The final chapter is a guide to intentional integration of K-12 STEM experiences within the priorities of regions. The chapter contains a summary of results from an authors' conference hosted by the IC$^2$ Institute at UT Austin, attended by most of this volume's lead authors. Based on conference results and on content in this volume, readers are provided a roadmap and tools to help apply the principles shared.

The author urges stakeholders in positions of influence to consider how the authentic societal concerns of their regions can help increase the quality of STEM education in their communities. The three cases presented in this chapter, and those presented in later chapters, demonstrate the potential for positive impact when STEM education is viewed as a concern larger than the schools themselves. Readers are encouraged to keep reading and to act in situations they deem promising.

## References

24th Air Force (2012). 24th Air Force fact sheet. Retrieved October 19, 2012 from http://www.24af.af.mil/library/factsheets/factsheet.asp?id=15663.

25th Air Force (2016). 25 AF brings San Antonio Chamber into the ISR world. Retrieved September 24, 2017 from http://www.25af.af.mil/News/Article-Display/Article/904966/25-af-brings-san-antonio-chamber-into-the-isr-world/.

Bardeen, M., & Cooke, M. P. (2011). Celebrating 30 years of K-12 educational programing at Fermilab. In *Proceedings of the DPF-2011 Conference, Providence, RI, August 8–13, 2011*.

Bhatnagar, J. (2008). Managing capabilities for talent engagement and pipeline development. *Industrial and Commercial Training, 40*(1), 19–28.

Christenson, J. (2011, November). Ramaley coined STEM term now used nationwide. Retrieved November 16, 2014 from http://www.winonadailynews.com/news/local/ramaley-coined-stem-term-now-used-nationwide/article_457afe3e-0db3-11e1-abe0-001cc4c03286.html.

Clary, E. G., & Snyder, M. (1999). The motivations to volunteer: Theoretical and practical considerations. *Current Directions in Psychological Science, 8*(5), 156–159.

CyberPatriot (2013a). What is CyberPatriot? Retrieved April 23, 2016 from https://www.uscyberpatriot.org/Pages/About/What-is-CyberPatriot.aspx.

CyberPatriot (2013b). Centers of excellence. Retrieved August 10, 2019 from https://www.uscyberpatriot.org/Pages/About/Centers-of-Excellence.aspx.

CyberPatriot (2016). CyberPatriot coaches' agreement. Retrieved March 14, 2016 from https://www.uscyberpatriot.org/Documents/Competition%20Terms.pdf.

Echeverri, E. (2014). Ruta n and science, technology and innovation plan, Medellín ST+i 2011–2021. Retrieved November 13, 2014 from http://www.upb.edu.co/pls/portal/docs/PAGE/GP_REPOSITORIO_IMAGENES/PG_RIMG_DEINTERES/PLAN-ST%2BI.PDF.

Etzkowitz, H., & Leydesdorff, L. (2000). The dynamics of innovation: From National Systems and "mode 2" to a triple Helix of university–industry–government relations. *Research Policy, 29*(2), 109–123.

FIRST (2019). At a glance. Retrieved August 10, 2019 from https://www.firstinspires.org/about/at-a-glance.

FIRST in Texas (2016a). FIRST in Texas. Retrieved April 24, 2016 from http://firstintexas.org.

FIRST in Texas (2016b). Texas Workforce Commission grant information | 2015–2016. Retrieved April 24, 2016 from http://firstintexas.org/twc/.

Forssen, A., Lauriski-Karriker, T., Harriger, A., & Moskal, B. (2011). Surprising possibilities imagined and realized through information technology: Encouraging high School Girls' interests in information technology. *Journal of STEM Education: Innovations and Research, 12*(5/6), 46.

Gamse, B. C., Martinez, A., Bozzi, L., & Didriksen, H. (2014). *Defining a research agenda for STEM corps: Working white paper*. Cambridge, MA: Abt Associates.

Gibson, D. V., & Conceição, P. (2003). Incubating and networking technology commercialization centers among emerging, developing, and mature technopoleis worldwide. In L. V. Shavinina (Ed.), *International handbook on innovation*. Oxford: Elsevier Science Ltd..

Gibson, D. V., Rhi-Perez, P., Cotrofeld, M., De Los Reyes, O., & Gipson, M. (2003). *Cameron County/Matamoros at the crossroads: Assets and challenges for accelerated regional and binational development*. Austin, TX: IC2 Institute.
Goonatilake, R., & Bachnak, R. A. (2012). Promoting engineering education among high school and middle school students. *Journal of STEM Education, 13*(1), 15.
Greater San Antonio Chamber of Commerce (2012). *Cyber City USA*. Retrieved 6/14/2013 from http://www.alamoaog.org/lunch/SACAP.pptx.
IEEE (2017). Smart cities. Retrieved September 24, 2017 from http://smartcities.ieee.org/about.html.
Jones, M. (2018). *A civic entrepreneur: The life of technology visionary George Kozmetsky*. Austin, TX: Dolph Briscoe Center for American History at the University of Texas at Austin.
Laursen, S., Liston, C., Thiry, H., & Graf, J. (2007). What good is a scientist in the classroom? Participant outcomes and program design features for a short-duration science outreach intervention in K–12 classrooms. *CBE-Life Sciences Education, 6*(1), 49–64.
Markham, T., Larmer, J., & Ravitz, J. (2003). *Project based learning handbook: A guide to standards-focused project based learning for middle and high school teachers*. Novato: Buck Institute for Education.
Melchior, A., Burack, C., Hoover, M., & Marcus, J. (2016). FIRST longitudinal study: Findings at follow-up (year 3 report). Waltham: The Center for Youth and Communities, Heller School for Social Policy and Management, Brandeis University.
National Science Foundation (2015). Awards advanced search. Retrieved June 1, 2015 from http://www.nsf.gov/awardsearch/advancedSearch.jsp.
NIST (2017). Global city teams challenge. Retrieved September 24, 2017 from http://pages.nist.gov/GCTC/.
Parque Explora (2016, November). La escuela se transforma: Encuentro e intercambio de experiencias educativas. Retrieved May 6, 2017 from http://www.parqueexplora.org/visitenos/noticias/la-escuela-se-transforma-encuentro-e-intercambio-de-experiencias-educativas/.
Porter, M. E. (1990). *The competitive advantage of nations*. New York: The Free Press.
Powell, A., Watson, J., Staley, P., Patrick, S., Horn, M., Fetzer, L., Hibbard, L., Oglesby, J., & Verma, S. (2015). Blending learning: The evolution of online and face-to-face education from 2008–2015. *Promising Practices in Blended and Online Learning Series*. International Association for K-12 Online Learning.
Reigeluth, C. M., & Carr-Chellman, A. A. (2009). *Instructional-design theories and models: Volume III: Building a common knowledge base*. New York: Routledge.
Reigeluth, C. M., Beatty, B. J., & Myers, R. D. (Eds.). (2016). *Instructional-design theories and models, volume IV: The learner-centered paradigm of education*. New York: Routledge.
Resta, P., & Laferrière, T. (2015). Digital equity and intercultural education. *Education and Information Technologies, 20*(4), 743–756.
RobotEvents.com (2016). Formosa Vex Starstruck 2016. Retrieved September 24, 2017 from https://www.robotevents.com/robot-competitions/vex-robotics-competition/RE-VRC-16-4937.html.
Scott, C. (2012). An investigation of science, technology, engineering and mathematics (STEM) focused high schools in the US. *Journal of STEM Education: Innovations and Research, 13*(5), 30–39.
Smilor, R. W., Gibson, D. V., & Kozmetsky, G. (1989). Creating the technopolis: High-technology development in Austin, Texas. *Journal of Business Venturing, 4*(1), 49–67.
Taipei American School (2017, May). Taipei American school robotics. Retrieved May 6, 2017 from https://www.facebook.com/TASRobotics/.
Team 6191 TFG x FRC (2017). Retrieved September 24, 2017 from https://www.facebook.com/tfgxfrcteam6191/.
Texas Education Agency (2014, August). TEA news releases online. Retrieved December 15, 2014 from http://tea.texas.gov/news_release.aspx?id=25769815392.
U.S. Department of Defense Office of Economic Adjustment (2016). Port San Antonio plans to target cybersecurity companies. Retrieved September 24, 2017 from http://www.oea.gov/port-san-antonio-plans-target-cybersecurity-companies.

U.S. National Center for Education Statistics (2017, May). Fast facts. Retrieved May 8, 2017 from https://nces.ed.gov/fastfacts/display.asp?id=372.

Vex Robotics (2013, November 15). Nearly everything you wanted to know about participating in VEX Robotics cdivompetitions. Retrieved March 15, 2016 from http://www.vexrobotics.com/wiki/Nearly_Everything_You_Wanted_to_Know_About_Participating_in_VEX_Robotics_Competitions.

Vygotsky, L. S. (1978). *Mind in society*. Cambridge: Harvard University Press.

Wilson, M. (2005). STC 383 session 5: Technology transfer policies and structures [PowerPoint slides].

Wu, H. K., Lee, S. W. Y., Chang, H. Y., & Liang, J. C. (2013). Current status, opportunities and challenges of augmented reality in education. *Computers & Education, 62*, 41–49.

Young, K. L. (2007, November). Recruiting future engineers through effective guest speaking in elementary school classrooms. In *Meeting the Growing Demand for Engineers and Their Educators 2010–2020 International Summit, 2007 IEEE* (Vol. 50, pp. 1–14). IEEE.

Zintgraff, C., Fletcher, C., Jordan-Kaszuba, J., & Webb, J. (2015). StemDev Medellín and an instrument proposal to assess regional STEM-economic development alignment. In J. Slovak & D. Gibson (Eds.), *Building Sustainable R&D Centers in emerging technology regions* (pp. 209–242). Brno: Masaryk University.

Zintgraff Jr., A. C. (2016). STEM professional volunteers in K-12 competition programs: Educator practices and impact on pedagogy (doctoral dissertation).

# Chapter 2
# Regional Industry Clusters: A STEM Center of Gravity for Educators, Industry, Government and Non-Profits

**Cliff Zintgraff, Songhee Han, and Richard V. Butler**

**Abstract** The authors argue that prominent industry clusters in a city or region play an oversized role in driving the most robust and forward-thinking K-12 STEM programs, and the authors encourage greater intentional adoption of a local-industry-cluster-centric strategy when developing K-12 STEM programs. Working from the premise that cities and regions have become the core economic unit of competition globally (Florida, Adler, and Mellander, Regional Studies, 51(1), 86–96, 2017), the authors logically connect the principles and activities of industry clusters with principles and activities of STEM education. They review cases drawn from the current volume in support of the ideas shared, and they propose a STEM Technopolis model. Using results from a survey of actors across sectors, they highlight differences between the views of K-12 educators versus professionals from other sectors, and they use these observations to highlight where actions can be taken to better leverage advantages found in the STEM Technopolis.

## 2.1 Introduction

A core assertion of the current volume is that industry clusters at the city or regional level should be leveraged to create great STEM education experiences for primary and secondary (K-12) students. The current chapter contains the argument for why the authors believe this is true. The authors see the role of industry clusters as essential. It is hoped this chapter will lead to greater mutual recognition between K-12 educators, industry professionals, and all those working in academia, government

C. Zintgraff (✉) · S. Han
The University of Texas at Austin, Austin, TX, USA
e-mail: cliff.zintgraff@utexas.edu; song9@utexas.edu

R. V. Butler
Trinity University, San Antonio, TX, USA
e-mail: rbutler@trinity.edu



C. Zintgraff et al. (eds.), *STEM in the Technopolis: The Power of STEM Education in Regional Technology Policy*,
https://doi.org/10.1007/978-3-030-39851-4_2

and non-profits, regarding all the reasons they should pool their talents around the priority industry clusters in a community. Greater vocal support, volunteers, mentors and funding are possible outcomes for K-12 educators, not to mention great STEM experiences for their students. A more robust talent development pipeline is the potential outcome for industry cluster stakeholders, and this leads to better quality-of-life for citizens across a community.

### 2.1.1 Research Lens: Critical Theory

This chapter is developed within the research philosophy of critical theory. Faced with the question of whether social research should focus on measurable and generalizable outcomes, or whether such research should give a large role to local context, critical theory strikes a balance. Critical theory acknowledges objective reality while also enabling strong influence of "social, political, cultural, economic, ethnic, and gender factors…crystallized (reified)" (Guba and Lincoln 1994, p. 110). In academic terms, critical theory is both ontologically real and epistemologically subjective. Figure 2.1 illustrates the critical theory research philosophy.

The lens of critical theory is consistent with the posture of the authors. The authors of this chapter have been active as researchers, practitioners and advocates of the kinds of systems described in this chapter. Critical theory acknowledges the role of the applied researcher, one striving for accuracy and rigor consistent with their own values and mission.

### 2.1.2 Why Cities and Regions? Their Preeminence in Global Competition

Does the national, state or city context have the greatest effect on STEM experiences for students? All levels play a role. In this volume, selected chapters describe at length the national and/or state factors that affect K-12 STEM education. Still, strong arguments have been made that cities are the preeminent unit of global competition. If this is correct, then the talent pipeline of a city takes on increased importance.

The academic literature supports the central role of cities and regions in global competition. Florida et al. (2017) argued that "the city is the ultimate enabler of innovation, entrepreneurship, and growth" (p. 91). Florida built on other research

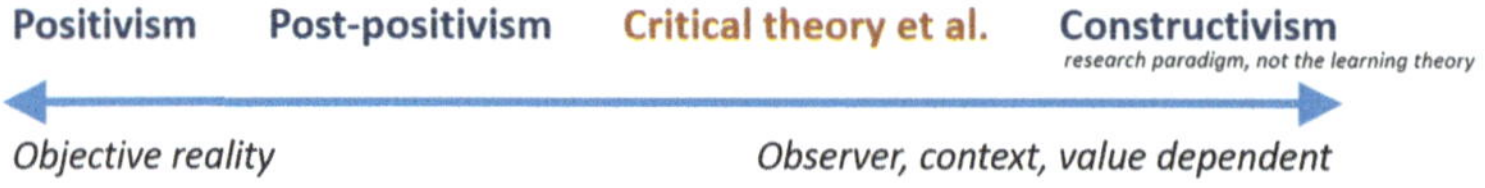


**Fig. 2.1** Critical theory on the positivist vs. non-positivist continuum (Zintgraff 2016)

placing cities at the center of the creative process (Duranton and Puga 2001; Almirall and Wareham 2011). The ideas presented are consistent with Porter's (1990) highlighting of the industry cluster social network as a key to its competitive advantage.

One can also see the competitions happening between cities. The recent bidding war for Amazon's second headquarters, HQ2, is a high-profile example (Garfield 2018). The presence of economic development foundations and routine competition between cities for business operations is a staple of municipal life. Cities address related policy issues to make themselves more competitive. For example, there is a growing push in the San Francisco Bay Area to pursue affirmative policies and initiatives to retain high caliber IT talent in the face of a high rate of professional exodus from San Francisco (Thomas 2018).

This phenomenon does not only happen in the American context. In South Korea, there was intense competition between municipalities to host SK Hynix's newly planned semiconductor cluster. This development will be supported by 120 trillion won ($106 billion) in South Korean government funding (Jun 2019).

The technopolis model that served as the framework for the development of Austin, Texas, U.S. similarly took a city-centric perspective on global competitiveness (Smilor et al. 1989). Meanwhile, and independent of assertions in the current volume, the availability of local workforce has long been a concern for industry (Mills et al. 2008). The authors argue that the city or city-sized regional perspective is preeminent in global competition, and further, that a qualified local workforce is fundamental to competitive success.

### *2.1.3 What Is an Industry Cluster?*

The term *industry cluster* was popularized in contemporary economic development by Michael Porter (1990). The same idea was explored as *geographical economics* by Paul Krugman (1991). The essential idea is that a collection of expertise in a region around a particular area of industry will create an advantage for all efforts within the cluster.

The industry cluster idea moves beyond the related notions of economy of scale and/or branding of a region, though all are related. Porter's (1990) conception focuses on knowledge sharing between organizations, innovation driven by focus and scale, and by a virtuous cycle that leads to the creation of even more new businesses in the cluster. The social networks (physical and virtual) associated with a cluster are the key element giving a cluster an advantage over competitors. While industry clusters are not fundamentally geographical, most are, and proximity is an advantage in the development of those social networks.

Some examples of industry clusters provide a clearer picture. Well-known industry clusters are software development in Silicon Valley and movie-making in Hollywood. Austin has become recognized globally for software development, information technology, creativity, and the Austin Model (Jones 2018).

Other examples in this volume demonstrate clusters well-known within their fields and/or more-immediate geographies. São Carlos, Brazil has a robust and mature agriculture sector, with many players across multiple sectors located within a nine-square-mile area (Gattaz, Falvo, and Cruvinel, this volume). San Antonio, Texas, U.S. claims the second highest number of certified information security professionals in the United States (Zintgraff 2016), with both local and global recognition for their cyber security capacity.

Industry clusters are not limited to large cities, and their size does not necessarily indicate their validity. In this volume, Aguiar and Pereira (this volume) describe how Fundão, Portugal, with a population of just under 30,000, is creating an Information and Communications Technology (ICT) cluster. Their city leaders and citizens are working in response to their specific challenges as a Portuguese rural community.

### 2.1.4 The Extended Industry Cluster Includes K-12

As it relates to the current discussion of STEM education, further technical definition is less relevant than viewing K-12 education as part of an extended industry cluster. It is useful to pause for a moment and consider the ways K-12 and industry interact in benefit of K-12 education. Schools bring in career speakers. They receive both funding and donations of equipment and materials from their industry partners. Some of these donations and materials are directly applied to the work of the cluster, but donors also provide general support. Special afterschool programs and competitions are created for local students, programs that use volunteers and mentors in their operation.

Those are the explicit connections, but there are also many tacit connections. The spouses, family members and friends of teachers also have jobs, and simply by chance, some will work in the local priority industry clusters. Teachers learn of happenings in those clusters and bring them into their classrooms in informal ways, and sometimes in formal ways. A small but meaningful percentage of teachers will have come from prior careers in these fields, meaning there is a core of industry cluster professionals already in schools.

From an economic development angle, Gibson et al. (2014) highlighted the importance of exposing tacit knowledge in innovation ecosystems. In a telling congruence, the STEM-compatible learning theories of situated learning (Brown et al. 1989) and situated cognition (Collins et al. 1989) bring specific focus to tacit knowledge.

The knowledge sharing, innovation, and new business development that Porter (1990) sees in clusters can also be seen in an extended cluster inclusive of K-12. Sharing of content knowledge and of applications in industry are two example of knowledge sharing. The creation of afterschool and competition programs is an example of innovation. It is also an example of new business creation, with the best ideas being the seeds for non-profits. Those non-profits are businesses that must operate with sustainable revenue models, and they fundamentally leverage the capacity of the cluster.

In short, K-12 is an actor in local industry clusters. They receive benefits, and they contribute to innovation and talent pipeline development. In the larger view, K-12 is a significant actor in the local technopolis.

### 2.1.5 Cases, Model, and Study Insights

The remainder of the chapter expands the argument through cases, a model, and insights from a brief study of the differing perspectives of K-12 educators versus actors from other sectors. The chapter begins with a brief overview of selected cases from this volume. The authors focus specifically on the industry-cluster-like attributes of those cases. A special discussion covers technical education, its relationship to STEM education, and its grounding in industry clusters. Based on the chapter's content, a *STEM Technopolis* model is proposed adapted from an existing model in the literature. Then, results of a study are shared that can help surface where perspectives different between K-12 educators and industry cluster actors from industry, academia, government, and non-profit stakeholders.

## 2.2 Cases: Seeing Industry Clusters in STEM Programs

The authors have observed that some of the strongest regional STEM programs are built around local priority industry clusters. These clusters tend to be high profile in their communities and accessible for all to learn about. K-12 teachers and administrators are themselves community members, and they see content and resources and bring them into the classroom by their own initiative. Local college programs often revolve around these clusters and/or use them for study. Policymakers are strongly motivated to support them. Non-profits build programs around them. To the extent these activities are actually happening, one should see the activities illustrated in real cases. The cases below demonstrate evidence of those activities.

### 2.2.1 San Antonio and CyberPatriot

Sánchez and Zintgraff (this volume) describe the CyberPatriot program built on the cyber security cluster in San Antonio, Texas, U. S. That cluster claims the second highest number of certified information security professionals in the United States, including military, federal, large company and startup company activities in the area (Zintgraff 2016). The chapter shares the history of U. S. Air Force security, IT and cyber missions dating back several decades, emphasizing the strength and deep roots of the cluster in the local community. Recent redevelopment efforts of the former Kelly Air Force Base include major focus on the cyber security cluster with accompanying policies for STEM education reflecting those advocated by this chapter and volume. Overall, the CyberPatriot case is a clear example of STEM education built on the strength of a local cluster.

The connection between the industry cluster, STEM activities, and the city's development can be seen in a set of complementary plans. The city's SA2020 plan, overseen by a non-profit organization of the same name, identifies cyber security as a priority cluster. The San Antonio Chamber of Commerce transformed its IT Committee into a Cyber Security Committee, and it pursued an action plan that resulted in a funded staff position leading cluster development. Port San Antonio, the organization driving redevelopment of the former Kelly Air Force Base, is pursuing a plan that makes cyber central to its mission, and the organization is providing major support to a science museum. That museum has STEM education at the core of its mission and is in close partnership with a STEM education non-profit.

CyberPatriot, a U. S. national and recently international program, is the basis for San Antonio's local program. The organization driving the local program, the CyberTexas Foundation, now has programs at the high school (grades 9–12), middle school (grades 6–8), and elementary school (grades 1–5). The local program includes robust training for teachers, mentors and students. The mentors are "industry professionals, students, academics, and otherwise IT-experienced adults who volunteer their time to teach cyber defense skills and cyber ethics to CyberPatriot teams" (CyberPatriot 2013, Technical Mentor(s), para. 1). Major support is received from industry. An international reader might recognize these organizations: Accenture, AT&T, Bank of America, Booz|Allen|Hamilton, Deloitte, General Dynamics IT, Rackspace, SAIC, and Symantec. U. S. and local readers might recognize Frost Bank, H.E.B. and USAA.

The support of the industry cluster has led to major successes. The city has the highest per capita program participation in the nation, with over 300 student teams. Successive mayoral administrations have supported the program through a yearly award, the Mayor's Cyber Cup. A 2012 team from the local Information Technology and Security Academy (ITSA) won the CyberPatriot national championship. That winning team came from a program started in 2001 that presaged what today is called an *early college high school*. In eight of nine years, a San Antonio team has advanced to the finals, with twenty San Antonio teams being finals participants.

The ITSA team is at the heart of one strand of narrative about the city's virtuous cycle. The San Antonio Technology Accelerator Initiative (SATAI) was an early 2000s effort to advance technology-based development in the city. ITSA grew in-part out of SATAI efforts. ITSA has been a successful venture and part of an award-winning model (Gonzalez 2016). ITSA and CyberPatriot have both benefited from the cluster and produced talent for its development. Influenced by local cluster development, the Air Force chose San Antonio for a large and high-profile new mission, the 24th Air Force (24 AF Office of History 2014). In turn, the cluster will grow, and more talent will be available for mentoring, to serve as volunteers, and to advocate for cyber security STEM education.

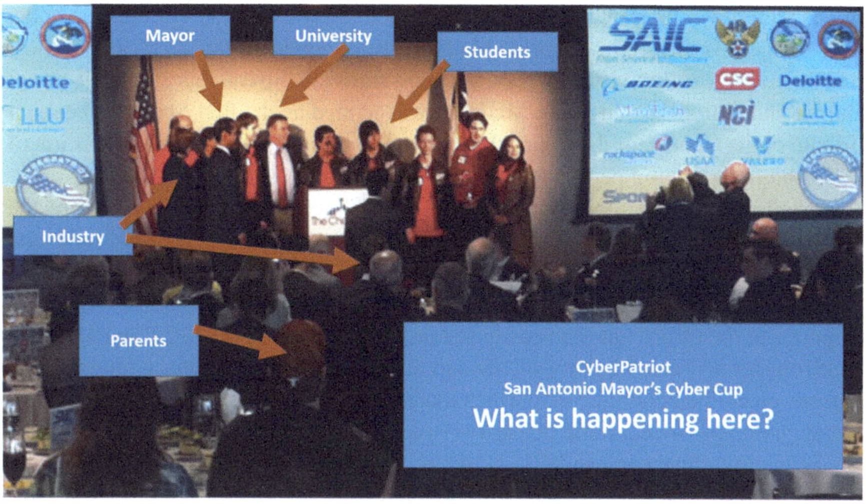


**Fig. 2.2** 2012 San Antonio CyberPatriot Mayor's cup winners; Eventual National Championship team

**Table 2.1** Attributes in San Antonio's cyber security cluster

| Clusters | Attributes |
|---|---|
| Cyber security | Deep local cluster history<br>Identified in multiple complementary formal plans<br>City government support<br>STEM/STEAM education integrated in plans<br>A robust list of supporters from the military, federal, private, startup, government and non-profit sectors<br>A champion local non-profit serving to organize the sectors |

Figure 2.2 is a picture of the 2012 national championship team receiving their Mayor's Cup *bomber flight jackets*. The main award was presented by then-Mayor and recent U. S. presidential candidate Julián Castro. The picture is a clear illustration of the sectors of the community in partnership around the program.

Table 2.1 lists the main industry attributes the authors observe in the case of CyberPatriot and the cyber security cluster in San Antonio.

## 2.2.2 *STEAM Programs and Medellín 2021*

Roldán (this volume) describes the fundamental integration of STEAM education in development of the Medellín technopolis. Development is driven by the Medellín 2021 Science, Technology and Innovation (ST + i) Plan (Echeverri Garcia 2014). The plan is a direct effort to shift Medellín to the posture of a knowledge economy, and in the process, to change the culture of the city, especially for its youth. This plan is well down its implementation path. Once known for drug culture and as the most dangerous city on the world, Medellín was recognized in 2012 as the most innovative city in the world by Citibank and the Wall Street Journal (Wall Street Journal and Citi 2013). In 2019, Medellín was named as the first Spanish-speaking affiliate of the Centre for the Fourth Industrial Revolution Network, a recognition of the leading work being performed in the city (Fourtané, 2019).

Medellín 2021 specifically identifies STEM/STEAM education as core to the development plan, and it identifies three technology-focused industry clusters, energy, ICT and health (Echeverri Garcia 2014). Medellín's leading industry foundation, Proantioquia (a reference to the surrounding state of Antioquia) is a strong supporter of the effort. One can easily see the connection between ICT and 4IR. The health cluster addresses quality-of-life and the well-being of citizens, a fundamental concern given the economic disparities than remain in the city, ones that reflect the global challenge of economic disparity (Roldán, this volume).

The energy cluster in Medellín is led by a dominant player, Empresas Publicas Medellín (EPM), the city's public utility company. EPM is owned by the municipality of Medellín and is a major contributor to the city's yearly budget. With 6000 employees (United Nations Global Compact n.d.), EPM is a major local employer of STEM professionals. EPM's business serves customers through Colombia and Latin America. EPM is a major supporter of Ruta N, Medellín's innovation agency and the driver of the Medellín 2021 plan. Une, the local telecommunications company and part of EPM's larger portfolio of companies, is also a major supporter of local development efforts.

The Medellín chapter (Roldán, in press) documents a new cluster that is forming as a result of the integration of STEAM education into the Medellín development plan. Through an intentional decision, educational technology companies, most of them startups, are taking the lessons learned from STEAM programs and developing solutions for schools, all as part of Ruta N's Interchange project. The result will be dissemination of lessons learned, dissemination of products, Spanish-language STEM products with a broad global market, increased culture change, and a small-but-growing cluster of educational technology companies.

Table 2.2 lists the main industry attributes the authors observe in the Medellín case.

**Table 2.2** Industry cluster attributes in the Medellín case

| Clusters | Attributes |
|---|---|
| Formally identified: | Driven by crisis-level problems of the city |
| Energy | Identified in formal plan |
| ICT | STEM/STEAM education integrated in plan |
| Health | Industry players included, including a dominant player and through the |
| New: | local industry foundation |
| Educational | Strong government support |
| technology | STEM-STEAM education led by same agency driving industry innovation |

### 2.2.3 Agriculture in São Carlos, Brazil

Gattaz, Falvo and Cruvinel (this volume) describe the STEM Technopolis in São Carlos, Brazil. Their focus is on the agriculture industry cluster in the region. The roots of the cluster go back more than 150 years, to efforts to increase coffee production, an important and profitable crop that led to a prestigious reputation for the city. Over time, the city and immediate region expanded its agricultural focus and offerings. Today, within a three-mile radius, multiple actors in the agricultural cluster educate university students, perform research, perform or manage industrial operations, develop agricultural innovations, and set policy for the region.

The city is known as a benchmark for providing high-quality agribusiness education, including a robust STEAM programs for students. Students between ages and 6 and 17 receive age-appropriate instruction around topics including nanotechnology, precision agriculture, rural basic sanitation, soil and water management and conservation, and more.. Programs are supported by teacher and counselor professional development, curriculum and material development, social mobilization regarding education, and systematic evaluation. Programs also leverage national initiatives encouraging STEM education experiences for students and professional development for teachers.

The agriculture programs do not operate in isolation. They receive strong support from city leaders and policymakers, within the context of Sanca Hub and local focus on research and innovation. Gattaz et al. (this volume) share a long list of actors active in the agriculture hub, with those actors crossing all industry sectors. Primary and secondary education are fundamentally part of the cluster's development. Recent focus is on developing mathematical and language literacy using the topics of agriculture industry noted above, all within an educational framework called psychopedagogy, addressing the cognitive, social and emotional needs of students.

Table 2.3 lists the main industry attributes the authors observe in the São Carlos STEM Technopolis focused on agriculture.

**Table 2.3** Industry cluster attributes in São Carlos, Brazil

| Clusters | Attributes |
|---|---|
| Agriculture | Long history of cluster (150+ years)<br>All technopolis actors represented within three-mile radius<br>Long list of players supporting program across all sectors<br>Strong government support<br>Rich educational content, specific to agriculture education, being used to teach mathematics and language literacy, and to teach cognitive, social and emotional skills needed by employers |

### 2.2.4 ICT in Fundão, Portugal

Aguiar and Pereira (this volume) authored this volume's chapter on the ICT cluster in Fundão, Portugal. Fundão's case is unique in this volume. The city, with a population of less than 30,000, represents the challenges faced by smaller cities and rural regions. In their chapter, Aguiar and Pereira picked up on the term preferred by local leaders: *desertification*. Like many similar areas around the globe, Fundão was seeing decreasing population and the exodus of young citizens and families to larger cities.

With the leadership of the Mayor, city and regional policymakers acted aggressively to address the challenge. Their Strategic Plan for Innovation addressed the needs of long-term, traditional clusters for which the region is already known (example: cherry farming), and it also embraced new technologies and envisioned an ICT cluster in the region. 2.5 million euros were raised to address the challenge, including European project funding. The city's vision has become reality. With two centers of gravity established under this plan, the Living Lab Cova da Beira, and Centro de Negócios e Serviços Partilhados (Business and Shared Services Center), the city has seen a reverse in net migration to positive numbers, and has seen 14 new companies start work in Fundão, including four multinationals, creating 500 quality jobs; 70 new startups; and over 200 privately-funded innovation projects. The work has received multiple European awards.

Fundamental to their plan is the development of coding academies. The goals of those academies are: (1) the re-training of adult workers, about half previously unemployed, to be ICT workers; and (2) the education of primary school students, using coding to develop technical and twenty-first century skills; and (3) to expand soon into secondary schools. The success of the primary school program has led to its adoption in schools around the country at high scale. Overall, Fundão is a strong example of a technopolis that has taken proactive steps to fundamentally integrate STEM education—in this case coding, as well as vocational education conducted using STEM principles—into their strategic development plan.

Table 2.4 lists the main industry attributes the authors observe in the Fundão, Portugal STEM Technopolis focused on ICT.

**Table 2.4** Industry cluster attributes in the Fundão Case

| Clusters | Attributes |
|---|---|
| ICT/ coding | Driven by major city developmental challenges<br>Championed by Mayor<br>Tight knit cross-sector leadership team<br>Actors supporting program across all sectors<br>Strong government support<br>Talent pipeline development from primary school through college and worker re-training (secondary school programs are imminent) |

### 2.2.5 *Summary of Industry Cluster Examples and Attributes*

The tables below summarize observations from the four cases above. Table 2.5 is a brief summary of the cases for reference. Table 2.6 notes common areas of direct interaction between the clusters (including industry, but other actors as well) and schools. Table 2.7 provides a summary of benefits by technopolis stakeholder, offering another angle on the benefit of industry clusters.

### 2.2.6 *Industry Clusters, STEM Education, and Technical Education*

The Fundão case opens a question worthy of discussion. One of Fundão's programs is vocational education, or as some say in the U.S., career and technical education. Technical education does not consist primarily of science, math, technology education or engineering classes, but rather is aimed directly at gaining employment for students in local jobs. Even experiential education, which is increasingly found in schools and exhibits a hands-on philosophy, is directed mainly at education for its implicit value, lacking the direct goal of finding employment. Most technical education requires STEM knowledge, acquired in an applied context. Is technical education also STEM education? What is the role of technical education in the STEM Technopolis?

Examples can help frame the question. Through technical education, one can become an aircraft mechanic. To become a certified mechanic in the U.S., a student must pass the U.S. Federal Aviation Administration licensing exam on which the applicant must demonstrate knowledge of trigonometry ("Basic Requirements," n.d.). Nursing also provides an example. To be a nurse, one must administer medicine safely, requiring mathematical literacy. Giving students the opportunity to apply academic knowledge in a real-world setting is a powerful way of cementing that knowledge and making it relevant to students.

The fundamental premise of the current volume is that STEM thrives when it is part of an ecosystem that includes not just educational institutions but also government, industry and other partners. Both government and industry have strong self-interests in a vibrant workforce development system. Industry depends on the availability of workers with skills specific to their field. Government success in economic development requires that a city/region make a convincing case to businesses that local workforce is available.

In San Antonio, Texas, U.S., the same partnership that developed the Information Technology and Security Academy (ITSA), first mentioned in the CyberPatriot case, also developed four other academies around industry clusters in the city. ITSA was the second academy developed using the *Alamo Academies* model developed with the Alamo Community College District in San Antonio ("Information Technology & Security Academy," n.d.). The Alamo Academies model brings

**Table 2.5** Summary of example cases

| Case / Location | Host type | Description |
|---|---|---|
| CyberPatriot<br>San Antonio, U.S. | Non-profit organization | Non-profit host partnered with a robust local industry cluster that includes the second highest number of certified information security professionals in the U.S.<br>Supported by multiple complementary planning and policy documents from the city government, a leading non-profit, chambers, and a redevelopment board.<br>A long and robust list of partners across all technopolis sectors.<br>New missions in San Antonio demonstrate the virtuous cycle. |
| Medellín 2021<br>Medellín, Colombia | Public joint venture serving as innovation agency for the city | Comprehensive development plan involving all sectors<br>Multiple fronts of effort: Innovation, company recruiting, startups, technology transfer, space planning, and more<br>All efforts driven by the city's innovation agency<br>Integrated within the plan, and also run by the innovation agency planning and deploying STEAM education experiences tied directly to city development goals and inspired by the needs of target clusters |
| Sanca Hub agriculture cluster<br>São Carlos, Brazil | Partnership among multiple actors | A long-time agriculture cluster in São Carlos, with multiple players located within a nine-square-mile radius<br>Developing language and mathematical literacy, and cognitive, social and emotional skills, in children between ages 6 and 17<br>Using agriculture-inspired content<br>Involves city government and all technopolis sectors<br>Focused on knowledge/skills needed by the agriculture cluster |
| Small city/rural strategic plan for innovation<br>Fundão, Portugal | City government | Effort of a city government working to avoid desertification of their small city/rural region<br>Programs for existing agricultural industry, and programs focused on new technologies, especially ICT<br>Coding training programs re-training adult workers, 50% of whom were unemployed, and educating primary school students on coding, with primary school programs expanding throughout the country<br>$2.5 million euros of support, including through European programs<br>Numerous awards, and a positive turnaround in net migration |

Other locations covered in this volume: Austin, U.S.; China; India; Querétaro, Mexico; Taipei, Taiwan

**Table 2.6** Most common interactions between schools and other technopolis sectors

| Case / Description | Mentoring | Workplace application | Talk about careers | Projects |
|---|---|---|---|---|
| San Antonio, U.S. / CyberPatriot | X | X | X | X |
| Medellín 2021 | X | X | X | X |
| São Carlos, Brazil / Agriculture | X | X | X | X |
| Fundão, Portugal ICT | Academia de Código curriculum does not mandate interactions, any activities would be arranged by individual teachers and schools. | | | |

**Table 2.7** Benefits of industry cluster focus, by Stakeholder

| Stakeholder | Benefits |
|---|---|
| Students | Relevant, engaging experiences |
| Families | Education opportunities; economic security |
| Teachers | Relevant content as a platform |
| Secondary schools | Content, mentors, funding |
| Colleges | A richer and deeper student pipeline |
| Employers | A richer and deeper talent pipeline |
| Cities and regions | Economic development; education and economic opportunity |

technical education into 11th and 12th grade classrooms, designed in a manner consistent with the requirements of Texas schools. Today in the U.S., the model might be called *early college high school*, with college instructors teaching dual-credit classes.

For this chapter, the important principle is that academy designs are a partnership among all sectors of the technopolis. The Academies began in the late 1990s as part of then-Mayor Howard Peak's vision that workforce development is a primary tool of economic development. The Academies adopted the principle that jobs follow workforce, a principle not widely understood at the time of Academies' first development. The Academies melded high school education, technical education, and industry cluster focus, all within the U.S. context, and all designed by robust industry cluster partnerships across sectors.

The Alamo Academies design is a platform for partnership development. Students attend their home high schools for half of each day, where they study the academic subjects required for high school graduation. Via school buses, they move to a community college site where they take the standard college technical courses, taught by college faculty. These courses carry dual credit, so the students simultaneously fulfill their high school requirements and earn a year's worth of college credit at no personal cost. Employers provide paid internships that students complete during the summer between their junior and senior years of high school (between 11th and 12th grade). Paid internships are an integral part of the Academies model. The paid internship design illustrates the type of win-win model found in sustainable partnerships, with employers contributing to the program, and with employers deriving immediate and long-term benefits.

The first academy supported the aerospace industry. The driving circumstance was risk of desertification in one area of the city. Kelly Air Force Base had been closed as part of national effort to reduce the number of military based in the U.S. Kelly's closing ended well over 10,000 jobs in the city ("History – Where," n.d.). Before closing, the base was a focal point for aircraft maintenance. During the base's transition to civilian status, the city attracted major aerospace companies, including Boeing, Lockheed Martin and Standard Aero, to begin operations in the former military facilities. These companies quickly discovered their workforce was close to retirement age, with few new qualified mechanics being trained to take the retirees' places.

The crisis of Kelly's closing was also an opportunity for innovation and development of new partnerships. The city, aerospace employers, the community college district and local public schools all had a stake in solving the problem. None could solve the problem alone, and the only option was to forge a cooperative effort. The Alamo Area Aerospace Academy opened its doors to students in August 2001.

ITSA and Advanced Manufacturing soon followed. More recently, Health Professions and Heavy Equipment have been added to the portfolio. The Academies have won national recognition, including a Bellwether Award in 2015 ("Achievements, Awards and Recognition," n.d.).

The Academies are a true partnership. Each interested party has a critical responsibility. The community college district provides facilities and instructors, and houses the Academies' professional staff. The operating expenses of the program, including professional staff salaries, are provided by the city as an economic development initiative. The school districts provide transportation to the community college instructional sites, and the participating employers pay their own interns.

The Alamo Academies are vivid examples of the regional technopolis approach to STEM education and workforce development.

- Academies are regional in scope. Students may come from any school in the San Antonio area, a feat complicated by the more than twenty government entities in the city responsible for childrens' education. Companies may be located anywhere in the region, and similarly their workforce will not be confined to the individual city in which they are located.
- The curriculum is employer-driven. Graduates emerge workforce-ready. Employers participate in the Academies for business reasons, not as occasional philanthropy.
- The occupations in question are inherently technical. Graduates must master not only specific job skills, but also a great deal of the underlying math, science and technology, and also the communication, collaboration, critical thinking and creativity skills (the *4 Cs*) valued by modern employers.
- For this reason, the Academies are also an effective model for STEM education. Students are more motivated to learn, for example, trigonometry and biology when they can see the immediate application of newly obtained knowledge.

Since college-level technical education requires facilities and instructors not commonly found in the secondary school system, the Academies are a force multiplier for school districts.
- Workforce development is part of core mission for community colleges. The opportunity to expand their intake of technical students by connecting to the secondary school system enhances community colleges' ability to carry out that mission.
- The availability of a skilled workforce is a vital component of successful economic development. The technopolis model is an effective way of thinking about engagement of other stakeholders in meeting this need.

## 2.3 Model: The STEM Technopolis Wheel

The principles in this chapter, and of this volume, converge in this volume's adapted version of the *technopolis model*, also known as the *technopolis wheel*. This adapted model is presented in three chapters of this volume: here, in the chapter discussing the place of secondary STEM education in the technopolis (Kidwell, Zintgraff, and Pogue, this volume), and in the final chapter of this volume providing a summary and guide to implementation in communities (Zintgraff, this volume). The original technopolis definition comes from work at the $IC^2$ Institute, The University of Texas at Austin, by Smilor et al. (1989).

The adaptations are related to edits, clarity, and to connecting an underappreciated element of the model strongly to the ideas of this volume. One proactive change is a dedicated spoke for primary and secondary education. In the model, we use the common U.S. shorthand, *K-12* (kindergarten through 12th grade). We argue that K-12 education must be an equal participant in a robust technopolis.

A second proactive edit is the inclusion of a dedicated spoke for technical colleges. These colleges are known by different names in different countries, with each having mild to moderate differences in focus. They are called community colleges in the U.S. They are called technical universities in many other countries. By explicitly acknowledging technical colleges, we give a dedicated place to colleges that focus mainly on workforce preparation. These colleges often deliver certifications, two-year degree programs, and adult continuing education. This chapter made the case for why technical education is a robust and common form of STEM education. The separate and special role of research universities is maintained in a separate spoke and is covered in depth in original technopolis publications (e.g., Smilor et al. 1989).

Third, we remove much of the detail from the spokes in favor of readability and clarity. The deeper purpose is accessibility to an audience that spans from researchers to policymakers to classroom practitioners. The authors wish the simple idea of the sectors working together to sharply come through. This change does not imply an undoing of the greater detail of the original model.

The fourth change addresses the underappreciated and often overlooked element of the model: that it is meant to be a wheel. More often than not, the wheel attribute is omitted from descriptions. Dr. Donna Kidwell asked the insightful question during a gathering of authors, "What generates torque for the wheel?" This question led us quickly to two energy sources, one being STEM learning, and the second being the industry clusters found in technopoles.

The STEM learning argument is made in the technopolis chapter (Kidwell, Pogue, and Zintgraff, this volume). As made clear in this chapter, the industry clusters do not consist solely of industry partners. The cluster includes academic institutions, K-12, government actors and non-profits who collaborate to make results greater than the sum of their parts. Figure 2.3 illustrates the STEM Technopolis Wheel.

## 2.4 Study: Examining Stakeholder Views, K-12 Versus Other Sectors

To complement the qualitative observations to this point in the chapter, and to shed light on how the views of K-12 educators compare to those from outside K-12 schools as it relates to industry clusters, a short study was devised. This section

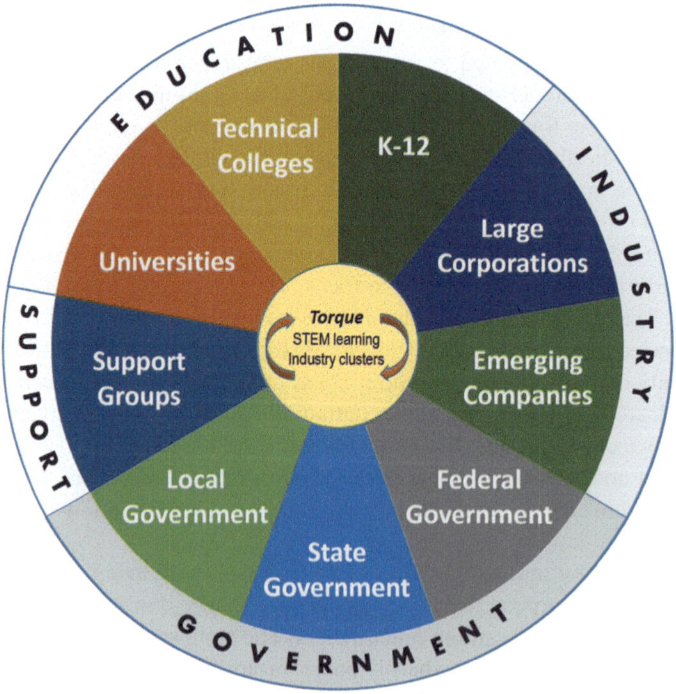


**Fig. 2.3** The STEM technopolis wheel: Industry clusters as Torque. Adapted from Smilor et al. (1989)

describes the study design, questions, execution and results. Analysis is offered in light of study results, the qualitative cases, and the experience of the authors.

### 2.4.1 Research Questions

The study pursued answers to these three research questions:

1. What role do industry clusters play for K-12 educators?
   - Are K-12 educators aware of local industry clusters?
   - What benefits do K-12 educators receive from local industry clusters?
   - What other benefits would K-12 educators like to receive from local industry clusters?
2. How do those outside K-12 see industry clusters helping K-12 programs?
   - How do stakeholders from other sectors interact with K-12?
   - In what other ways would stakeholders from other sectors (university, industry, government, non-profit organizations) like to interact with K-12?
3. What are the successes in K-12 collaboration with other sectors, and where are the challenges and barriers?

### 2.4.2 Survey

Two surveys were designed to address the research questions, one for K-12 educators, and a second for other stakeholders. The survey consisted of a series of Likert-scale questions and relevant open-ended questions. Given the varied backgrounds of participants, there was risk that many questions would be outside the expertise of respondents. Therefore, Likert questions offered an option to pass on individual questions, avoiding noise from those not qualified to answer. Table 2.8 lists the impact categories offered in both surveys, and Table 2.9 lists the related-but-separate questions asked of K-12 educators versus other respondents. Appendix A contains a fuller description of the surveys.

### 2.4.3 Execution

The survey was administered in both paper-based and web-based forms to reach a larger and more diverse sample population. The authors ran the paper-based survey at the Greater Austin STEM Community Convening, June 6, 2019, at the Austin Chamber of Commerce in Austin, Texas. This community meeting was focused on

**Table 2.8** Categories of possible K-12 STEM program benefits from area industry clusters

| Categories (Zintgraff 2016) | | | | | |
|---|---|---|---|---|---|
| Mentoring for students | Speaking to students about careers | Helping educators understand workplace applications | Helping students with projects | Directly teaching content to students | Helping educators secure STEM program resources |
| Suggesting content or/and developing curriculum | Set up, configuration, and maintenance of equipment | Delivering or/ and hosting professional development | Hosting field trips | General, non-specific assistance | |

**Table 2.9** Summary of survey questions, by Respondent type

| K-12 Educators | Other stakeholders |
|---|---|
| Are they aware of industry clusters in their city or region? List them.<br>What additional interactions or benefits would K-12 STEM educators like to receive from industry clusters in their area? | How do industry clusters in their area interact with K-12 STEM programs?<br>What other interactions or benefits do they wish area industry clusters could provide to K-12 STEM programs? |

the Central Texas STEM workforce, pursuing deeper collaboration across educators, industry, government, and non-profit professionals to ensure STEM programming is accessible to all students throughout Greater Austin. As such, the authors presumed the majority of the participants to be stakeholders in K-12 STEM education.

The web-based survey was created through Google Forms and distributed through the authors' social media channels including Facebook and Twitter from May 6th to July 5th, 2019. Because of the authors' background in teaching K-12 STEM education at schools and volunteering for K-12 STEM-events hosted by non-profits, K-12 teacher or administrator colleagues expressed interest in this survey and helped redistribute it through their own social media channels. Questions were included to collect demographics from respondents.

### 2.4.4 Descriptive Results

There were 30 total survey respondents with a fairly equal distribution between K-12 and other sectors.

- There were 14 K-12 educator respondents: 1 student, 9 K-12 teachers, & 4 K-12 administrators.
- There were 16 other stakeholder respondents: 4 university administrators, 2 industry, 2 government, & 8 non-profit professionals.

All respondents reside in Texas: 25 in the Greater Austin Area, 4 in San Antonio, and 1 in Houston. The mean score for each question was calculated, and then the mean difference of each group's scores were compared to identify similarities and divergence in the two group's views of how industry clusters relate to K-12 STEM programs.

## *2.4.5 Key Findings*

The authors present the findings in summarized form, rather than per research question, focusing on the key learnings apparent from study results. Responses were not sufficient in number for analytical statistics; therefore, descriptive statistics are shared and interpreted. Table 2.10 summarizes the gaps found between educator's desires for outside help, versus the kinds of help stakeholders think they should deliver.

### 2.4.5.1 Finding 1: K-12 Educators' Vague Understanding of Industry Clusters

The most obvious finding from the survey was educators' vague understanding of the concept of an industry cluster.

- 9 of 14 K-12 respondents indicated they knew at least one industry cluster in their city/region active in K-12 STEM programs.
- Of the 9, only 3 accurately identified a cluster.
- The other 6 identified geographic regions, STEM program hosts, or the programs themselves, which are not industry clusters.

### 2.4.5.2 Finding 2: Gap in Benefits Received and Desired

Part 1: Expert help for students and teachers wanted; general help and field trips less so.

**Table 2.10** The educator versus outside stakeholder collaboration gap

| K-12 Educators | Other stakeholders |
|---|---|
| Want more than stakeholders think:<br>• Career information<br>• Expert knowledge | Think K-12 wants, or wants more of:<br>• General assistance<br>• Field trips |
| Want less than stakeholders think:<br>• General assistance<br>• Field trip opportunities | |

There was a significant gap in what benefits K-12 educators wished to receive from other stakeholders versus what other stakeholders wished to provide for K-12 STEM programs. This could be seen in both the ratings of current benefits received, and in ratings of what additional benefits were desired by K-12, or could be given by other stakeholders.

- K-12 educators most wished to receive the following benefits from other stakeholders:
- Speaking to students about careers (3.50, highest score: 5)
- Mentoring for students (3.08)
- Directly teaching content to students (3.08).

Other stakeholders most wished to provide:

- General, non-specific assistance (3.38)
- Hosting field trips (3.27).

This result from educators is consistent with Zintgraff's (2016) study of STEM professional volunteers in K-12 competition programs. K-12 educators generally want outside professionals to provide specialized information or assistance that is outside of educators' ability to deliver to students. K-12 educators also find field trips useful, but relatively difficult to execute.

The current study adds new information to Zintgraff's (2016) findings, which only considered the perspectives of educators. It shows that other stakeholders do not share the educators' perspectives, but rather think general help for STEM programs is most needed. K-12 educators have low interest in 'general, non-specific assistance' (2.55), ranked fourth from the bottom. Their interest in field trips is higher, but not high (2.83) due to administration challenges, schedule changes, and student security concerns outside of school.

Part 2: Inform me, but let me do the teaching professional work.

K-12 educators responded that 'suggesting content or/and developing curriculum' (2.33) was one the least favorite benefits they wished to receive despite other stakeholders generally willingness to provide it (2.93). Likewise, 'delivering or/and hosting professional development' category scored only 2.64 from K-12 educators but 3.00 from other stakeholders, which was above the average score.

At first review, this might seem in conflict with educators' wish to receive expert help. However, the authors see a difference between (a) receiving expert information from outside the school, versus (b) having external stakeholders generate curriculum and professional development deliverables, which external stakeholders are not trained to do, while educators are. The authors also speculate that educators' professional development calendars, which are quite full, leave little time for new professional development experiences.

Reinforcing that observation, K-12 educators notably wished to receive help in understanding workplace applications (helping educators understand workplace applications, 3.00) instead. This reinforced the earlier observation that K-12 educators were most interested in technical knowledge and workforce experience which educators do not possess.

Part 3: Career speakers: High value, and keeping it simple.

'Speaking to students about careers' was ranked high from K-12 educators (3.50) and other stakeholders (3.20). It seemed 'speaking to students about careers' was perceived to be one of the most convenient and efficient ways for both K-12 educators and other stakeholders to follow through STEM programs.

Zintgraff (2016) identified four mental constructs as dominating the minds of teachers when they think about bringing STEM professional volunteers into their teaching: practicality, pedagogy, knowledge and skills, and rapport. Rapport is likely most related to long-term mentoring in programs like FIRST Robotics and CyberPatriot. For shorter term considerations, career speakers are practical, they introduce real-world knowledge and skills, and they bring the real world into the classroom (pedagogy).

#### 2.4.5.3 Finding 3: How Common Is Partnering?

Earlier sections might leave the impression that partnering among technopolis sectors is common, including between K-12 and other sectors. Clearly, there are cases in this volume that exhibit strong cross-sector partnerships. Zintgraff (2016) reported many other examples; for example, in U.S. engineering education.

While this study lacks the sample sizes and broad geography needed to generalize, study results suggest more partnering is possible. The differences in perceptions about how K-12 should collaborate with actors outside schools, and the general lack of awareness among responding educators of what an industry cluster is and how it operates, makes clear there is more opportunity.

This state of affairs is not a surprise to the authors. In fact, it is consistent with the message of the current volume. This volume has been written to share principles of partnership, supported by high quality cases from around the world. Even for those cases, each case author could readily share opportunities in their region to increase partnerships and their effectiveness. The authors assert these opportunities are available for harvesting in every community. Meanwhile, study findings include the raising of two questions that might be worthy of further research: How common is partnering? And how common is effective partnering?

## 2.5 Conclusions

The current volume, STEM in the Technopolis, advocates using the particular and priority industry clusters of a city or region as a foundation for great STEM experiences for students. It is argued that, at the intersection of industry clusters, STEM pedagogy and the technopolis model, there is rich content, and there are supportive people, volunteers, mentors, funding, and the mix of skills needed to educate students for twenty-first century jobs, all to the benefit of local communities.

In this chapter, the authors examined in depth the role of industry clusters. The cases covered in this chapter and volume demonstrate the benefits that can result when a priority cluster in a city or region puts its mind to helping K-12 students. These examples are seen from around the world. Industry clusters, especially strong priority ones, inject energy into the system; they are torque to turn the technopolis wheel.

However, a Texas-based study indicates gaps between (a) the help educators want, and (b) the help industry, academia, government and non-profits think they should provide. The authors do not claim generalizable results; still, the observation notes important potential barriers. K-12 educators and other technopolis stakeholders cannot maximize their collaborative results if they do not understand one another. Further exploration of the mutual perceptions of K-12 educators versus technopolis stakeholders is indicated.

The education and workforce needs of the future will demand regional partnerships, keyed to the economic advantages of the region, providing the technical skills and STEM knowledge required by the twenty-first century economy. Cities' and regions' economic futures, and quality-of-life for citizens, will hinge in no small part on bringing educators, employers, policymakers, economic developers, and other stakeholders together in common cause. Together, they are a strong team for building meaningful educational experiences and an appropriately skilled workforce.

## Appendix A: Summary of Ecosystem Stakeholder Surveys

STEM Ecosystem Stakeholder Surveys: Concise Summary

1. Your relationship to K-12 STEM programs? Choose one from below.

| | |
|---|---|
| K-12: | Student/ Parent/ K-12 teachers/ K-12 administrator |
| Other sector: | University faculty/ University administrator/ Industry professional/ Government or related policymaker. Non-profit professional |

2. (K-12) Are you aware of at least one industry cluster in your area that is active in K-12 STEM programs? / (Other Sectors) How do industry clusters in your area interact with K-12 STEM programs?
3. List the industry clusters you know are active in area K-12 STEM programs.
4. (K-12) What other interactions or benefits would K-12 STEM programs like to receive from industry clusters in your area? / (Other Sectors) Other interactions or benefits you wish area industry clusters provided to K-12 STEM programs?
5. Select the degree (5,4,3,2,1, *I don't know) to which K-12 STEM programs in your area receive benefits from area industry clusters, along the categories indicated, to the best of your knowledge.

| Mentoring for students | Speaking to students about careers |
|---|---|
| Helping educators understand workplace applications | Helping students with projects |
| Directly teaching content to students | Helping educators secure STEM program resources |
| Suggesting content or/and developing curriculum | Set up, configuration, and maintenance of equipment |
| Delivering or/and hosting professional development | Hosting field trips |
| General, non-specific assistance | |

6. Your city or region?
7. Any particular experiences that inform your answers to this survey?

*For a digital copy of this survey, contact the corresponding author.*

## References


Achievements, Awards and Recognition (n.d.). Retrieved August 7, 2019 from https://alamoacademies.com/awards/.

Almirall, E., & Wareham, J. (2011). Living labs: Arbiters of mid- and ground-level innovation. *Technology Analysis & Strategic Management, 23*(1), 87–102.

Basic requirements to become an aircraft mechanic (n.d.). Retrieved August 7, 2019 from https://www.faa.gov/mechanics/become/basic/.

Brown, J. S., Collins, A., & Duguid, P. (1989). Situated cognition and the culture of learning. *Educational Researcher, 18*(1), 32–42. https://doi.org/10.3102/0013189X018001032.

Collins, A., Brown, J. S., & Newman, S. E. (1989). Cognitive apprenticeship: Teaching the crafts of reading, writing, and mathematics. In L. B. Resnick (Ed.), *Knowing, learning, and instruction: Essays in honor of Robert Glaser* (pp. 453–494). Hillsdale: Lawrence Erlbaum Associates, Inc.

CyberPatriot (2013). Team organization. Retrieved August 7, 2019 from https://www.uscyberpatriot.org/competition/Competition-Overview/team-organization.

Duranton, G., & Puga, D. (2001). Nursery cities: Urban diversity, process innovation, and the life cycle of products. *American Economic Review, 91*(5), 1454–1477. https://doi.org/10.1257/aer.91.5.1454.

Echeverri Garcia, E. (2014). Ruta n and science, technology and innovation plan, Medellín ST+i 2011–2021. Retrieved November 13, 2014 from http://www.upb.edu.co/pls/portal/docs/page/gp_repositorio_imagenes/pg_rimg_deinteres/plan-st%2bi.pdf.

Florida, R., Adler, P., & Mellander, C. (2017). The city as innovation machine. *Regional Studies, 51*(1), 86–96. https://doi.org/10.1080/00343404.2016.1255324.

Fortané, S. (2019, February). Medellin becomes a fourth industrial revolution network's affiliate center. Retrieved July 31, 2019 from https://interestingengineering.com/medellin-becomes-a-fourth-industrial-revolution-networks-affiliate-center.

Fourtané, S. (2019, February). Medellin becomes a Fourth Industrial Revolution Network's Affiliate Center. Retrieved February 15, 2020 from https://interestingengineering.com/medellin-becomes-a-fourth-industrial-revolution-networks-affiliate-center.

Garfield, L. (2018, January 18). Amazon has announced the top contenders in its $5 billion bidding war for HQ2-here are their craziest proposals. Business Insider. Retrieved August 7, 2019 from https://www.businessinsider.com/amazon-headquarters-city-proposals-hq2-2017-10.

Gibson, D., Foss, L., & Hodgson, R. (2014). Institutional perspectives in innovation ecosystem development. In *Moderne Konzepte des organisationalen marketing*. Wiesbaden: Springer Gabler.

Gonzalez, I. (2016, February). San Antonio produces top CyberPatriot teams. Rivard Report. Retrieved August 7, 2019 from https://therivardreport.com/san-antonio-produces-competitive-cyberpatriot-teams/.

Guba, E. G., & Lincoln, Y. S. (1994). Competing paradigms in qualitative research. In N. K. Denzin & Y. S. Lincoln (Eds.), *Handbook of qualitative research* (pp. 105–117). Thousand Oaks, CA: Sage Publications, Inc.

History – Where it All Started. (n.d.). Retrieved August 7, 2019 from https://alamoacademies.com/history/

Information Technology & Security Academy. (n.d.). Retrieved August 7, 2019 from https://alamoacademies.com/information-technology-security-academy/.

Jones, M. (2018). *A civic entrepreneur: The life of technology visionary George Kozmetsky*. Austin: Dolph Briscoe Center for American History at the University of Texas at Austin.

Jun, J. (2019, July). Municipalities vying to host SK Hynix-led chip cluster. The Korea Times. Retrieved August 7, 2019 from https://www.koreatimes.co.kr/www/tech/2019/02/133_263836.html.

Krugman, P. (1991). *Geography and trade*. Cambridge: MIT Press.

Mills, K. G., Reynolds, E. B., & Reamer, A. (2008). *Clusters and competitiveness: A new federal role for stimulating regional economies*. Washington, DC: Brookings Institution.

Porter, M. E. (1990). *The competitive advantage of nations*. New York: The Free Press.

Smilor, R. W., Gibson, D. V., & Kozmetsky, G. (1989). Creating the technopolis: High-technology development in Austin, Texas. *Journal of Business Venturing, 4*(1), 49–67.

Thomas, A. (2018, November 30). 5 major reasons people are leaving San Francisco [Newsletter]. Retrieved August 7, 2019 from https://www.inc.com/andrew-thomas/5-major-reasons-people-are-leaving-san-francisco.html.

24 AF Office of History (2014). History of HQ twenty-fourth Air Force and 624th operations center (pamphlet). [PDF file]. Retrieved August 7, 2019 from https://www.afcyber.af.mil/Portals/11/documents/About_Us/AFD-140429-035.pdf?ver=2016-04-26-113101-810.

United Nations Global Compact (n.d.). Empresas Publicas de Medellin. Retrieved August 7, 2019 from https://www.unglobalcompact.org/what-is-gc/participants/3319-Empresas-Publicas-de-Medellin.

Wall Street Journal and Citi Announce Medellín Wins 'City of the Year' Global Competition (2013). Retrieved August 7, 2019 from https://www.citigroup.com/citi/news/2013/130301a.htm.

Zintgraff Jr., A. C. (2016). STEM professional volunteers in K-12 competition programs: Educator practices and impact on pedagogy (Doctoral dissertation). [PDF file]. Retrieved August 7, 2019 from https://digital.library.unt.edu/ark:/67531/metadc955031/m2/1/high_res_d/ZINTGRAFF-DISSERTATION-2016.pdf.

# Chapter 3
# The Education Philosophy, Theories and Models That Enable STEM Policy Integration

**Anthony J. Petrosino, Maximilian K. Sherard, and Sneha A. Tharayil**

**Abstract** As federal and state policies continue to emphasize the need for STEM Education reform, it is important to understand how a collaboration between industry, academia, governments, nonprofits, and K-12 schools, can bolster this effort. However, any worthwhile attempts at this must arguably first be steeped in an understanding of the STEM Education movement as well as the deep lineage of learning theories which underscore what has now come to be understood as evidence-based best practices in bolstering STEM teaching and learning. As such, the ensuing chapter recounts a brief history of the STEM Education movement, discussing the impact of pivotal documents, such as *A Nation at Risk*, which fanned the flames of education reform. It then presents the various perspectives as to what "STEM Education" means, focusing on the separated perspective versus the integrated perspective. This is followed by a synopsis of major learning theories and paradigms, such as behaviorism, constructivism, social constructivism, and constructionism, which have informed and continue to underscore education research. Finally, the chapter concludes with a thick description of authentic, situated inquiry-based pedagogies like Problem- and Project-Based Learning, as well as their potential to foster connections and partnerships between industry clusters and K-12 schools.

A. J. Petrosino (✉)
Southern Methodist University, Dallas, TX, USA
e-mail: apetrosino@smu.edu

M. K. Sherard · S. A. Tharayil
The University of Texas at Austin, Austin, TX, USA
e-mail: mksherard@utmail.utexas.edu; sneha.tharayil@utexas.edu



C. Zintgraff et al. (eds.), *STEM in the Technopolis: The Power of STEM Education in Regional Technology Policy*,
https://doi.org/10.1007/978-3-030-39851-4_3

## 3.1 Introduction

The current volume argues for the development of STEM education programs based on the needs of cities and regions, with an emphasis on the collaborative benefits when schools work with industry, academia, government and nonprofits using topics of local importance (examples: transportation, equity, cyber security, or whatever is found in the local development plan). One method of achieving cooperation between industry, academia (K-16 education) and government is through authentic implementation of problem- and project-based learning in K-16 schooling. In this chapter, the authors explain how STEM education research supports this claim. A brief history of reform in STEM education is provided to contextualize the demand for and types of STEM education that have been conceived. Then, a discussion of emerged and emerging STEM education philosophies – namely constructivism, social constructivism, and constructionism – are provided to contextualize problem- and project-based instruction in their philosophical roots. Finally, a thick description of problem- and project-based instruction are provided. Examples are provided regarding how problem- and project-based instruction open schools to volunteers, mentors, and other organizations operating in a technopolis, and how the model enables connections to the local needs in a city or region.

## 3.2 A Brief History of Science, Technology, Engineering, and Mathematics Education

Science, Technology, Engineering, and Mathematics—or STEM as it is commonly known—has a relatively short but expensive and explosive history in United States education reform. This history must be understood to effectively explain how regional industry clusters can become *meaningfully* involved with K-16 (kindergarten to fourth-year college) STEM education. This section will outline a brief history of STEM education, compare the varying conceptions of STEM that exist, and discuss the role of *integrated* STEM education.

### *3.2.1 Panic Over Science and Mathematics Education*

To understand where the term STEM comes from and where it has gone (and is going), we could look back to many points in history; however, this chapter will begin in 1983 with the publication of *A Nation at Risk*. This (in)famous report produced by the U. S. National Commission on Excellence in Education stirred fear in the hearts of citizens with its warning of the "rising tide of mediocrity that threatens our very future" (1983, p. 1) related to student performance in science and mathematics in American schooling. To remedy the nation's slipping prominence on the global stage of innovation, the report recommended "a longer school year, more homework, tighter university admission standards, more testing for students as indicators of proficiency, higher standards for becoming a teacher, an 11-month professional year, and market-sensitive and performance-based pay." (Mehta 2015, p. 21). The nation at risk responded swiftly. In the year after the release of the report, six million copies of the report were sold, 250 state task forces were created, and the role of the federal government in education reform began expanding (Mehta 2015). The panic from *A Nation at Risk*, whether real or manufactured, resulted in an expansion of government bodies involved in improving science and mathematics education. In 1989 the American Association for the Advancement of Science (AAAS) created project 2061, which created a national set of goals for science, mathematics, and technology education in a report titled *Science for All Americans*. The National Council of Teachers of Mathematics (NCTM) and the National Science Teachers Association (NSTA) emerged in the late 1980's to bring practitioners to the table with regards to education reform, and created a number of documents which would become early national efforts to standardize science and mathematics education. In the two decades following *A Nation at Risk*, efforts to improve education in the United States would largely be focused on individual improvements to mathematics and science, separately. It wouldn't be until the early 1990's when the idea of integrating science, technology, engineering, and mathematics would come into the national discourse.

### 3.2.2 The Emergence of STEM

Before STEM, there was SMET. In the early 1990's the National Science Foundation used "SMET" to describe efforts to improve science-mathematics-engineering-technology education; however, the acronym caused dissonance amongst users and was converted to STEM (Sanders 2009). The STEM brand was quickly swept up by NSF, K-16 educators, educational researchers, and a variety of other stakeholders. Federal funding for STEM education surged, with $3.7 billion being allocated in 2011 and an additional $4.3 billion set aside for the *Race to the Top* initiative (Breiner et al. 2012). In the K-12 sector, much of these funds went to the creation of STEM academies; with a goal of having 1000 STEM academies by 2020 (President's Council of Advisors on Science and Technology, 2010). STEM academies are a loosely organized group of schooling strategies that intend to place STEM disciplines at the center of the school's curriculum. However, critiques of STEM academies discuss their role in gentrification of inner-city schools and furthering the gap in opportunities between dominant and non-dominant students in cities which utilize them (Bullock 2017).

Discussing the application of STEM funding, creation of STEM academies, and efforts to improve STEM education can seem like putting the cart before the horse. Understanding the many conceptualizations of or perspectives on STEM education are necessary to identify high leverage "hooks" for industry clusters to become involved. STEM education can be considered from a variety of angles depending on the stakeholder. Some of the many stakeholders are: governing bodies, industries, practitioners, the general public, and educational researchers (Breiner et al. 2012). However, efforts to improve STEM education usually fall into one of two camps: (a) separated perspectives on STEM; or (b) integrated perspectives on STEM.

Separated perspectives on STEM are when each subject is taught separately with the hope that the synthesis of disciplinary knowledge will be applied. This may be referred to as STEM being taught as "silos." This tradition often manifests itself in efforts target each individual letter of STEM; such as, efforts to increase STEM course enrollment, STEM degree attainment, or professional development for teachers in teaching content-specific courses. However, isolating the letters in STEM, or even isolating the disciplines within each letter (for example, chemistry and biology remaining separate courses), does not prepare learners to engage in the complex, interdisciplinary, and pernicious problems that face the twenty-first century. As evidence of this, Dr. Jo Boaler published a recent article in *Scientific American* highlighting the differing achievement amongst students who take the triennial test delivered by the *Program for International Student Assessment* (PISA). Results demonstrated that 15-yearolds in the United States students performed beneath other developed nations in mathematics. When analyzed to understand how students approach problems, students who had been taught to memorize rote strategies for particular types of mathematics performed the lowest; whereas, students who learned about the relationship between types of mathematics and engaged in complex problems performed higher.

At the other end of the spectrum, integrated STEM (iSTEM) seeks to bring the practices and structure STEM education closer to those of STEM professionals. In iSTEM, the principles of science and the analysis of mathematics are combined with the design process of technology and engineering in the classroom. The boundaries between traditional sciences, mathematics, and engineering or technology fields is intentionally blurred. iSTEM engages learners in complex and ill-defined problems to give learners a chance to identify their prior knowledge, understand the gaps in their knowledge, and iteratively engage in problem-solving and evaluation to constantly refine solutions. In iSTEM, learners engage authentically in real world problems, which lack the clear definitions of knowledge necessary to generate solutions, as might be portrayed in siloed approaches to STEM. Some key outcomes of iSTEM, presented to the National Academy of Engineering and National Research Council – Board of Science Education (Petrosino 2012) are:

- Using an interdisciplinary or integrated STEM curriculum provides opportunities for more relevant, less fragmented, and more stimulating experiences for learners.
- When done properly, integration of STEM brings together overlapping concepts and principles in a meaningful way and enriches the learning context.
- Learning situated in such enriched (macro) contexts often lead to meaningful learning experiences.

As noted in Chap. 1, this volume adopts the integrated view of STEM. In other chapters, classes operating in isolation are often referred to as "STEM-related"—for example, an individual mathematics course, or a traditional science course. STEM-related classes are still important, but STEM-related classes are not STEM, and in isolation, they do not come close to addressing STEM-related goals. If there is a pathway to connecting education, governing bodies, and industry; it is through integrating STEM in classroom learning, rather than separating the domains into their silos. The remainder of this chapter describes two models for classroom instruction, Problem- and Project-based instruction, which integrate STEM. The underlying educational philosophies will be described in brief detail. The chapter will culminate with a discussion of how the tenets of Problem- and Project-based instruction provide a context for the involvement of local and regional industry clusters.

## 3.3 Education Philosophies for Integrated STEM Learning

If asked to conjure an image of the classic classroom, the following might appear: rows of classroom desks, a chalk board at the front and center, a teacher standing at the front of the room, and students seated, silent, and facing the teacher. While some might describe this as a caricature, these fixtures still persist in modern classrooms around the world. Furthermore, these fixtures are relics of an education philosophy and design theory which dominated classroom teaching and learning for much of the past century: behaviorism and instructionism. The behaviorist perspective is most represented through the work of B. F. Skinner. One basic tenet of the behaviorist perspective is that all individuals have equal potential and the knowledge is equally accessible to all people. However, behaviorism purports that learning must be organized in a fashion that is strictly hierarchical, with the curriculum progressing from easy to difficult. The resulting education-design strategy, instructionism, leans heavily upon incremental tasks which build upon each other, repetition of practice, and the notion that one task must be mastered completely before conceiving of the next. While this philosophy and resulting design strategy isn't altogether useless for education; modern perspectives on knowing and learning suggest that learning is collaborative, non-linear, and relies heavily upon individuals and groups prior knowledge and experiences. While instructionist designs and behaviorist philosophies of learning have taken up considerable space in the minds of educators for some time, problem- and project-based instruction are historical examples of instructional models which leverage the social and constructed aspects of learning.

Problem- and Project-based instruction (also known as problem- and project-based learning or inquiry) are research verified models for instruction which open learners up to engaging in meaningful experiences with real world problems and stakeholders. While similar in lineage, Problem- and Project-based instruction have slight differences in their utility; however, both are deeply steeped in constructivist, sociocultural, and constructionist principles. Furthermore, neither instructional model is completely novel, rather, both can be traced back to over a century ago (Holm 2011; Buck Institute for Education 2007; NEA 2017; Kilpatrick 1918). Before diving into the tenets of both instructional models, we offer a brief history of constructivist, sociocultural, and constructionist education philosophies and discuss their relevance to the purpose of this book.

### *3.3.1 Constructivism in Education*

Problem- and project-based learning are best illustrated in the constructivist model of knowing and learning. Stemming primarily from John Dewey, Jean Piaget, and Jerome Bruner, constructivist perspectives on learning claim that understanding is actively pieced together as learners connect new knowledge from educational situations to prior experiences and understandings of the world. Distinct from behaviorism, constructivism places learners' prior knowledge and experience at the center of instruction. Some features of instructional design organized with constructivist principles in mind, are: (a) assessing student understanding of concepts before beginning instruction; (b) allowing students to encounter misconceptions and engage in experimentation to desettle ideas; and, (c) iterative assessment and reflection to understand how ideas are changing at the individual and class scale. The emphasis of process over product is found in constructivist-designed learning environments. Thinking and working with a defined scheme and understanding are valued over rules and correct answers. Instruction with a constructivist lens tends to be non-linear and often incorporates big ideas or major concepts into the curriculum before presenting details or heuristics. The use of discordant events to illuminate misconceptions to the students followed by cognitive disequilibrium promotes conceptual change on behalf of the student. As the student attempts to strike a balance between assimilation and accommodation, the role of the teacher is a guide or facilitator.

### *3.3.2 Social Constructivism in Education*

Lev Vygotsky, a Russian psychologist, is often credited with sociocultural theories of learning; however, much of his work was expanded upon post-mortem by scholars following in the sociocultural tradition. Sociocultural perspectives on learning highlight the role of cultural knowledge, cultural tools, and social interaction in constructing knowledge for individuals and groups. One of the perspective's central concepts, the zone of proximal development, attempts to explain the range of tasks that a group of learners can complete, and how this range can be expanded by the careful and directed use of scaffolding and social interaction during a learning experience. In instruction, as students demonstrate a capacity for more detailed information to proceed through the project, the teacher may provide relevant information on a need to know basis, so as to not overwhelm the student but to provide information when it will be perceived by the student as necessary. It is the classrooms' perception and involvement and construction of their own knowledge in the problem- or project-based instructional methods that ties into the constructivist and social constructivist theories of education.

Problem- and Project-based instruction lean heavily upon the group-situated nature of knowing and learning. Typical classrooms engaged in these instructional models will leverage interaction within small groups and between small groups. In this perspective, knowledge of the group is considered just as important as knowledge of the individuals.

### *3.3.3 Constructionism in Education*

Seymour Papert, a student of Jean Piaget's, is credited with the expansion of constructivist philosophies through the creation of a philosophy called *constructionism*. Constructionist perspectives on education are similar to constructivist perspectives in that learning takes place when groups of learners engage in collective meaning using personal experience and new information to construct mental representations of phenomena. However, it differs in that constructionist perspectives heavily value the creation of some tangible product (whether it be a working circuit board or a theoretical model of a phenomena) as evidence of one's learning. In this perspective, the physical manifestation of learning recycles to become a tool to engage with. For example, a learner can generate a simulation of how they believe an ecosystem works, but then use the simulation to test new ideas and evaluate their thinking.

Problem- and Project-Based Instruction lean heavily upon the creation of some artifact of to demonstrate the process and product of one's learning. Typical projects completed by students can be designs to solve some problem, prototypes of a technology, or models of how a phenomenon works.

## 3.4 Problem-Based Instruction

Though Project-Based Learning and Problem-Based Learning are distinct, the two have a shared history; with the former being borne out of the latter. Though problem-based learning as an instructional model first emerged during the 1950s and 1960s (Prince and Felder 2006), it rose to prominence during the 1980s and 1990s through the work of physician and scholar, Howard S. Barrows. During his tenure as a medical education professor at McMaster University in Ontario, Canada, Barrows documented his use and the effectiveness of "ill-structured problems" to help medical students develop their diagnostic skills (for a more thorough account of Barrows work and his principles of Problem-Based Learning, refer to Barrows (1980, 1985,

1986, 1996). Since then, Problem-Based Learning has garnered some prevalence in higher education, particularly within the fields of law, medicine, and engineering, among others.

As noted above, Problem-Based Learning is characterized by the use of "an open-ended, ill-structured, authentic (real-world) problem and [students] work in teams to identify learning needs and develop a viable solution, with instructors acting as facilitators rather than primary sources of information" (Prince and Felder 2006, p. 128). Thus, a core tenet of Problem-Based Learning is the use of realistic problem scenarios or cases that may emerge within the practice of a certain field or discipline as anchor for the curriculum. In so doing, the Problem-Based Learning pedagogy aims to fulfill the following educational objectives:

1. Structuring of knowledge for use in clinical (practical) contexts
2. The developing of an effective clinical (practical) reasoning process
3. The development of effective self-directed skills
4. Increased motivation for learning (Barrows 1986)

In his 1986 article in the *Medical Education* journal, Barrows identified a taxonomy of Problem-Based Learning models of instruction that vary in the degree to which the anchor problem is simulated or left open-ended, as well as the degree to which the curriculum is teacher-directed or self-directed. According to Barrow's taxonomy, the more student-driven and the greater the opportunity for free-inquiry the Problem-Based Learning unit is, the deeper the aforementioned educational objectives are addressed (Barrows 1986). Thus, the truest sense of Problem-Based Learning allows students to engage in the genuine inquiry of solving a realistic problem, encouraging them to employ the critical thinking and reasoning skills necessary to undergo the problem-solving process which typically might include the following stages: *naming* (fully defining the problem and the main issues within it); *framing* (establishing the boundaries and scope of the problem); *moving* (taking steps to begin solving the problem); and, *reflecting* (analyzing and evaluating steps or solutions attempted to solve the problem) (Prince and Felder 2006). An example of a problem-based scenario is presented. This scenario is found by the Case It! Project, a National Science Foundation (NSF)-funded project from the University of Wisconsin-River Falls to develop case-based scenarios for molecular biology learning. The example provided below[1] presents a case on infectious diseases, specifically HIV/AIDS (Berland 2018). Examples such as these illustrate how authentic and situated problem scenarios can be used to elicit meaningful inquiry and develop sophisticated problem-solving and reasoning skills at almost any developmental stage.

> Infectious diseases are caused by bacteria, viruses, or other pathogenic agents. Diagnosis may involve detecting the presence of proteins or nucleic acid from the suspected pathogen using ELISA or PCR, respectively. In some cases, the patient's blood will be tested for the presence of antibodies specific for a pathogen, as an indication that the person was previously infected with that agent.

[1] Source: This excerpt of a problem statement is taken from the Case It! Project. For more information, visit http://www.caseitproject.org or see the full citation for Berland (2018) in the References page at the end of this chapter.

*Background*: Human immunodeficiency virus (HIV) causes the disease Acquired Immunodeficiency Syndrome (AIDS). AIDS is characterized by the inability to mount an effective immune response to bacteria and other pathogens, resulting in a variety of life-threatening infections. The virus is spread when bodily fluids, such as blood and semen, from an infected person directly enter the bloodstream or tissue fluids of an uninfected person. For example, unprotected sexual intercourse and sharing needles during injected drug use can spread the virus. Once in the body, HIV infects and destroys certain white blood cells (called CD4 cells) and impairs the immune system. It may take years after the initial HIV infection for the symptoms of AIDS to appear. HIV infection is routinely detected indirectly, via tests which measure whether a person's blood contains antibodies against HIV; if so, they must have been previously infected with the virus. PCR can also be used to directly measure the amount of HIV in a person's blood or lymph nodes. For additional information about HIV/AIDS-related disease, detection methods, treatment, and prevention, visit the CDC's web site, www.cdc.gov/hiv.

These case scenarios are based on real people infected with HIV. Each case includes a video, accessed from the Case It web site, showing that person discussing their experience with the disease. As you study the cases, here are some general questions you might consider:

1. How did this person become infected with HIV?
2. Have others also been infected? Who else should be tested?
3. How reliable are the tests?
4. How often should someone be tested?
5. Why do people engage in risky behaviors?

## 3.5 Project-Based Instruction

Two seminal works related to Project-Based Instruction inform this chapter's conception of the pedagogy. The first is an early article in the *Journal of the Learning Sciences* written by Brigid Barron and colleagues from Vanderbilt University titled *Doing with Understanding: Lessons from Research on Problem- and Project-based Learning* (Barron 1998). The second is a chapter from the *Cambridge Handbook for the Learning Sciences* written by Joseph Krajcik and Phyllis Blumenfeld titled *Project-based Instruction* (Krajcik and Blumenfeld 2006). The formation of Project-Based Instruction has involved many changes and adaptations to academic curricula. As such, there have been numerous iterations of definitions of Project-Based Instruction. Nevertheless, a survey of the literature on Project-Based Instruction tend to point to at least seven essential elements that characterize Project-Based Instruction (Krajcik and Shin 2014; Marshall et al. 2010; Hasni et al. 2016):

- *A driving question or anchor*: A question, problem statement, or phenomenon that organizes and drives the project activities, providing a meaningful context through which students can explore and use disciplinary content and practices, and achieve learning goals. It should inherently encompass relevant content, mirror real-world situations, and provide continuity and coherence throughout the duration of the project (Krajcik and Blumenfeld 2006).

- *Focus on learning goals*: The Project-Based Learning unit is integral and central to the curriculum and is not only aligned with, but deeply integrates the core learning goals and content standards appropriate for the students' learning needs and trajectories (Thomas 2000).
- *Student participation in disciplinary practices*: Students engage in the same types of processes and authentic practices of experts in the domain or discipline of knowledge pertinent to the project context or the content (Krajcik and Shin 2014).
- *Collaboration and/or cooperation*: Students, teachers, and community partners work together to solve complex problems, mirroring the innately social aspect of expert problem-solving (Krajcik and Shin 2014).
- *Scaffolding*: Instructors and facilitators design and employ strategies, tools, and progressions to help students participate in activities that may be beyond their immediate abilities, but with appropriate support (scaffolding) are accessible and attainable (Reiser and Tabak 2014).
- *Use of cognitive or learning tools and technologies*: Cognitive tools refers to the use of visualizations, representations, and technologies which may facilitate learning and understanding (Krajcik and Shin 2014). A well-designed Project-Based Learning unit will not only provide opportunities for, but also inherently require the use of cognitive tools and technologies
- *A tangible, final product or artifact*: The culmination of a Project-Based Learning unit is marked by the production or deliverance of a tangible final product or artifact. The final artifact is appropriate and relevant to addressing the driving question or anchor phenomenon and often mimics the types of products or artifacts produced by experts or professionals in the discipline. These artifacts should be shared and publicly accessible and should be representative of student learning (Krajcik and Shin 2014).

Although not necessarily identifiable as distinct elements, some other overarching defining qualities of Project-Based Instruction include student-centeredness and authenticity of the project context (Blumenfeld et al. 2000; Thomas 2000; Kokotsaki et al. 2016). That is, project contexts should mimic realistic or worthwhile problems, challenges, or scenarios that allow students to actively and meaningfully engage in understanding, planning, questioning, learning, and addressing the driving question or problem.

## 3.6 Connections to Industry Clusters

Before discussing how industry clusters may partake in and promote STEM Education reform efforts, it is useful to first define the term "industry clusters." Coined by Michael Porter (1990), *industry clusters* refer to businesses, suppliers, and other similar institutions which function in a particular field, service, or discipline and which all occupy and/or operate within the same geographic region.

According to Porter, the clustering of industry players stimulates competition by increasing productivity of the companies in the region, driving innovation, and encouraging the development of new businesses and organizations.

These effects of industry clusters thus have some implications for partnerships with K-12 schools, especially in light of Problem- and Project- Based Instruction. Problem- and Project-Based Instruction lean heavily upon students actively constructing an understanding of a complex problem, defining parameters for working, and generating a solution. Central to both instructional models is the use of an anchoring phenomena or problem. How one selects these anchoring events is often up to the teacher, but herein lies a meaningful way for industry leaders and all those involved in development of local priority industry clusters to become involved in K-16 STEM education. When the actors working to understand and solve locally or regionally relevant problems were to engage K-16 educators and students, one may see an increase in interest and engagement while simultaneously enriching the learning environment and better preparing students for futures in STEM careers. Furthermore, practitioner-oriented literature on Project-Based Instruction typically discusses the role of an external audience for students to design for. For example, if students are engaging in a Project-Based Instruction unit to understand traffic in a city, it is useful and authentic to present their propositions to some *actual* stakeholders, like a city council member or civil engineering firm.

The cases in this volume present multiple proposed examples. In the language of this chapter, the cases from Medellín, Colombia are a mix of Problem- and Project-Based STEM+Arts (STEAM) Instruction. The case of CyberPatriot from San Antonio illustrates a Problem-Based competition operating at city-wide scale, and it introduces the idea of mentor-based learning, reflecting the depth to which industry cluster mentors are integrated in programming. The Fourth Industrial Revolution case from Taiwan literally engages high school teachers and students in the development and execution of national strategy. Constructionism is explicitly seen in the cases of Medellín; Taiwan; São Carlos, Brazil; Fundão, Portugal; Querétaro, Mexico, and China, and is likely present in all cases in the volume.

## 3.7 Conclusion

In summary, Problem- and Project-Based Instruction are existing and well researched instructional models which support collaboration between governing bodies, industry, non-profits and education. Making connections between these spheres and engaging learners in locally or regionally relevant problems provides a platform for integrated STEM education and builds capacity for the future generation of the STEM workforce.

**Acknowledgement** This work was partially supported by awards CNS 1837687 and DRL 1615207 from the National Science Foundation. We thank our colleagues Cliff Zintgraff and Paul Resta from The University of Texas and Walter Stroup from the University of Massachusetts-

Darthmouth who provided insight and expertise that assisted in the preparation of this chapter. While appreciative of the expertise and guidance of our colleagues, the interpretations and conclusions of this paper remain our own.

## References

Barron, B. J. S. (1998). Doing with understanding. *The Journal of Learning Sciences, 7*(3&4), 271–311.

Barrows, H. S. (1985). *How to Design a Problem-based Curriculum for the Preclinical Years*. New York: Springer.

Barrows, H. S. (1986). A taxonomy of problem-based learning methods. *Medical Education, 20*(6), 481–486.

Barrows, H. S. (1996). Problem-based learning in medicine and beyond: A brief overview. In L. Wilkerson & W. H. Gijselaers (Eds.), *Bringing problem-based learning to higher education: Theory and practice: New directions for teaching and learning* (Vol. 68, pp. 3–12). San Francisco: Jossey-Bass.

Barrows, H. S., & Robyn M. Tamblyn, B. S. N. (1980). *Problem-Based Learning: An Approach to Medical Education*. Springer Publishing Company.

Berland, M. (2018, May). Resource manual for Case It! Version 7. Retrieved August 24, 2019 from http://www.caseitproject.org/wp-content/uploads/2012/07/RM2017.html.

Breiner, J. M., Johnson, C. C., Sheats-Harkness, S., & Koehler, C. M. (2012). What is STEM? A discussion about conceptions of STEM in education and partnerships. *School Science and Mathematics, 112*(1), 3–11.

Blumenfeld, P., Fishman, B. J., Krajcik, J., Marx, R. W., & Soloway, E. (2000). Creating usable innovations in systemic reform: Scaling-up technology-embeded project-based science in urban schools. *Educational Psychologist, 35*, 149–164.

Buck Institute for Education (2007). *Handbook: Introduction to project-based learning*. Retrieved July 24, 2018 from http://www.bie.org/images/uploads/general/20fa7d42c216e2ec171a212e97fd4a9e.pdf.

Bullock, E. C. (2017). Only STEM can save us? Examining race, place, and STEM education as property. *Educational Studies, 53*(6), 628-641.

Hasni, A., Bousadra, F., Belletête, V., Benabdallah, A., Nicole, M.-C., & Dumais, N. (2016). Trends in research on project-based science and technology teaching and learning at K–12 levels: A systematic review. *Studies in Science Education, 52*(2), 199–231. https://doi.org/10.1080/03057267.2016.1226573.

Holm, M. (2011). Project-Based Instruction: A review of the literature on effectiveness in prekindergarten. *River Academic Journal, 7*(2), 1–13.

Kokotsaki, D., Menzies, V., & Wiggins, A. (2016). Project-based learning: A review of the literature. *Improving Schools, 19*(3), 267–277.

Krajcik, J., & Blumenfeld, P. (2006). Project-based learning. In R. K. Sawyer (Ed.), *The Cambridge handbook of learning sciences* (1st ed., pp. 317–333). New York: Cambridge University Press.

Krajcik, J. S., & Shin, N. (2014). Project-based learning. In R. K. Sawyer (Ed.), *The Cambridge handbook of learning sciences* (2nd ed., pp. 275–297). New York: Cambridge University Press.

Kilpatrick, W. H. (1918). The project method: The use of the purposeful act in the education process. *Teachers College Record, 19*, 319–335.

Marshall, J., Petrosino, A., & Martin, T. (2010). Preservice teachers' conceptions and enactments of project-based instruction. *Journal of Science Education and Technology, 19*(4), 370–386. Retrieved August 24, 2019 from http://www.jstor.org/stable/40864010.

Mehta, J. (2015). Escaping the shadow: "A nation at risk" and its far-reaching influence. *American Educator, 39*(2), 20–26.

National Education Agency (2017). "Research spotlight on project-based learning." *National Education Agency: Great Public Schools for Every Student*. Retrieved July 24, 2018 from http://www.nea.org/tools/16963.htm.

Petrosino, A. J. (2012). *iSTEM and learning outcomes. Guest lecture to the committee on integrated STEM education*. Washington, DC: National Academy of Engineering.

Porter, M. E. (1990). *The competitive advantage of nations*. New York: The Free Press.

Prince, M., & Felder, R. M. (2006). Inductive teaching and learning methods: Definitions, comparisons and research bases. *Journal of Engineering Education, 95*(2), 123–138.

Reiser, B. J., & Tabak, I. (2014). Scaffolding. In R. K. Sawyer (Ed.), *The Cambridge handbook of learning sciences* (2nd ed., pp. 44–62). New York: Cambridge University Press.

Sanders, M. (2009). STEM, STEM education, STEMmania. *Technology Teacher, 68*(4), 20–26.

Thomas, J. W. (2000). *A review of research on project-based learning*. San Rafael: The Autodesk Foundation. Retrieved August 24, 2019 from http://www.bobpearlman.org/BestPractices/PBL_Research.pdf.

United States. National Commission on Excellence in Education. (1983). *A nation at risk: The imperative for educational reform: A report to the Nation and the Secretary of Education, United States Department of Education*. Washington, DC: The Commission: [Supt. of Docs., U.S. G.P.O. distributor].

# Chapter 4
# The STEM Technopolis Wheel: In Motion Through STEM Learning

**Donna K. Kidwell, Cliff Zintgraff, and Gregory P. Pogue**

**Abstract** This chapter explores the history, current state and possible futures for STEM learning, starting in K-12 schools, and throughout a lifetime, in the context of the technopolis model (Smilor, Gibson and Kozmetsky, *J Bus Ventur, 4*(1):49–67, 1989), and in the context of the current volume, STEM in the Technopolis: The Power of STEM Education in Regional Technology Policy. This chapter does not focus on a specific geography. Instead, 31 years after publication of the technopolis model/wheel, this chapter reconsiders the original technopolis wheel against trends in the future of work and learning. We celebrate the wheel as a mechanism for moving regions forward, and as one that anticipated the role of STEM before the term existed, and one that planted seeds for primary and secondary schools (K-12) in economic development strategy. The authors suggest that STEM learning is torque for the technopolis wheel. A STEM Technopolis Model is presented that reflects the role of all STEM learning in city/region-based technology and economic development.

## 4.1 Background: The Technopolis Model/Wheel

The original technopolis model (Smilor et al. 1989), illustrated in Fig. 4.1, considered the role of the research university, the need for purposeful and proactive policies by all levels of government, the catalytic capability of large companies, the vibrant need for local companies, and the need for all of these actors to collaborate in a coordinated approach to developing strong economic regions. The model recognized the need for research universities to provide talent through graduate students and scholars and to provide the seeds for spin-out companies that could foster

D. K. Kidwell (✉)
Arizona State University, Phoenix, AZ, USA
e-mail: Donna.Kidwell@asu.edu

C. Zintgraff · G. P. Pogue
The University of Texas at Austin, Austin, TX, USA
e-mail: cliff.zintgraff@utexas.edu; gpogue@ic2.utexas.edu

 
C. Zintgraff et al. (eds.), *STEM in the Technopolis: The Power of STEM Education in Regional Technology Policy*,
https://doi.org/10.1007/978-3-030-39851-4_4

**Fig. 4.1** The technopolis model/wheel (Smilor et al. 1989)

economic opportunity through job creation, wealth generation, and economic diversification. It encouraged the concept that these companies start their revenue engines off with fledgling universities contracts and encouraged an ongoing relationship with the university.

On the education front, the model recognized that government at the local, state and federal levels had a responsibility to help nurture programs. One sub-spoke of the wheel, State Government-Educational Support, acknowledges U.S. state governments' responsibility for education. Although the 1989 article does not deeply explore that sub-spoke, it is not far reaching to imagine education and training as part of the economic development system envisioned, including K-12 (primary and secondary[1]) schools, community colleges and workforce educational initiatives.

In fact, we see this inclusion of K-12 in the biography of George Kozmetsky, founder of the $IC^2$ Institute at The University of Texas at Austin. The $IC^2$ Institute is an economic development think tank, and was the laboratory for development of the technopolis model. George Kozmetsky himself was an active innovator in creating

[1] K-12 is a common shorthand in the U.S. for grades kindergarten through 12th grade. The term is used in this volume as a shorthand for primary and secondary schools in any location, U.S. or around the world.

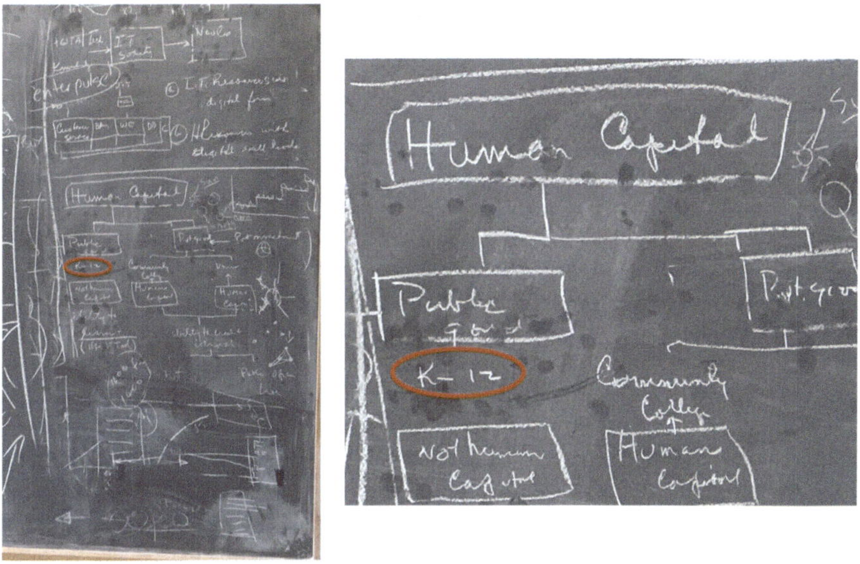


**Fig. 4.2** Kozmetsky's inclusion of K-12 in economic development

new educational models that would provide torque for the technopolis wheel. He established a considerable number of catalytic programs within the IC$^2$ Institute that pushed economic development forward in Austin, essentially twisting the wheel to provide forward motion that resulted in impressive economic impact.

In his biography of Kozmetsky, Jones (2018) documented examples of high school entrepreneurship training and job training being pursued actively by the IC$^2$ Institute. In addition, the Cross Border Institute for Regional Development (CBIRD) (Gibson et al. 2003), an effort overseen by Kozmetsky in the early 2000s, included a full section on the design of secondary school programs. Most telling, a preserved example of Dr. Kozmetsky's famous white board sessions makes clear that he saw K-12 as a standard part of the community's economic development engine; in other words, as part of the technopolis. Figure 4.2 shows Kozmetsky's preserved drawing, still present in one of his old IC$^2$ Institute offices.

However, the IC$^2$ Institute did not focus consistently on K-12 integration into the technopolis. They yielded to others with greater K-12 knowledge and capacity. Based on available documentation, they only revisited these ideas seriously in the ELT workforce training program and CBIRD project, both projects late in Kozmetsky's life. It was during or shortly after these projects that Judith Ramaley, a Director in the National Science Foundation Division of Education and Human Resources, re-cast the acronym SMET (science, mathematics, engineering and technology) as *STEM*, more pleasing to the ear, and capturing "the relationship among the subject areas… science and math serve as bookends for technology and engineering…STEM weaves those elements…into all aspects of education" (Christenson 2011, para. 6).

Further exploration sheds more light on the kinds of learning that need to happen in a technopolis. Below are more examples from K-12, from workforce training, and of education experiences for undergraduate and graduate students.

## 4.2 Entrepreneurship Education in K-12

Jones (2018) documented Kozmetsky's early support for high school entrepreneurship education. Again Kozmetsky was prescient, as entrepreneurship education has grown dramatically in K-12 settings: 42 states have standards, guidelines or proficiencies in entrepreneurial education, and 18 states require entrepreneurship education in high school (Junior Achievement n.d.). Thomas Friedman provided a complementary perspective, noting traditional jobs are declining, and that schools need to produce students who are ready for innovation (Friedman 2013). Students must develop critical thinking, collaboration and communication skills that are necessary for graduates to create their own jobs, often throughout their careers.

Can students receive an effective entrepreneurship education experience without the involvement of industry and other professionals from outside schools? Students can learn about entrepreneurship, but they cannot do entrepreneurship or deeply understand the experience without the involvement of experienced entrepreneurs. To the extent students are actually trying to start a business, either for the experience of it or for real, collaboration with other technopolis sectors is required. Students in these classes, and especially entrepreneurially minded students, should be linked to professionals who can fill gaps in management, market access, development and other skills.

## 4.3 Workforce Learning in the Technopolis: E-Learning and Training Labs

In 2002 the IC$^2$ Institute embarked on a new model for educating students in the growing Austin ecosystem, trying to help all Austin residents participate in the region's growing economy. To drive this new model, the institute launched E-Learning & Training Labs (ELT Labs) (Harris 2002). ELT Labs assembled transdisciplinary teams from academia, business and all levels of government to research and develop leading-edge workforce development, training and learning programs. They focused on programs designed to primarily help at-risk youth by teaching employability skills such as responsibility, dependability, positive work attitudes, teamwork and communications skills that result in higher productivity and retention rates. Jones (2018) reported that the program was deployed first for adult job training, and then in area high schools.

The transdisciplinary approach was almost always the preferred approach in IC$^2$ Institute programs that were launching into major new areas. The term was understood to mean something much more than multidisciplinary (each group providing a part) or even interdisciplinary (cross-functional teams working together), but rather that a new mini-discipline of sorts was being created at the intersection of the fields involved (Zintgraff et al. 2013). The mini-discipline might require new theories and practices for success. High respect was maintained for existing disciplines, but no dogma existed preventing the adoption of whatever philosophies, theories and practices were deemed necessary. The transdisciplinary idea helps one think beyond common perceptions about the degree to which K-12 should be working side-by-side with other technopolis actors.

## 4.4 University Student Learning in the Technopolis: Aspiring Women Entrepreneurs

In an interconnected world, learning the 4 Cs—communication, collaboration, critical thinking and creativity—may involve these activities across country borders. Language, context and culture add a serious new dimension to the idea of learning to work in a team. The 4 Cs cannot be learned via lecture; likewise, international team skills can only be learned through experience.

A recent IC$^2$ Institute program exercised this approach. As a spinoff of the institute's Nexus Startup Hub@American Center in New Delhi, India, a new program was created, Aspiring Women Entrepreneurs (AWE) (IC$^2$ Institute 2018). Students were selected from three locations: India, Sweden, and Austin. The students from India came from Delhi, Chennai, Hyderabad, Pune, Kolkata, Ahmedabad, and Jaipur, India. They represented 15 universities. The student age range was 19–31 with nine pursuing bachelor's degrees, five master's degrees and five pursuing Ph.D. degrees. The innovations spanned multiple target markets, including consumer products, clean air and water technologies, and apps for educational and social engagement.

Students completed two phases. First, students completed the Innovation Readiness Series™ (IR) program, an eLearning course that provides the necessary steps for entrepreneurs, engineers, researchers, and scientists to evolve their technology-based concepts into a viable product or service. The course includes 10 modules on key topics in commercialization with activities and a 70+ page eBook. Second, students traveled as a team to the other countries, formed small teams, and researched products that might be sold in the country's market.

The AWE program represents the kind of deep learning that can be transformational for students, with principles learned and then applied in practice and in real context. The program also highlights the administrative complexity and general expense of deep learning programs. Not every program requires a traveling international team, but as a general rule, rich experiences also require rich administration and budget. How to bring programs like these to scale, or to find a balance between deep results and scalability, is a key question that almost always arises for transdisciplinary teams from a technopolis. Still, it is this kind of learning that turns the technopolis wheel.

## 4.5 Graduate University Learning: MS in Science and Technology Commercialization

Perhaps the best example of technopolis learning is embodied by the $IC^2$ Institute's Masters of Science in Science and Technology Commercialization (MSSTC) program. Launched in 1996, the program focused on graduate entrepreneurship education, and it also focused on all the factors that encourage commercialization in an ecosystem. The technopolis model was a core model used in the program. The program drew from multiple colleges on campus, business, engineering and law. The program was thirteen months in calendar length, running every other weekend in a format suited to the needs of learners already engaged in careers. It was also one of the first programs that welcomed online learners, supporting a truly global cohort.

The program had innovative curricula built through the collaboration of faculty from across colleges. It was designed with a project- and team-based approach that modeled the entrepreneurial and commercialization journey, the same journey for which students expected to become proficient. Two of the current authors completed MSSTC. The program used a forward thinking assessment model, where students took a total of two traditional examinations during the course. All other coursework was project-based, most team-based, some with international colleagues, and all requiring the 4 Cs skills in abundance.

## 4.6 The Future of Learning and Work

Deep learning has long been part of technopolis-oriented experiences. To the extent these experiences involves the STEM fields, learning happens in a cross-disciplinary (or greater) way, and in an integrated way. This kind of learning is not new. But what is the future of learning and work?

The nature of work is changing rapidly. Across learning at all levels, the rapid pace of technology change is straining the traditional models of education, both academic and workforce training, as well as the policies supporting those systems. In the US, debates on the tension between the relevance of a Bachelors degree versus the cost of tuition are topics of news and policy debates. Moody's Investor's Service (Moody's 2018) forecasts that 2019 will continue a negative financial outlook as tuition growth is greatly outpaced by expenses on campuses.

Meanwhile, we are seeing massive calls around the need for a talented and skilled workforce. The European Commission estimates 756,000 unfilled jobs in the European ICT sector by 2020 (Palotai 2017). The pace of new technologies creates a tumult of new job categories, with new skill combinations and new roles that are not readily anticipated by the traditional pace of educational and training curriculum.

The nature of work is changing rapidly, as employers find new ways to broker between talent and business needs. This has led to the rise of freelance economy demonstrated in the 57 million freelance Americans (Upwork 2017) and the 3 million employed globally by Uber (2019). Alongside this, we are seeing an increasing awareness of the need to consider the frameworks and support needed for a citizen whose future will require a lifelong ability to learn.

Harvard University, University of Washington and others have engaged in a deliberate effort to examine and create models for a 60 Year Curriculum (Branon 2018). They recognize that longer lifespans and quickly changing jobs will require thinking in a broader notion about learning over a lifetime. This conversation is emerging alongside efforts across the US to blend pathways between high schools, community colleges and universities. It's also driving conversations around the unbundling of traditional curricula into models that can be stackable, more transportable across institutions, and more readily converted into credentials that learners can use in the workforce as quickly as they acquire them.

At Arizona State University, President Michael Crow is leading a campus and community spanning effort to design a new model for a learner where they can learn anything, any time, and anywhere (Faller 2018). In keeping with ASU's charter, the campus is being asked to rethink the curriculum, processes, technology and policies that were designed for a model that no longer serves the modern learner and the workforce they are engaged in. ASU's experiments with deep company partnerships are efforts to co-design this future with companies. As of early 2019, over 10,000 Starbuck's employees are getting their degrees through their partnership with ASU, with over 2,000 graduates since June of 2014.

Uber's partnership with ASU allows drivers to pursue a degree while driving, or alternately, drivers can choose to share that educational offer with a family member. Uber is particularly interested in helping learners that often fall between the cracks of traditional systems. For example, some current students are immigrants whose engineering degrees are not commonly recognized in the U.S. While driving with Uber, they can transfer to ASU, complete an American degree, and engage their talents to full capacity. Uber recognizes that they can address the challenges of diversity and ongoing demand of a technology enabled workforce through unique partnerships with universities such as ASU.

In both cases, the companies are dedicating significant internal efforts to the partnership. More than just identify a pathway to an education, these partnerships are designed with a deep understanding of the day to day life of the learner alongside the workforce and human capital needs within the company.

One can also see these types of customization and pipeline trends at the K-12 level and with a focus on industry clusters of local importance. An industry cluster is a collection of companies, vertically or horizontally integrated, all working in the same fundamental field and serving the same overall market (Porter 1990). The Hollywood film industry and Silicon Valley computer industry are examples of industry clusters.

Through the lens of the technopolis model, the industry clusters also include the academic, government, and non-profit players who educate, set policy, and provide advocacy and support for the cluster. Most current K-12 programs are not built around clusters. However, many of the best examples in this volume are. Among the examples in this volume are agriculture in Brazil (Gattaz, Falvo and Cruvinel, this volume), cyber security in San Antonio (Sánchez and Zintgraff, this volume), advanced manufacturing in Taiwan (Yang, Yang, Chou, Wei, Chen and Kuo, this volume), and ICT in Fundão, Portugal (Aguiar and Pereira, this volume). In those programs, the future of learning and connection to work is embedded in designs that leverage close relationships between K-12 schools and local priority industry clusters.

## 4.7 Fully in the Cluster: K-12 as Development Collaborators for Learning Products

Kozmetsky was a thought leader in understanding the potential for education to provide torque to the technopolis wheel. Various initiatives have focused on new models for pedagogies and delivery modalities to highly actionable learning experiences. Kozmetsky reflected that:

> …education is knowledge an individual must convert to value; whereas, training is knowledge with immediate market value. ($IC^2$ Institute, 2019)

He recognized that a combination of skills alongside a deep and comprehensive education enabled talent to immediately contribute to new technology firms. As Kozmetsky tended to do, he saw not only a linear progression from K-12 student to employee, but a system where K-12 schools are full collaborators in technopolis development. To understand his perspective, it is important to understand what Kozmetsky (1990) saw as some of the key relationships in any community advancing technology-based development:

1. Universities working with private and public institutions to move innovations toward markets.
2. Aspiring entrepreneurs getting connected to capital, mentorship, markets, etc.
3. Giving entrepreneurs and their partners help—entrepreneurship training and support--in making market-ready products.
4. Monitoring outcomes, including economic performance, market access and job creation.

Earlier, the authors mentioned some of the efforts made by the $IC^2$ Institute in working with schools. Kozmetsky viewed these efforts through a systems lens. He viewed both K-12 schools and vocational schools as consumers and collaborators (Jones

2018, p. 309). Schools should train and educate students, and they should work side-by-side with industry and other partners to integrate technology, develop new products, and play a role in making those products market ready.

In fact, this case can be seen in the current volume. In Medellín, Colombia, the Ruta N Innovation Agency, the agency responsible for implementation of the city's science, technology and innovation plan, is doing exactly the collaboration described (Roldán, this volume). The agency has facilitated development of a small cluster of educational technology companies. These companies are fully integrated with K-12 staff and agency supporters who are building STEAM programs for the community. The companies are developing related products to sustain the development efforts, and in the process, new entrepreneurial activity is taking root in the community.

Thus, the concept of innovation diffusion from educational entities that Kozmetsky championed in the 1970s (Jones 2018) has extended from primarily university settings to that of K-12 education. One can take this thought even further. In this volume's chapter on industry clusters, Zintgraff, Han and Butler (this volume) argue that K-12 can legitimately be viewed as more than a collaborator with industry clusters, but as a part of industry clusters. How is that possible? What does K-12 have to do with private industry?

But in the technopolis view of the world, an industry cluster includes stakeholders from all sectors. If K-12 is in close partnership with industry, with knowledge diffusing both directions, and with collaborative development of in-school and out-of-school products and services underway, with those products and services developing the talent pipeline that fuels the cluster, then K-12 is important to the cluster in the same manner as government policymakers or supportive non-profits.

For sure, this is not often the case. But in several cases in this volume, and notably in Roldán's (this volume) chapter on Medellín, Colombia, this is exactly the case. Regardless of how one might debate the definitions, these cases show K-12 in tight alignment with technopolis goals, working with other stakeholders in a transdisciplinary way.

## 4.8 The STEM Technopolis Model/Wheel

### *4.8.1 Illustration*

The ideas of this chapter are modeled in the STEM Technopolis Wheel. The modified model was introduced in Zintgraff, Han and Butler (this volume), and is shown in Figure 4.3 with an additional *flywheel* adaptation. The figure is otherwise a simplified and adapted version of the original technopolis model/wheel from Smilor et al. (1989). This model is also introduced in this volume's chapter on industry clusters. The following changes have been made to the model.

- A dedicated spoke has been added for primary and secondary education. The authors argue that K-12 education must be an equal participant in a robust technopolis.

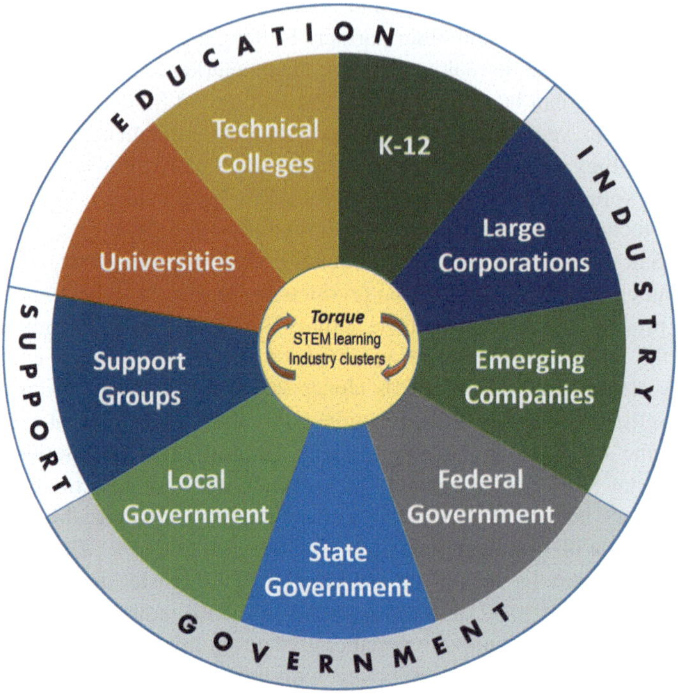


**Fig. 4.3** The STEM technopolis wheel: STEM learning as torque. Adapted from Smilor et al. (1989)

- A dedicated spoke has been added for technical colleges and the STEM education that is part of their generally workforce-driven efforts. The argument for this spoke is made in Zintgraff, Han and Butler (this volume).
- Detail has been removed from the spokes in favor of readability and clarity. The deeper purpose is accessibility to an audience that spans from researchers to policymakers to classroom practitioners.
- Sources of torque have been added to the wheel: industry clusters, and STEM learning. This change addresses the underappreciated and often overlooked element of the model: that it is meant to be a wheel.

### *4.8.2 Extending the Metaphor: A Flywheel for Economic Development*

Readers may be familiar with the workings of a flywheel. For those who are not, imagine the simple example of a spinning top. The energy one imparts by twisting of fingers, hand and wrist give momentum to the top. In a vacuum, with no friction, the top maintains momentum permanently. STEM learning, along with grounding STEM experiences in priority industry clusters, imparts momentum to the flywheel.

**Fig. 4.4** The STEM technopolis flywheel metaphor: Momentum for the technopolis

This metaphor, illustrated in Fig. 4.4, is another way to understand the STEM Technopolis virtuous cycle in action.

## 4.9 Conclusion

The concept of the technopolis wheel has served as a strong model for regional economic development. The model has provided a thoughtful way to envision partnerships in a technopolis, and it helps stakeholders see opportunities across sector boundaries. The authors suggest that by adding K-12 as a spoke (sector), the technopolis wheel is augmented to reflect three important ideas.

1. K-12 can work with universities, all three levels of government, and industry to create the talent needed to drive forward to the future. Rather than expecting K-12 to feed the system while operating in a silo, a holistic system is at work. Young people are exposed to the opportunities and challenges of industry, in age-appropriate ways, from their very first formative experiences.
2. Institutions dedicated to learning are worth a 25% slice of the technopolis wheel, and learning continues throughout a person's lifetime.
3. Learning is foundational to a vibrant economic community and is becoming more critical as the nature and future of work changes. Educational policy should ensure opportunities are open across sector boundaries and that roadblocks are removed for interaction between K-12, universities, and other sectors. When policies are well conceived, educational institutions can work effectively with industry to create engaging learning experiences. Through those experiences, students actively participate in a region's growth.

The authors also tried to re-capture the idea of the *turning* technopolis wheel. The rotation of the wheel is a wonderful metaphor, and the metaphor of a flywheel helps technopolis actors remember that coordinated effort imparts momentum that lasts beyond immediate actions. The release of stored energy from regional assets sustains economic progress. Two energy sources have been identified: priority industry clusters from a region's areas of excellence, and STEM learning. Thoughtfully designed STEM programs engage many spokes of the wheel, mobilizing regional assets to collaborate and create ongoing motion and impact. Such programs create great learning experiences for students of all ages. The inevitable result of great learning experiences is forward motion in a technopolis.

## References


Branon, R. (2018, November). Learning for a lifetime Retrieved July 28, 2019 from https://www.insidehighered.com/views/2018/11/16/why-longer-lives-require-relevant-accessible-curricula-throughout-long-careers.

Christenson, J. (2011, November). Ramaley coined STEM term now used nationwide. Retrieved July 28, 2019 from http://www.winonadailynews.com/news/local/ramaley-coined-stem-term-now-used-nationwide/article_457afe3e-0db3-11e1-abe0-001cc4c03286.html.

Faller, M. B. (2018, March). ASU's vision of future: Learning across lifespan — anytime, anywhere, any age Retrieved July 28, 2019 from https://asunow.asu.edu/20180301-creativity-asu-crow-community-conversation-lifelong-learning-future.

Friedman, T. (2013, March). Need a job? Invent it Retrieved July 28, 2019 from https://www.nytimes.com/2013/03/31/opinion/sunday/friedman-need-a-job-invent-it.html.

Gibson, D. V., Rhi-Perez, P., Cotrofeld, M., De Los Reyes, O., & Gipson, M. (2003). *Cameron County/Matamoros at the crossroads: Assets and challenges for accelerated regional and binational development*. Austin: IC2 Institute.

Harris, B. (2002, May). IC$^2$ Institute launches e-learning and training collaboration Retrieved July 28, 2019 from https://www.govtech.com/e-government/IC2-Institute-Launches-E-Learning-and-Training.html.

IC$^2$ Institute. (2018, May). Aspiring women entrepreneurs program kicks off in New Delhi Retrieved July 28, 2019 from https://ic2.utexas.edu/news/aspiring-women-entrepreneurs-kickoff/.

IC$^2$ Institute. (2019). George Kozmetsky Retrieved July 28, 2019 from http://ic2.utexas.edu/about/mission-and-history/george-kozmetsky/.

Jones, M. (2018). *A civic entrepreneur: The life of technology visionary George Kozmetsky*. Austin: Dolph Briscoe Center for American History at the University of Texas at Austin.

Junior Achievement. (n.d.). The states of entrepreneurship education in America Retrieved July 28, 2019 from https://www.juniorachievement.org/documents/20009/20652/Entrepreneurship+standards+by+state.pdf/494b5b34-42a2-4662-8270-55d306381e64

Kozmetsky, G. (1990). Economic growth through technology: A new framework for technology commercialization. In D. V. Gibson, G. Kozmetsky, & R. W. Smilor (Eds.), *The technopolis phenomenon*. Austin: IC2 Institute.

Moody's. (2018, December). Research announcement: Moody's – US higher education outlook remains negative on low tuition revenue growth Retrieved July 28, 2019 from https://www.moodys.com/research/Moodys-US-higher-education-outlook-remains-negative-on-low-tuition%2D%2DPBM_1152326.

Palotai, P. (2017, May). E-skills and jobs in the digital age. European Commission Retrieved July 28, 2019 from https://ec.europa.eu/epale/fr/content/e-skills-and-jobs-digital-age.

Porter, M. E. (1990). *The competitive advantage of nations*. New York: The Free Press.

Smilor, R. W., Gibson, D. V., & Kozmetsky, G. (1989). Creating the technopolis: High-technology development in Austin, Texas. *Journal of Business Venturing, 4*(1), 49–67.

Uber. (2019). Company info Retrieved July 28, 2019 from https://www.uber.com/en-PK/newsroom/company-info/.

Upwork. (2017, October). Freelancers predicted to become the U.S. workforce majority within a decade, with nearly 50% of millennial workers already freelancing, annual "freelancing in America" study finds. Retrieved July 28, 2019 from https://www.upwork.com/press/2017/10/17/freelancing-in-america-2017/.

Zintgraff, C., Green, C. W., & Carbone, J. N. (2013). A regional and transdisciplinary approach to educating secondary and college students in cyber-physical systems. In S. C. Suh, U. J. Tanik, J. N. Carbone, & A. Eroglu (Eds.), *Applied cyber physical systems*. New York: Springer.

# Chapter 5
# Moving Toward Digital Equity in the Technopolis

**Paul E. Resta**

**Abstract** While the vast majority of residents in cities have access to the internet at home, many remain under-connected due to limited data or speed, cost barriers, insufficient digital devices or lack of tech support or digital skills. This negatively impacts the economic mobility, academic performance and social inclusion of low-income students and families. This chapter provides examples of how six cities in the United States have successfully addressed these challenges. All six cities meet the criteria of a technopolis, and each leveraged the strength offered by a technopolis through the development of innovative local policies and collaborative community action. The chapter also describes how policies developed at the state and national level may support technopolis cities in their efforts to move toward digital equity.

## 5.1 Introduction

The technopolis represents a major force for innovation and development in today's technology and knowledge-based global society. Across the globe, we have witnessed the power of information and communication technologies (ICT) to accelerate science and industry but we have seen that it can also contribute to growing disparities and the digital exclusion of peoples in society. Although efforts have been made to increase access to technology in schools and communities, today we find that progress has plateaued in the United States (Pew Report). These disparities are often most marked in the technopolis where the ICT plays a dominant role in generating wealth and prosperity for those with digital knowledge, skills and access to ever advancing digital technology. In contrast many low income and minority families in the technopolis are without access to a device and broadband in the home. As noted in the Industry Trade report (Molla, 2017), over 62 million urban Americans don't have access to or cannot afford broadband and Internet.

P. E. Resta (✉)
College of Education, The University of Texas at Austin, Austin, TX, USA
e-mail: resta@austin.utexas.edu

 
C. Zintgraff et al. (eds.), *STEM in the Technopolis: The Power of STEM Education in Regional Technology Policy*,
https://doi.org/10.1007/978-3-030-39851-4_5

The economic and educational implications for the low-income families in these cities are profound in that without digital access and skill, parents can't find, apply or qualify for even low-paying jobs. The children are also disadvantaged in their lack of access to knowledge resources and digital tools to support their learning. Today over 70% of teachers in our schools assign homework assignments that draw upon Web resources (McLaughlin, 2016). Students without digital access face what is known as "the homework gap." This gap disadvantages students in all subject areas but may be most disadvantaged related to courses in science, technology, engineering and mathematics.

In the United States, there is growing recognition that the country is falling behind other countries in digital access. At present the U.S. ranks 25th among countries in broadband speed (DSL Reports 2015). The recognition that the U.S. is falling behind other countries has led to both increased awareness as well as efforts at the federal, state and local level to move toward digital equity. Such efforts, however, will require new policies and initiatives at the national, state and local level.

This chapter is focused on understanding what constitutes the digital divide and the issues and challenges confronting policy-makers in addressing the significant differences in the access and use of technology based on race and socioeconomic status. It identifies initiatives that have or are being taken at the federal, state and local level to address the digital divide and offers recommendations that may help the technopolis in moving toward the goal of digital equity.

## 5.2 What Is Digital Divide?

Many consider the digital divide simply in terms of lack of access to a device and connectivity. However, it is increasingly recognized that this is too narrow a definition (Gorski 2009). A more current view is that the digital divide exists when a group's access to digital technologies and resources differs based on a group's race, socioeconomic status, or national identity (Light 2001). Access to a device and connectivity represents the first level divide. But access without the digital skill to use these resources represents an even more significant barrier confronting many low income families and elders and deprives them of many of the benefits in the use of ICTs and the internet. For example, research demonstrates that those with the highest education levels are able to derive greater benefits from ICTs and the internet compared to those with low levels of education. (Van Deursen and Helsper 2015). All of these factors contribute to the digital exclusion of a group and efforts to achieve digital inclusion of these groups represent a more complex problem than simply access to "boxes and wires." The "T" in STEM represents not only a critical element of STEM but increasingly those with limited digital access and skill have less access and resources to develop their knowledge in the other aspects of STEM. To overcome these barriers involves moving toward digital equity.

## 5.3 What Is Digital Equity?

Digital equity means ensuring that, independent of race, socioeconomic status, national identity, or disability, students have access to:

- *up-to-date hardware, software and connectivity to the internet.* Local STEM challenges attract local industry funding, and therefore helps address this challenge.
- *meaningful, high quality, culturally responsive content.* Local challenges have culturally responsive content at some level. Open education resources help bridge this need.
- *creating, sharing, and exchanging digital content.* Having students move beyond using technology for drill and practice and focus on higher order thinking skills and knowledge creation and sharing.
- *educators who know how to use digital tools and resources effectively.* Locally important industry support thier staff to work with students and teachers on projects and offering professional development and industry-based internships and experiences.
- *high-quality research on the application of digital technologies to enhance learning.* Industry cluster-focused programs benefit from resources that enable research.

To accomplish these goals represents a complex undertaking and involves leadership and collaboration at the national, state and local level in which industry clusters may play a major role.

## 5.4 Past Efforts to Move Toward Digital Equity

At the federal level, Title I of the Elementary and Secondary Education Act (ESEA) provided funding that enabled schools serving disadvantaged students to acquire technology. Often this funding was used to acquire technology systems based on the promise that it would result in significant gains in students test scores. In addition the E-Rate program has helped provide funding that enabled many schools serving low income students to have access to the internet. The U.S. Department of Education's *Preparing Tomorrow's Teacher to Use Technology* (PT3) program helped colleges of education across the country to better address the need to prepare future teachers with the knowledge and skills to use current and future generations of learning tools to enhance student learning. Another program, *twenty-first Century Learning Centers* enabled schools in urban and rural areas to remain open longer to provide homework centers and technology and other academic enrichment programs. Regional laboratories and educational service centers also provided professional development to help educators make better use of technology resources to enhance learning. Unfortunately, there has been limited evidence thus far that these efforts resulted in significant gains to student achievement on standardized tests (Resta and McLaughlin 2003) and despite these efforts, there continue to be disparities between schools both in access and effective use of technology resources for learning. It appears that existing policies at the national, state and local level may need to be assessed in terms of their impact and to consider what policies both within the technopolis as well at the state and national level may help move toward digital equity. Since the technopolis exists at the local level, this mainly refers to local policies.

## 5.5 The Digital Equity Act of 2019

The Digital Equity Act of 2019 establishes two grant programs to be administered by the National Telecommunications and Information Administration (NTIA) to promote digital equity nationwide:

- *Building Capacity within States through Formula Grants:* The legislation creates an annual $120 million formula grant program for all 50 States, the District of Columbia, and Puerto Rico to fund the creation and implementation of comprehensive digital equity plans in each State
- *Spurring Targeted Action through Competitive Grants:* The legislation also creates an annual $120 million competitive grant program to support digital equity projects undertaken by individual groups, coalitions, such as the technopolis innovative approaches to using technology to enhance STEM education.
- *Supporting Research and Evidence-Based Policymaking*: The legislation tasks NTIA with evaluating digital equity projects and providing policymakers at the local, state, and federal levels with detailed information about the effectiveness of these projects.

This legislation will provide substantial nationwide help in addressing the digital divide (NCDE 2019).

To assess the effectiveness of this act, as well as other federal programs to help move toward digital equity, it will be important for policy-makers to continue to consider the following questions:

- What are the differences in the access and use of technologies between schools in low-income and high-income areas of the technopolis? Schools serving low-income students often have them working with obsolete technology that is not able use current software and learning tools. Industry grants and provision of refurbished computers to schools and low income families help to address this need.
- What are the differences in the frequency and types of use of technology experienced by low and high achieving students? Previous studies have shown that low-income and minority students typically use technology tools and resources for drill and practice as compared to other students who use technology-based

programs focused on higher order cognitive skills (Kitchen and Berk 2016). This trend continues despite the fact that other research studies (Pogrow 1988) demonstrate that having low income and minority students engage with technology programs focused on higher order thinking skills actually improves their self-confidence and academic performance. In contrast, technopolis programs are generally focused on the higher order skills that use digital tools to prepare students for a rapidly changing technology and knowledge-based global society.
- What is the teacher readiness required to use state-of-the art technologies and pedagogies? For technology to be used in ways that enhance learning, the teachers must be skilled in how to effectively integrate technology into their teaching practices. Unfortunately, teachers in schools serving low-income children are less likely to have access to high quality professional development opportunities than teachers from wealthy school districts. Industry clusters often help address this need by sharing staff expertise with the schools and providing teacher summer internships.
- What content is available to address the needs of diverse learners? There often remains a lack of high quality, culturally responsive digital content to address the needs of minority students with limited proficiency in English. The growing quality and diversity of open education resources in the technopolis helps address this need

The above questions are important for schools, business, industry and government within the technopolis to determine what actions and policies need to be taken to prevent or reduce the digital exclusion of students as well as other members of low-income families.

As noted in other chapters, the strategies and focus of the technopolis on STEM K-12 education programs may also be expanded to helping to close the digital divide. The technopolis initiatives described in other Chap. 1 and other chapters provide examples of the efficacy of a technopolis-centric approach to engage students, teachers, business and industry to enhance K-12 STEM experiences and other critical educational needs. This approach can also be applied to closing the digital divide.

## 5.6 Technopolis Cities Address the Digital Divide

Even in the technopolis, a not uncommon situation is that the schools serving low income students in the community often have both less access to technology resources as well as to educators who are skilled in their use to enhance learning compared to schools serving high income areas of the city. To close this divide requires a concerted and coordinated effort among schools, industry and city government. It was through such concerted and collaborative efforts that technopolis cities such as Austin, Seattle and Portland have made substantial progress in enabling low income families and students to have access to:

- *Devices and connectivity*. High tech industries have made grants to schools serving low income families to enhance digital access. For example, in Austin, over 20 years ago, IBM provided multiple computers for each classroom in the schools in East Austin serving low income students. More recently, Google also worked to provide very low-cost, broadband access to low income families in a significant portion of the city.
- *High quality, culturally responsive content*. Schools have been encouraged to use open educational resources to provide the high quality digital content needed in the curriculum.
- *Educators who know how to use digital tools and resources effectively.* Engineers, scientists and technology experts from industry have played a key role in working with teachers in the school to address their STEM knowledge and skill gaps. Teachers also benefit from having these experts meet directly with their students and work with them on projects. Although the teacher may facilitate these experiences, they also increase their STEM knowledge as well. In addition, industries within the technopolis may offer teacher summer internships to help them understand current trends and developments in STEM fields.
- *Visionary leadership.* In all of the above examples, helping to close the digital divide within the technopolis was accomplished by industry clusters, schools, foundations, and city government working closely together in planning and implementing efforts to close the divide

There are a number of examples in the U.S. of the technopolis taking action in planning and implementing initiatives to address the digital divide. For example, the cities of Austin, Portland, and Seattle have been successful in addressing the digital divide through collaborative efforts with local industry coordinated by the city government. Google Fiber cities such as Kansas City, San Francisco, Austin and San Antonio have leveraged the industry relationship to make broadband access more widely available. These efforts are helpful in marshalling the resources and bringing together a critical mass or organizations and city leaders to address the divide.

In 2011, Portland developed and released a document entitled *Connecting to Our Future: Portland's Broadband Strategic Plan* This enabled the Portland Office for Community Technology (OCT) to coordinate resources from multiple city departments to engage all sectors of the community in the development of a plan to eliminate digital exclusion in the city. The city held a Digital Inclusion Summit in 2014 that was attended by city and county government leaders as well as business and community leaders. This collective commitment and action resulted in the successful development and implementation of the plan (Resta and McLaughlin 2003).

Similar actions were taken by Seattle and Austin in response to gaps in digital technology use and literacy. In Portland, the Mayor's Office, the Department of Information Technology and the Office of Civil Rights worked together to develop a plan to measure the gaps in digital equity in the city. They then worked with the community in designing and implementing a plan to move toward digital equity in the technopolis. In Austin the city's Innovation Office played a leadership role in the development of the digital equity strategic plan. Implementation of the plan involved the collective action of other city departments and locally important industry, businesses, public health agencies, the libraries, nonprofits and philanthropic organizations and neighborhood groups (Rhinesmith. C. 2016).

Other cities taking action to close the digital divide include San Francisco and Kansas City. San Francisco is establishing a citywide fiber project that will be a public-private operation offering free service to any resident living below the federal poverty line in the city (Descant 2018). In Kansas City, the Connecting for Good organization, the city government, HUD and Google Fiber are working together to provide frcc broadband access to its public housing residents.

## 5.7 Recommendations at the Local Level

The above examples demonstrate that leadership and collective action in the technopolis can help successfully address the digital divide. The following recommendations for local action may also prove helpful:

- Successful planning and execution of a digital equity plan in the technopolis requires that the city government take a leadership role in facilitating awareness of the digital divide and building support for collective action to close the gap. As noted above, different departments within the city government may take the leadership role in facilitating collaboration with locally important industry, business, banking, school, university and community leaders to facilitate their working together in development of a plan to move toward digital equity.
- The federal Community Reinvestment Act requires that banks must invest a portion of their profits back into the community they serve. It is now possible for banks to use a portion of these funds to address the digital divide. Guidelines for how schools and communities may work with banks in supporting digital equity efforts are available from the National Collaborative for Digital Equity (NCDE). The organization may also provide assistance to communities in planning digital equity summits such as the one initiated by Portland. For example, during 2018, NCDE held digital equity summits in Boston, Massachusetts and Manchester, New Hampshire (www.digitalequity.us).

The technopolis-centric approach discussed in other chapters demonstrate how locally important industries play and important role in the technopolis in helping to support STEM education in the schools. This approach is also essential in helping to close the digital divide within the technopolis. Table 5.1 lists the digital divide barriers discussed in this chapter as well as the technopolis-centric strategies using locally important industry clusters to help address each of these barriers.

**Table 5.1** Technopolis digital equity barriers and strategies

| Barriers | Strategies |
|---|---|
| Lack of access to current hardware and software in schools serving low income students | Industrial clusters provide technology grants and refurbished computers to schools (business and industry replace high capability computers every 3 years that are still viable for schools) |
| Lack of broadband access in low income family homes | Business and industry work with city and state government and telecommunications organizations to identify the areas lacking broadband access and develop plans to fill gaps. They work with cable providers to encourage them to provide deep discounts for broadband access by low income families |
| Lack of access to high quality, culturally relevant content in schools | Industry clusters provide consultant assistance to schools to assure STEM curriculum meets current and emerging needs; assist in identifying high quality open educational resources that address STEM needs |
| Lack of educators who know how to use digital tools and resources for learning | Industry clusters encourage their staff to work with students and teachers on STEM projects and activities. They provide teacher STEM professional development and resources and also offer summer internships for teachers enabling them to stay current on STEM developments and applications |
| Lack of high-quality research on the application of digital technologies to enhance STEM learning | Industry clusters collaborate with and fund university research on applications of digital technologies to enhance STEM learning |

## 5.8 Policy Recommendations at the State Level

At the state level, there is increasing recognition that low income and minority families in the technopolis may lack access to broadband. In response to this need, Missouri initiated a state-wide initiative to bring broadband to 95% of the state's population. Ohio is revamping the states existing 1850-mile fiber optic network to expand s capacity to 100 GPS. A similar project was initiated in California in which a consortium was formed to extend a 100 GPS network from Los Angeles to Seattle. New York is allocating funding to extend broadband access to the Adirondacks. Georgia has enacted legislation titled: Achieving Connectivity Everywhere Act (SB402) designed to help the state close the digital divide (Georgia Department of Community Affairs 2018). These initiatives demonstrate that states can and must play a central role in moving toward digital equity.

The above examples provide evidence of the actions that can be taken at the state level to move toward digital equity. In the U.S., states have a primary responsibility for funding schools, monitoring their performance and assuring they have access to technology tools and digital resources. In addition, they are responsible for teacher certification standards and the approval of teacher preparation programs to assure that teachers are competent in the use of the digital tools and resources to improve learning.

Recommendations for the state level include:

- Develop standards for technology access in schools. Although there is wide variance in state funding formulas for technology in schools, it is helpful for states to develop minimum standards for access to technology.
- Provide access to digital devices and high-quality digital resources through state purchasing agreements and contracts with vendors to achieve discounted pricing.
- Develop standards for both student and teacher fluency in the use of digital tools and resources for learning. The national educational technology standards developed by the International Society for Technology in Education is used by many states as well as countries across the globe (ISTE 2019). States will need to continually review and update their standards based on the exponential growth of digital technologies. Teacher certification standards should also reflect the national standards for teacher fluency.

## 5.9 Policy Recommendations at the National Level

As noted at the beginning of this chapter, the U.S. federal government has initiated a number of policies and programs to help address the digital divide. Three other chapters in this volume, those on India, China and Brazil, also cover policies undertaken at the national level. The following are recommendations for continued and expanded leadership at the U.S. national level to move toward digital equity.

- Initiate programs such as *Preparing Tomorrow's Teachers to Use Technology* (U.S. Department of Education 2015), to help teacher education programs to more effectively prepare teachers with the high level of skills and knowledge to effectively integrate technology into their teaching practices.
- Continue to provide leadership to provide disadvantaged schools with access to current technology tools and resources for learning through programs such as Title 1.
- Continue to support research on best innovative practices in the use of technology for learning through the U.S. Department of Education and the National Science Foundation.
- Conduct national studies to monitor the impact and effectiveness of the Digital Equity Act of 2019 on moving toward digital equity in schools, homes and the workplace. In addition, the studies may identify the need for other programs or initiatives to address remaining barriers in moving toward digital equity at home and in school.
- Assure accurate information on broadband access across the country. Microsoft indicates that the estimate that 25 million people lack access to broadband is greatly understated and their data indicates that over 162 million people lack access to broadband speeds (Lohr 2018). It is recommended that more accurate data be acquired to provide a clear picture of the lack of access to broadband across the country.

From the perspective of an individual technopolis, the question becomes how a city or region leverages these policies for maximum benefit. Many cities have equity or inclusion initiatives. These initiatives, which are broader than K-12, should very intentionally include K-12. Local initiatives should measure broadband access and inequities, find ways to connect teachers to training opportunities, and take the lead in ensuring local school governing bodies are taking maximum advantage of national policies and resources.

In addition to these recommendations that generally address digital equity at the city or regional levels, there are specific ways of bringing a city/regional (technopolis) focus to STEM K-12 education programs can address digital equity. Chapter 1 of this volume argued for a technopolis-centric approach to creating K-12 STEM experiences—it recommended using the particular challenges of cities to engage students, teachers, and other STEM education stakeholders. This volume has numerous examples of this being done using locally-important industries: computers and software in Austin; cyber security in San Antonio; ICT, energy and health in Medellín; advanced maufacturing the Fourth Industrial Revolution in Taiwan; the agriculture industry in São Carlos, Brazil; ICT in the small City of Fundão in Portugal. Using programs of high local importance can be a strategy for addressing specific barriers identified in digital equity research. Table 5.1 listed some of the barriers previously identified in this chapters, and it lists ways a technopolis-centric strategy using locally important industry clusters can address digital equity barriers.

The above recommendations may help the technopolis take a leadership role in addressing the digital divide within their city and region.

## 5.10 Summary

The digital divide within the technopolis is one in which low-income and minority families may experience digital exclusion that adversely affect their educational and economic opportunities. Actions taken at the national and state level can help bridge the divide through effective policy action and digital equity initiatives. At the local level, closing the divide can be achieved through collective efforts of the city government, business and industry, community organizations, schools and universities. More specifically, collaborations among the stakeholders in the technopolis can directly addressed detailed barriers identified in digital equity research. Such actions will be essential to:

- End the "homework gap" and enable all students to benefit from access to high quality resources to support their learning.
- Enable members of these families to find, apply and to qualify for jobs.
- Support the educational and economic development of the technopolis.

## References

Descant, S. (2018). San Francisco aims to close the digital divide with citywide fiber project. Accessed at: https://www.govtech.com/network/San-Franciso-Aims-to-Close-the-Digital-Divide-with-Citywide-Fiber-Project.html

DSL Reports. (2015) CWA: U.S Ranks 25th In broadband speed. Accessed at: http://www.dslreports.com/shownews/CWA-US-Ranks-25th-In-Broadband-Speed-111866

Georgia Department of Community Affairs (2018). Georgia Broadband Deployment Initiative. Accessed 2020, Febuary 20 at https://broadband.georgia.gov/about.

Gorski, P. (2009). Insisting on digital equity: Reframing the dominant discourse on multicultural education and technology. *Urban Education, 44*(3), 348–364. (2008, May 19).

ISTE. (2019) ISTE National standards for students (Accessed 2019, July 22 at https://www.iste.org/standards/for-students).

Kitchen, R., & Berk, S. (2016). Research commentary: Education technology: An equity challenge to the common core. *Journal for Research in Mathematics Education, 47*(1), 3–16.

Light, J. (2001). Rethinking the digital divide. *Harvard Educational Review, 71*(4), 709–734.

Lohr, S. (2018, December). Digital divide Is wider than we think, study says. Accessed 2020, February 20 at https://www.nytimes.com/2018/12/04/technology/digital-divide-us-fcc-microsoft.html.

Molla, R. (2017, June). More than 60 million urban Americans don't have access to or can't afford broadband internet. Access at: https://www.vox.com/2017/6/20/15839626/disparity-between-urban-rural-internet-access-major-economies.

McLaughlin, C. (2016, April). The homework gap: The 'cruelest part of the digital divide'. Accessed 2020, February 20 at http://neatoday.org/2016/04/20/thehomework-gap/.

NCDE. (2019). National and state summits on digital equity (Accessed 2019, July 22 at http://digitalequity.us/index.html).

Pogrow, S. (1988). Teaching thinking to at-risk students. *Educational Leadership, 45*(7), 79.

Resta, P., & McLaughlin, R. (2003). Policy implications of moving toward digital equity. Toward Digital Equity: Bridging the Divide in Education, 211–228.

Van Deursen, A. J., & Helsper, E. J. (2015). The third-level digital divide: Who benefits most from being online?. *Communication and Information Technologies Annual: Digital Distinctions and Inequalities: Studies in Media and Communications, 10*, 29–53.

# Chapter 6
# The Quantitative View: How to Measure STEM in the Technopolis

**Bruce Kellison, Ravae Villafranca Shaeffer, and Cliff Zintgraff**

**Abstract** This chapter presents ideas for quantitatively measuring the operations and impacts of a STEM ecosystem in a city or regional technopolis. The measurements are organized by the different stakeholders within such STEM ecosystems. The authors draw from ideas and actual practices described within the cases of this volume, and also from the STEM Ecosystems Initiative already present in three of this volume's case locations. Beyond stakeholder measures, the authors also share measures consistent with a systems view of the STEM community and argue for viewing the ecosystem holistically and synergistically. The chapter weighs the merits of *count* vs. *quality*; discusses the importance of properly defining measurement goals; and suggests measurements for a wide variety of STEM ecosystem partners in the technopolis.

## 6.1 Introduction and Background

The current chapter is part of the volume *STEM in the Technopolis: The Power of STEM Education in Regional Technology Policy*. The volume makes an argument for building STEM education experiences around the particular strategies and challenges of a city or region. The volume especially focuses on the priority industry clusters in a region and how alignment with those clusters can lead to robust support from local industries and workforce champions. The volume explicitly includes primary and secondary education (known in the U.S. as "K-12") in the ecosystem, and

---

Education Service Center-Region 20 is one of twenty education service/support centers for public school teachers in Texas. Region 20 serves the area in and around San Antonio, Texas, U.S.

---

B. Kellison (✉) · C. Zintgraff
The University of Texas at Austin, Austin, TX, USA
e-mail: bkellison@ic2.utexas.edu; cliffz@utexas.edu

R. V. Shaeffer
Education Service Center, Region 20, San Antonio, TX, USA
e-mail: Ravae.Shaeffer@esc20.net

 
C. Zintgraff et al. (eds.), *STEM in the Technopolis: The Power of STEM Education in Regional Technology Policy*,
https://doi.org/10.1007/978-3-030-39851-4_6

in fact, is mostly focused on K-12's role, while also covering cases that include higher education and all stakeholders in a community.

Imagine that you like the ideas in the overall volume and would like to implement relevant ones in your community. How might you measure progress and outcomes? This chapter is dedicated to the topic of quantitative measures. The chapter lists the stakeholders involved in building a STEM Technopolis and shares ideas about what measures can be implemented. The list includes formative measures that reflect on implementation and quality, and summative measures of outcomes.

No rigorous claims are made for the rightness of these measures or their appropriateness in general. The measures simply reflect the experience of the authors, including their experience writing for this volume. The chapter provides a structure and starting point for any community wishing to think about measurement.

### 6.1.1 Sources of Ideas for Measures

The measurement ideas in this chapter reflect two sources. The first source is the foundational chapters and case studies in this volume. The current volume includes cases from Medellín, Colombia; San Antonio, U.S.; Taipei, Taiwan; Austin, U.S.; Querétaro, Mexico; São Carlos, Brazil; Fundão, Portugal; India; and China. These cases include information about measurements, and those real examples inform the current chapter.

The second source is the STEM Ecosystems Initiative, a U.S. and now global initiative supported by the STEM Funders Network (SFN) (STEM Ecosystems 2019b). The initiative consists of 84 communities that have "demonstrated cross-sector collaborations to deliver rigorous, effective preK-16 instruction in STEM learning" (STEM Ecosystems 2019b, para. 3). The Austin case in this volume is significantly based on the initiative; the Mexico case is also related; and while the particular San Antonio case (the city's CyberPatriot program) well predates it, the San Antonio community is a recent addition to the Ecosystem network and was selected as the host for one of two 2020 network convenings.

### 6.1.2 Who Are the Stakeholders? The STEM Technopolis Model

Kidwell, Pogue and Zintgraff, and Zintgraff, Han and Butler, in this volume, build on the technopolis model (Smilor et al. 1989), developed from the experience of Austin's economic transformation. The model made an early case for the close collaboration of stakeholders in the community for driving technology-based economic development. Kidwell et al. and Zintgraff et al. argue the case for including K-12 as an equal partner. Regardless, the model delineates stakeholders involved in this

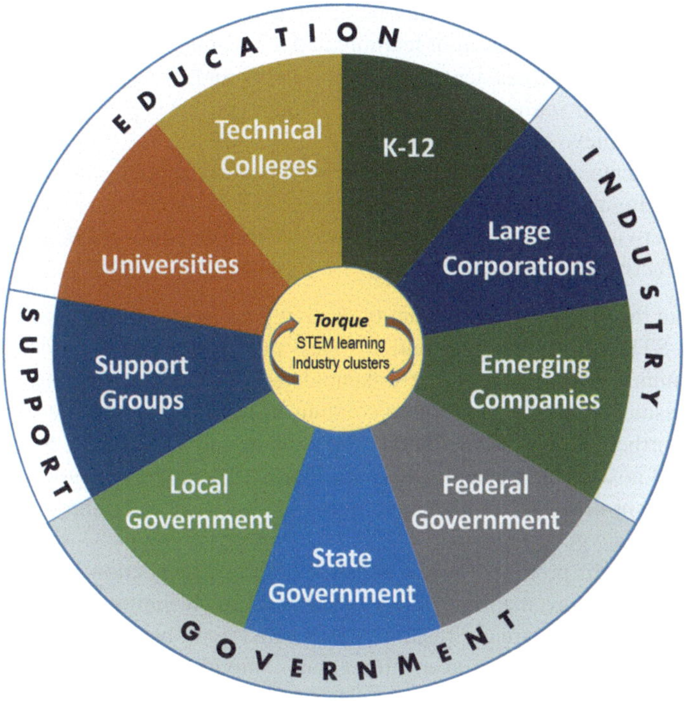


**Fig. 6.1** The STEM technopolis model (Kidwell, Pogue and Zintgraff; and Zintgraff, Han and Butler; both this volume, adapted from Smilor et al. 1989)

community collaboration. Figure 6.1 is an illustration of the suggested revision of the model. The original model included a sub-spoke called *educational support*. Zintgraff, Han and Butler, in this volume, provide a longer description of early work within the original technopolis concept that involved K-12 actors.

### 6.1.3 More About STEM Ecosystems

The STEM Ecosystem Initiative, incubated by the STEM Funders Network with support from the Teaching Institute for Excellence in STEM (TIES), is a focused effort to connect communities across the globe to a national (now international) Community of Practice, with the goal of scaling high-yield practices that increase access to science, technology, engineering and math for all students (STEM Ecosystems 2019a). The initiative provides expert technical support, peer-to-peer coaching, and community of practice convenings for the 84 designated communities. Each community receives two years of support that it likely could not afford alone. At its foundation, the STEM Funders Network is aligned around four principles (STEM Ecosystems 2019b, para. 4):

1. The STEM disciplines should not be in separate silos; rather they must be integrated through a continuum of learning and a dynamic framework.
2. STEM literacy emerges across learning settings and throughout one's lifetime. No one sector or setting can provide the comprehensive experiences and content needed to achieve STEM literacy.
3. Advancements and innovations in STEM education should connect directly to the changing needs of the STEM workforce.
4. SFN's activities and projects are informed by research and will be evaluated for their effectiveness and impact.

Throughout the process of a community receiving a designation (becoming a member of the STEM Ecosystems Initiative), TIES provides technical assistance. They help communities engage multiple stakeholders. To begin the process, TIES provides assessments and tools to support community reflections and to develop a baseline for further investigation. These tools invite the voices of STEM-expert institutions, postsecondary and other institutions of higher education, business/industry, parent-serving organizations, public and private philanthropy, and local, regional and state government.

A major benefit provided by the network is direct connection to international, federal and state conversations, policy discussions, convenings, and other coordinated efforts that yield greater results to each ecosystem than possible to access in isolation. For example, the conversation about computer science exhibits common themes across communities. Demand for talent with computer science training and experience is generally on the rise. PreK-16 generally struggles to increase computer science-certified teachers that could help train more talent. While the details of policy, practice and resources may differ among locations, underlying challenges are common, and communities can take approaches used elsewhere, select ones that fit locally, and adapt them as needed.[1]

In summary, the STEM Ecosystems Initiative is helping ecosystems align resources and effort to maximize results for students, parents and teachers and to "create systemic change in how the community prepares students for success through STEM" (STEM Ecosystems 2019c, para. 7). Some common themes in the initiative's baseline data for each community are (a) methods of career exploration for students; (b) family engagement in STEM; (c) equity in access to STEM; and (d) alignment between efforts with out-of-school time and in-school time STEM learning.

[1] There are numerous examples of this principle in the current volume, for example: (1) Use of project-based learning, but with adaptations; (2) leveraging different industry clusters; (3) selection and application of STEM workforce training models; (4) adopting STEM collaboration principles while adjusting for national policies and different industry clusters; (5) selection and application of STEM workforce training models; (6) adopting STEM collaboration principles while adjusting for national policies.

### 6.1.4 A Common Theme: The Systems View

The technopolis model and the STEM Ecosystems Initiative both bring focus to a systems view of a community. What are the main issues related to measurements of a STEM Technopolis system? What are stakeholders in the system trying to accomplish? When there is effective collaboration, different and generally better outcomes result. Duplication in the system is minimized. Win-win opportunities are identified. Fidelity of implementation to intended designs is more common. Knowledge transfer happens between K-12 stakeholders and their partners, and outputs from the system as a whole are improved.

#### 6.1.4.1 Reducing Duplication

In every ecosystem, there are actors who work independently of other actors, those that work collaboratively with others, and those that work in deep partnership. When actors are not in effective collaboration, the result is often duplicated efforts that serve small populations in selected areas of a community. Funding sources often drive this duplication; for example, grants that require service to at-risk populations or specific opportunity zones—highly worthy efforts, and in any system of independent actors, some duplication is inevitable. Still, in the end, one sees duplicate offerings, duplicate audiences, and/or multiple events offered on the same dates that compete for audience rather than creating access to programs.

On the other hand, actors who work collaboratively or in deep partnership create win-win opportunities that maximize limited resources and increase access. In communities with multiple non-profit partners offering hands-on STEM experiences, coalitions can share marketing resources, outreach expenses, coordination overhead and measurement of impact. If such resource sharing is not practical, actors may use ecosystem information to self-organize and deconflict, choosing to focus on areas where they have advantage in delivery, improving overall system results.

#### 6.1.4.2 Fidelity of Implementation

There is significant research into the problem of innovative STEM education designs being implemented in traditional ways, or simply not in accordance with the intention of the designer (e.g., Spector 2014). In STEM education, different actors are often working from different definitions, including differences about the construct of STEM itself (Zintgraff and Hirumi in press). Part of the ecosystem process is work by the community to establish definitions (explicit or implicit) for their aligned work. As a community builds common definitions that inform goals, all stakeholders can build consensus around what a good STEM design looks like, and around measures related to fidelity of implementation. The result can be an increase in fidelity. At a minimum, actors can become aware of differing assumptions, and the *small laboratories* of the ecosystem can work until consensus approaches emerge that can operate at scale.

#### 6.1.4.3 Knowledge Transfer

While the traditional model of PreK-16 education institutions was built on the transfer of information from experts to students, changes in workforce needs have created in turn a need for systemic change. The workforce of the future expects its talent to think creatively to solve problems, think critically about a problem from multiple perspectives, and collaborate effectively with a team to design multiple, viable solutions to problems. This workforce expectation is less about a transfer of knowledge and skills in PreK-16 education, and more about inspiring lifelong learners who persist to seek answers, information, and creative solutions.

In an ecosystem, when barriers come down between PreK-12 and other sectors, exchange of information is inevitable. Zintgraff, Han and Butler (this volume) argue that when non-lecture STEM pedagogies open the door for STEM professionals to deeply engage with students, knowledge transfer (KT) happens from employers into PreK-12, and vice versa as employers learn about their future workforce. It is the same type of KT common in a highly functional technopolis (Conceição et al. 1997). Well-designed systems encourage knowledge transfer for the benefit of the system as a whole.

#### 6.1.4.4 System-Level Outputs

If a system is focused on aligning resources to maximize results for a community, then how are those results measured? One important measurement is the alignment of STEM talent produced to STEM workforce demand. Within that category, one can look for STEM job forecasts, improvement (reduction) of the STEM talent pipeline gap, indicators of commitment by students to STEM course sequences, STEM persistence from secondary to postsecondary education, and system-wide measures of student and teacher efficacy in STEM. PreK-16 education providers, formal and informal, already engage in continuous improvement processes. They respond to and proactively design for change. Ecosystems should leverage this existing activity that is already aligned with the ideas in this chapter.

### *6.1.5 Counts Versus Quality*

Readers might relate to the following scenario. You are reading a report about the impact of a STEM program. Large impact numbers are reported regarding the number of people served. Question: Do you find that report inspiring? Or does it sometimes leave you wondering about the depth of impact? As participants in ecosystems, the authors know how easily numbers can be counted up, the potential difficulty in verifying them post-project, and the difference between activity and impact. Counts can be good, and usually well intentioned, but they do not tell the whole story.

Broadly speaking, this chapter highlights two types of measures. Some measures are related to counting participation. Examples include:

- How many students are in STEM-related K-12 programs?
- How many students are in STEM-specific (interdisciplinary) classes?
- How many students graduate with STEM-related degrees?

Other measures are related to the quality of activities being performed in a community. Examples of measures include:

- Evaluations of teacher professional development opportunities
- Completion of surveys about how much constructivist style or project-based type teaching is actually happening in classrooms, in the view of teachers and students
- Student grades

### *6.1.6 Having a Common Community Vision*

The context that a community sets for its technopolis/ecosystem is essential to executing measures in an effective way, and to having those measures count in the minds of stakeholders when they are eventually shared back to the ecosystem. Some actions that a community can take to strengthen the quality of data and future impact of that data, at a systems level, are listed below. Below is a roadmap to developing a common community vision:

1. *Convene.* Convene STEM leaders across sectors in the technopolis, and ask for volunteers to lead or steer an effort to develop a community vision for STEM. Determine who will be the backbone institution for convening and communicating throughout the process.
2. *Scan.* Scan the community for stakeholder groups: networks, councils, collective impact networks, and groups who are focused on STEM. Establish communication networks to provide ongoing invitations to participate in building a vision.
3. *Vision.* Draft a vision statement for what is possible. Invite stakeholder leaders and community groups from the scan to review with their networks.
4. *Survey stakeholder groups.* Invite all members of the community to provide input to who, what and how stakeholders are or should be involved in the STEM community.
5. *Establish norms for collaboration.* Invite stakeholder leaders to spend time establishing norms for collaboration; review/edit vision; craft a mission statement; brainstorm and prioritize goals.
6. *Establish common definitions.* Establish common definitions so that cross-sector stakeholders are working with a defined knowledge base.
7. *Community compact.* Invite leaders from all stakeholder groups to publicly agree to collaborate and support the community vision established by stakeholder leaders. One example is an official designation process to become a STEM Ecosystem as a community.

8. *Establish community goals.* Convene stakeholder leaders to revisit community goals for the purpose of designing metrics and a one-, three- and five-year roadmap to reach the goals.
9. *Ongoing communication.* Communicate with transparency at every stage of building a community vision. Invite participation from new partners at every step or convening. Develop a cadence for communication on a regular basis–weekly, monthly, quarterly–to fit the needs of the stakeholder groups.

### 6.1.7 How to Use Measurements

In thinking about what measurements can be taken, it is useful to consider how measurements will be used. Berwick (1996) wrote a primer about the improvement of systems in health care. Commenting on the use of measurement data, he wrote: "Measurement is best used for learning rather than for selection, reward, or punishment" (p. 619). In a system of stakeholders who are cooperating voluntarily, the authors consider this an important concept.

For example, imagine that an asset mapping effort of the region discovers that two programs are duplicating a service. A less wise response is to try and engineer an end to one service by one of the providers. A wiser choice is to share information and allow voluntary cooperation. For example, the providers might determine on their own that they best serve different segments of a market, or that their coordinated effort might serve a larger share of the market if provided at different times in the year.

In general, actors in the system should be cautious about responses that, taken collectively, look like punishment to another actor in the system. For example, a coordinated effort to adopt one partner over another will likely be obvious. The effort might prompt the less fortunate partner to fully withdraw from the system, or to engage in other behavior clearly in their legitimate best interest but not in the interest of the ecosystem.

Most of the time, the best use of measurements is for sharing with all stakeholders, and also for the collective decision-making process that happens at the system level among stakeholders. Actors can commit to their measured contribution to a goal and receive ongoing data from the ecosystem about progress toward the goal.

### 6.1.8 Measurement Goals and Driving Questions

In light of these ideas, the remainder of this chapter strives to answer the following questions:

1. Who are the stakeholders in a STEM Technopolis?

2. What does success look like (what are individual measures) for individual stakeholders?
3. What are significant measures about a STEM Technopolis as a whole?
4. How are measures communicated back to stakeholders to motivate reduced duplication, increase win-win collaboration, improve fidelity, improve knowledge transfer, and improve system outputs?

## 6.2 Measurements by Stakeholder

This section shares measurement ideas organized by stakeholder. For the reader not familiar with research terminology, *formative* refers to measures during implementation, providing insight into how a program element was implemented (how it was formed). *Summative* refers to outcomes of a step or of a full program. The following technopolis/ecosystem stakeholders are covered:

- Students (K-12, post-secondary)
- Teachers
- Schools
- School districts
- Colleges and universities (2-year, 4-year, post-graduate)
- Industry
- Non-profits
- Government (state and local)
- Ecosystem

### *6.2.1 Students*

The performance of students, their beliefs, intent, and participation in STEM education are fundamental to a STEM Ecosystem and to the long-term health of any technopolis. Table 6.1 contains student-related measures that might be part of a STEM Technopolis.

Two validated instruments might be used to address some of these measures.

- *STEM Semantics Survey* (Tyler-Wood et al. 2010). This survey is a validated instrument for measuring student interest in STEM topics and includes general measurement of STEM career interest.
- *STEM Career Interest Survey* (Tyler-Wood et al. 2010). This survey is more specific to career interest in particular STEM fields.

**Table 6.1** Possible measurements for students in a STEM technopolis

| Description/Source of measure for selected items | Formative or summative | Count or quality | Productive ways to report (selected) |
|---|---|---|---|
| Enrollment in traditional, STEM-related classes | Formative | Count | Public |
| Enrollment in STEM classes (integrating across 2+ subjects) | Formative | Count | Public |
| Measures of general student interest in STEM | Formative | Quality | Public |
| Student commitment to STEM course sequences | Formative | Count | Public |
| High school completion for students enrolled in STEM/STEM-related classes | Summative | Count | Public |
| Standardized test results in STEM/ STEM-related classes | Summative | Quality | Students, Family |
| Time with mentors / San Antonio, CyberPatriot) | Formative | Quality | Public |
| Measures of self-efficacy | Formative | Quality | Students, Family |
| Impact on life decisions / Medellín | Summative | Quality | Selected qualitative reporting (pro and con cases) |
| Number of students role-playing a future career | Formative | Count | Students, Family, Teacher, Counselor, School |
| Number of students performing outreach to other students (ie. Presentations, mentoring, coaching) | Formative | Count | Student, Family, Teacher, Counselor, School, District |
| Percent represented by underrepresented groups (women, minorities, etc.) | Formative | Count | Public |
| Percent and number of students succeeding in competitions | Summative | Count and quality | Public |
| Time on STEM tasks (ex. digital badging platform, Hour of Code) | Formative | Count | Student, Family, Teacher, Counselor, School, District |

## 6.2.2 *Teachers*

The commitment of teachers to professional growth and their implementation of proven STEM practices is critical to supporting a culture of lifelong learning in the STEM Technopolis/ecosystem. The technopolis is further strengthened with the implementation of proven practices that are informed by higher education and workforce expectations for STEM knowledge. Table 6.2 contains teacher-related measures that might be implemented.

A number of well-researched models and validated instruments might be used to implement some of the measures above.

**Table 6.2** Possible measurements for teachers in a STEM technopolis

| Description/Source of measure for selected items | Formative or summative | Count or quality | Productive ways to report (selected) |
|---|---|---|---|
| Teacher confidence with proven practices for STEM teaching and learning | Formative | Count | Teacher, Campus leadership |
| Teacher professional growth plan | Formative | Quality | Teacher, Campus, District |
| Teacher engagement with industry (externships, internships) | Summative | Count | Teacher, Campus, district |
| Teacher engagement with higher education | Summative | Count | Teacher, Campus, District |
| Workforce aligned units of instruction (PBL lessons or units aligned with industry expectations or advisory) | Formative | Quality | Teacher, Industry |
| Workforce-aligned instructional activities | Formative | Count | Teacher, Industry |
| Industry engagement in classrooms with students | Summative | Count | Public |

- *Concerns-Based Adoption Model (CBAM)* (Hall and Hord 1987). Under the belief that educators must be committed to novel classroom practices for those practices to be successful at scale (Cuban 1986; Hall and Hord 1987; Harris et al. 2009), CBAM provides tools for tracking educator adoption in the context of teacher beliefs.
- *Technological Pedagogical Content Knowledge (TPACK)* (Mishra and Koehler 2006). TPACK is built on the assumption that technology, pedagogy and content must be present and aligned for curriculum to be successful, and that this task is far from straightforward for novel curricula.
- *Constructivist Learning Environment Survey (CLES)* (Taylor, Fraser, & White, 1994). CLES is a well-validated instrument that can help determine if a classroom is actually using constructivist principles of instruction.
- *STEM Semantics Survey* (Tyler-Wood et al. 2010). Measures interest in STEM and is specifically noted as being appropriate for teachers.

### *6.2.3 Schools*

How well a particular primary or secondary school, two-year or four-year college, or graduate program is delivering a STEM curriculum is an important metric in a local STEM ecosystem. Table 6.3 lists possible metrics for schools.

The models and instruments listed for students and teachers also apply for overall school measurement. Measurement of individuals should be kept private and used for individual professional development. Aggregated measures using any of these instruments can be used to understand the STEM status of a school.

**Table 6.3** Possible measurements for schools in a STEM technopolis

| Description/Source of measure for selected items | Formative or summative | Count or quality | Productive ways to report (selected) |
|---|---|---|---|
| Number of teachers receiving professional development on STEM best practices, by count, or hours. | Formative | Count | Public |
| The degree to which STEM classes are taught using constructivist principles | Formative | Quality | Campus; public aggregated |
| Results of fidelity of implementation studies | Formative | Quality | Campus |
| Numbers of students taking STEM courses | Summative | Count | Public |
| Numbers of students engaged directly with industry (internships, mentorship with capstone projects, or job shadowing experiences, field trips) | Summative | Count | Public |
| Number of STEM-focused partners (ie. Mentors involved in STEM programs) | Summative | Count | Public |
| Number of district supported STEM activities | Summative | Count | Public |
| Number of STEM activities supported by partners | Summative | Count | Public |
| Student engagement in out of school time programs focused on STEM (after school, in partnership with other actors) | Summative | Count | Campus, Public |
| Aggregated measures of student interest in STEM and careers, as a measure of school effectiveness | Summative | Quality | Campus, Public |
| Number of events with at least nn participants running for at least mm years | Formative | Count | Public |
| School rank among peer schools in region | Summative | Count and quality | Public |

### 6.2.4 *School Districts*

The culture of STEM, when measured at school district level, can offer a robust view of a systemic approach to STEM education for a PreK-12 system. Table 6.4 suggests school district-level measures.

### 6.2.5 *Colleges and Universities (2-Year, 4-Year, Post-Graduate)*

The colleges and universities are critical partners in the technopolis/ecosystem as certifications and degrees are awarded by these institutions. Table 6.5 offers higher education measures for STEM education.

**Table 6.4** Possible measurements for school districts in a STEM technopolis

| Description/Source of measure for selected items | Formative or summative | Count or quality | Productive ways to report (selected) |
|---|---|---|---|
| Aggregated: Number of teachers receiving professional development on STEM best practices, by count, or hours. | Formative | Count | Public |
| Aggregated: The degree to which STEM classes are taught using constructivist principles | Formative | Quality | Campus; Public aggregated |
| Number of schools in a district that are offering STEM programs, and change over time | Summative | Count | District, Public |
| Number of STEM programs of study, as defined by a coherent sequence of 3 or 4 courses offered to achieve coherent studies in engineering, cybersecurity, or computer science, for example | Summative | Count | District, Public |
| Number of partnerships coordinated at district level | Summative | Count | District, Campus, Public |
| Total number of partnerships coordinated at each campus | Summative | Count | District, Campus, Public |
| School district and charter school partnerships yielding additional STEM courses or programs of study | Summative | Count | District, Charter Partner, Campus, Public |
| Aggregate number of students participating in STEM programs across district | Summative | Count | Public |
| Aggregated measures of student interest in STEM and careers, as a measure of school effectiveness | Summative | Quality | Campus, Public |
| Number of events with at least nn participants running for at least mm years | Formative | Count | Public |
| District rank among peer districts in region /state | Summative | Count and quality | Public |

### 6.2.6 *Industry*

The level of engagement of industry with other actors in the technopolis is a critical area to measure. Such measurement might begin with the development of an asset map of local STEM employers, and also of employers not traditionally viewed as in STEM fields, but employing a high number of STEM workers (e.g., IT departments in large companies). Table 6.6 outlines measures of industry engagement with other actors in the technopolis to impact the STEM workforce demand.

For industry measures, tools and techniques from standard economic development analysis can be deployed and integrated into a larger technopolis/ecosystem measurement design. Direct economic output of firms in STEM industry clusters is a straightforward, common, and quantifiable metric of impact. Off-the-shelf economic impact software packages such as IMPLAN or REMI are widely employed to analyze and model the impacts of firm-level or industry-level spending and would be quite useful and easily understood in the context of quantifying STEM industry

**Table 6.5** Possible measurements for colleges and universities in a STEM technopolis

| Description/Source of measure for selected items | Formative or summative | Count or quality | Productive ways to report (selected) |
|---|---|---|---|
| Number of students enrolled in interdisciplinary courses | Formative | Count | Public |
| Number of MOUs for STEM dual credit with school districts and colleges | Formative | Count | Public |
| Number of majors and graduates in STEM and STEM-related disciplines | Summative | Count | Public |
| Number of students with internships or apprenticeships in local firms | Summative | Count | Public |
| Number of interdisciplinary courses offered | Summative | Count | Public |
| Number of graduates and majors that lead to STEM jobs in high demand in the labor market | Summative | Count | Public |
| Percent of students graduating with an advanced STEM or STEM-related degree | Summative | Count | Public |
| Number of articulated agreements from community colleges to universities offering efficient time to degree earned in STEM | Summative | Count | Public |
| Number of students completing STEM dual credit | Summative | Count | Public |
| Post-graduation student evaluations | Summative | Quality | Campus |
| Aggregated employer ratings of graduates | Summative | Quality | Campus |

**Table 6.6** Possible measurements for industry in a STEM technopolis

| Description/Source of measure for selected items | Formative or summative | Count or quality | Productive ways to report (selected) |
|---|---|---|---|
| Number of STEM industry firms in an ecosystem asset map | Formative | Count | Public |
| Number of internships for high school or college students | Formative | Count | Industry, Public |
| Number of people employed in STEM-related firms or jobs | Summative | Count | Public |
| Direct economic output of STEM industries | Summative | Count | Public |
| Spending by industry STEM cluster | Summative | Count | Industry, Public |
| Indirect effects of spending by STEM industry cluster (B2B, personal income, salaries) | Summative | Count, Quality | Public |
| Longitudinal study of economic trends for STEM industries | Summative | Quality | Public |
| Total or per capita funding provided for STEM programs | Formative | Count | Industry, Public |
| Number of senior executives serving on STEM-education-specific boards | Formative | Count | Industry, Public |

impacts. (Direct impacts are firm-level spending for salaries, supplies, raw materials, and operating expenses.) These packages can also help quantify indirect and induced economic impact, as well as the fiscal benefits to local and state government of the spending done by firms in a STEM cluster. (Indirect effects are the sum of the increased business-to-business activity caused by the direct spending; induced effects are the quantifiable personal income increases caused by the direct and indirect effects.)

These tools can be applied in two ways: static, and longitudinal. Static application is made for a one-time measure. Static measures can be important in demonstrating impact, confirming the anecdotal beliefs of stakeholders in an industry cluster. Longitudinal studies of economic trends are even more useful to policymakers. An asset map can be created, and ecosystem assessment performed that includes measurement of economic impact. The assessment should be designed to be easily repeatable. Armed with empirical data on multi-year impact trends, STEM advocates can influence policy debates around community investments in STEM ecosystems.

Richard Butler, co-author of this volume's chapter on regional industry clusters, has performed such assessments for San Antonio, U.S., relative to numerous industry clusters, including the IT/cyber cluster that has benefited from the cyber security education initiatives described in this volume. Multiple studies have produced longitudinal data over time (e.g., Butler and Stefl 2010), with that data summarized in reports that highlight the importance of local cyber security education efforts. When combined with an aerospace study noted by Butler and Stefl, the community has static and longitudinal data across multiple industry clusters fed by Alamo Academies, an *early college high school* type program started in the early 2000s and one consistent with the principles of this volume. Among the data uncovered relative to the Alamo Academies program: (1) the aerospace cluster contributed $3.4 billion annually to the economy, with 10,000 direct jobs in the city, and an average annual wage of $78,000, far above the local average wage; (2) the cyber security cluster contains about 40 private firms that engage with both the private sector and the military and employ about 50,000 employees who classify as technology workers; (3) yearly economic impact of $10 billion (Butler and Stefl 2014); (4) estimated cyber security impact of between $1.75 and $3.5 billion (Sánchez and Zintgraff, this volume).

### 6.2.7 *Non-Profits*

Non-profit partners in an ecosystem serve as STEM education advocates, conveners and providers, and especially as providers of out-of-school-time experiences. Foundations and selected non-profits may serve as funders. Table 6.7 outlines measures for non-profit actors in the technopolis.

**Table 6.7** Possible measurements for non-profits in a STEM technopolis

| Description/Source of measure for selected items | Formative or summative | Count or quality | Productive ways to report (selected) |
|---|---|---|---|
| Capacity for STEM funding in a community | Formative | Count | Public |
| Number of services that can be provided to students | Formative | Count | Public |
| Number of legislative mandates requiring a non-profit partner, such as school turnaround partners | Formative | Count | Public |
| Number of legislative mandates that fund a non-profit partner, such as communities in schools | Summative | Count | Public |
| Number of students receiving STEM services | Summative | Count | Public |
| Total number or percent of students served in STEM | Summative | Count | Public |
| Measures of student STEM and career interest before/after services | Summative | Quality | Public |
| Statistical review of non-profit services on student grades, assessments, behavior (UP Partnership, n.d.) | Summative | Quality | Public |

### 6.2.8 Government (State and Local)

Local and state governments are critical actors in the technopolis that fund and often lead educational efforts, and that can influence sustainability or growth through legislative advocacy. Table 6.8 includes measures for local and state government in a STEM Technopolis.

### 6.2.9 Ecosystem

The power of measures at the ecosystem level are valuable to the technopolis in aggregate and disaggregated for the needs of groups of actors in the ecosystem. Table 6.9 includes measures for ecosystem-level impacts.

One of the quantifiable measures of STEM is labor force alignment with the needs of local industry. Such analysis is often done in the course of regional assessments and/or asset mapping, but the goal is to match the labor force needs of local firms with talent being produced locally or attracted to the region. There is a mismatch when local educational assets (K-12, college, and post-graduate) are not producing new workers with the skills demanded by local firms, or not producing enough of them. A healthy technopolis is marked by the close connection among assets in the STEM ecosystem: good communication about needs and resources, agreement among assets on regional economic development goals, and a workable

**Table 6.8** Possible measurements for local and state government in a STEM technopolis

| Description/Source of measure for selected items | Formative or summative | Count or quality | Productive ways to report (selected) |
|---|---|---|---|
| Number of policy agenda items supportive of STEM industry or STEM education | Formative | Count | Public |
| Number of articulated backyard strategy agendas (strategies involving local chambers, etc.) by industry for education partnership | Formative | Quality | Public |
| Number of articulated goals at local or state government focused on STEM | Formative | Count | Public |
| Number of articulated goals at local or state government focused on STEM, that were informed by education or workforce partners | Formative | Count | Public |
| Funding for traditional, STEM-related education programs | Summative | Count | Public |
| Funding for integrated STEM education programs | Summative | Count | Public |
| Aggregated versions of all student, teacher, school and district measures | Summative | Count | Public |
| Economic impact on STEM programs aligned to STEM industry clusters | Summative | Count | Public |
| Total or per capita funding provided for STEM programs (local, county, state, national) | Formative | Count | Public |

**Table 6.9** Possible measurements for an ecosystem in a STEM technopolis

| Description/Source of measure for selected items | Formative or summative | Count or quality | Productive ways to report (selected) |
|---|---|---|---|
| Comparison of workforce needs to talent being produced locally or regionally | Formative | Count | Public |
| Map of STEM assets in the ecosystem | Formative | Quality | Public |
| Strength of collaboration among actors in the ecosystem | Formative | Quality | Public |
| Community of practice connections with other ecosystems | Formative | Quality | Public |
| Number of industry engagements with other actors in the ecosystem | Summative | Count | Public |
| Aggregated versions of all student, teacher, school and district measures | Summative | Count | Public |
| Longitudinal study of industry engagement with actors in the ecosystem over time | Summative | Quality | Public |
| Changes in quality-of-life metrics reasonably attributable in part to educational outcomes | Summative | Quality | Public |
| Number of programs with stakeholders from all sectors: K-12, higher education, industry, government, non-profit | Formative | Count | Public |
| Number of defined career pathways from secondary school to careers | Summative | Count | Public |
| Number of STEM projects with outside-the-community funding sources | Formative | Count | Public |
| Number of industry clusters represented in the ecosystem | Summative | Count | Public |
| Distribution of STEM career choices of students in the ecosystem | Summative | Count | Public |

roadmap to achieve those goals. One of the characteristic benefits of a STEM ecosystem is more labor force alignment with local industry than would normally be expected without such a targeted ecosystem.

One measure suggested is connections to other ecosystems or the community of practice framework (Wenger et al. 2002). The Ecosystem to Ecosystem (E2E) concept, or community of practice (COP), that is supported by the Teaching Institute for Excellence in STEM (TIES) and the STEM Funders Network are international efforts to bring communities of practice together (virtually and face-to-face) to learn from one another, share resources and experiences, and provide communities of support across the globe. By understanding the assets and needs across communities, TIES is able to connect resources to networks with economies of scale.

### 6.2.10 Conclusion

This chapter has highlighted a range of measurements that partners in a STEM ecosystem can quantify to begin to articulate the educational and economic impacts of STEM programs in a technopolis. The goal was not to be exhaustive, but suggestive; many of these elements can and should be adapted or expanded by local advocates, depending on the local context.

The authors argued for considering these measures by technopolis/ecosystem stakeholder; for considering formative and summative measures; for considering count and quality measures; and for considering measures within a larger vision. Organizations should interpret measures within that vision, and they should be wise when sharing results, using results for improvement and coordination, not for explicit or implicit punishment of those cooperating voluntarily in an ecosystem. Readers can see examples of many of these measures in the cases of the volume. Readers are encouraged not to assume impact in their STEM ecosystems. In the most basic sense, this chapter's message is the following: Demonstrate the educational and economic impact of STEM education in your cities and regions.

## References

Berwick, D. M. (1996). A primer on leading the improvement of systems. In *First Annual European Forum on Quality Improvement in Health Care, London, 9 March 1996* (pp. 619–622).

Butler, R. & Stefl, M. (2010). The aerospace industry in San Antonio: Economic impact in 2010. Report accessed on 2019, August 13 from http://www.aerospacesanantonio.org/uploads/Aerospace_Brochure_2011_Final.pdf.

Butler, R., & Stefl, M. (2014). The San Antonio IT economy: 2014 economic impact. Retrieved August 24, 2019 from http://www.tinyurl.com/sa-it-impact.

Conceição, P., Gibson, D., Heitor, M. V., & Shariq, S. (1997). Towards a research agenda for knowledge policies and management. *Journal of Knowledge Management, 1*(2), 129–141.

Cuban, L. (1986). *Teachers and machines: The classroom use of technology since 1920*. New York: Teachers College Press.

Hall, G. E., & Hord, S. M. (1987). *Change in schools: Facilitating the process*. Albany: SUNY Press.

Harris, J., Mishra, P., & Koehler, M. (2009). Teachers' technological pedagogical content knowledge and learning activity types: Curriculum-based technology integration reframed. *Journal of Research on Technology in Education, 41*(4), 393–416.

Mishra, P., & Koehler, M. (2006). Technological pedagogical content knowledge: A framework for teacher knowledge. *The Teachers College Record, 108*(6), 1017–1054.

STEM Ecosystems. (2019a). STEM Learning ecosystems overview. Retrieved 2019, August 3 from https://stemecosystems.org/about/.

STEM Ecosystems. (2019b). About the STEM Funders network. Retrieved 2019, August 4 from https://stemecosystems.org/about-the-stem-funders-network/.

STEM Ecosystems. (2019c). Frequently asked questions about the STEM learning ecosystems initiative. Retrieved 2019 August 4 from https://stemecosystems.org/faqs/.

Smilor, R. W., Gibson, D. V., & Kozmetsky, G. (1989). Creating the technopolis: High-technology development in Austin, Texas. *Journal of Business Venturing, 4*(1), 49–67.

Spector, J. M. (2014). Program and project evaluation. In J. M. Spector (Ed.), *Handbook of research on educational communications and technology* (pp. 195–201). New York: Springer.

Taylor, P. C., White, L. R. and Fraser, B. J. (1994). A classroom environment questionnaire for science educators interested in the constructivist reform of school science. In: Annual Meeting of the National Association for Research in Science Teaching (NARST), March 1994, Anaheim, California.

Tyler-Wood, T., Knezek, G., & Christensen, R. (2010). Instruments for assessing interest in STEM content and careers. *Journal of Technology and Teacher Education, 18*(2), 345–368.

UP Partnership. (n.d.). Our approach & team. Retrieved 2019, August 12 from https://uppartnership.org/about-up-partnership-san-antonio/.

Wenger, E., McDermott, R., & Snyder, W. M. (2002). *Cultivating communities of practice: A guide to managing knowledge*. Boston: Harvard Business School Press.

Zintgraff, C., & Hirumi, A. (in press). Aligning learner-centered design philosophy, theory, research and practice. In J. M. Spector, B. B. Lockee, & M. Childress (Eds.), *Learning, design, and technology: An international compendium of theory, research, practice, and policy*. New York: Springer.

# Part II
# Cases at City or Regional Level

# Chapter 7
# Medellín, A Case of Self-STEAM (Esteem)

**Alejandro Roldán Bernal**

*This is the base of the argument: STEAM as a framework in science, technology, engineering, arts and mathematics, with the word esteem, meaning self-confidence—it is about the new pathway for many kids away from narcotics trafficking to STEAM careers.*

**Abstract** Medellín, a case of self-steam (Esteem), is a short journey through the last five years in which several projects have impacted the quality of education, the vocational orientation for STEAM careers, but mostly the mindset of students in the city. Transformations in mobility, social inclusion, infrastructure and security have been key factors, but as it relates to students' belief in their ability to solve real problems in their neighborhoods, the STEAM projects have provided the framework and the intellectual adventure for students. This chapter describes the projects and their impact. One goal is to highlight how the STEAM model, within the context of Medellín's city development plan, has yielded results. A second goal is to spotlight the urgency to be more focused on talent development. This topic must be the core of the public agenda, to increase the connection between education and economic growth. During the five years, STEAM has been the right path to strengthen the bond. This path should continue, increasing talent in the city and creating the new face of the Medellín to the world. The Ruta N Innovation Agency and allies started the movement, with programs now adopted by the city's Secretary of Education, private schools and universities who believe in the importance of a knowledge economy for Medellín in the XXI century.

A. R. Bernal (✉)
Ruta N Innovation Agency, Medellín, Antioquia, Colombia
e-mail: alejandro.roldan@rutan.co



C. Zintgraff et al. (eds.), *STEM in the Technopolis: The Power of STEM Education in Regional Technology Policy*,
https://doi.org/10.1007/978-3-030-39851-4_7

*By sharing my story, I saw my first steps reflected in the faces of our young Colombians, I saw my curiosity reflected in the incessant questions received, and I saw my desire to continue forward instilled in the paisa upbringing of my interlocutors.*[1]
*Juliana Garcia*
*Ford Foundation Fellow*
*NSF Graduate Research Fellow.*

## 7.1 Introduction

This chapter describes some of the experiences in STEAM education in Medellín, Colombia in order to present how Medellín's STEAM education framework helps to provide thousands of students with hands-on activities based on cutting edge technologies. The main goal of this framework is to inspire students and connect them with the opportunities of the XXI century labor market.

For decades, Medellín's communities were dominated by the role model of warriors, drug dealers, and violent people, leaving a deep impact (a cultural imprint) in the mindset of the young people. Although this mindset seemed difficult to change, Medellín figured out how to reinvent itself: violence decreased, social and economic inclusion indicators improved, and the city—leaders and citizens—decided to orient their urban/economic development towards generation of knowledge and high tech products and services.

## 7.2 About the Case

Medellín, the second largest city in Colombia with 2.5 million of people, arguably has been the city most deeply affected by the illicit drug trade. 70% of the total area of the city have a rural vocation, but just only the 30% of the population lives there and the 70% in the urban area. Closing the gap between rural and urban, rich and poor in the city has been the challenge of the last two decades. Designing programs against the inequity and violence, has been an important emphasis during various administrations periods, but because it's economy vocation as an industrial hub attract a lot of people from outside of the city that has increasing the gap. In order to present the city, is important to begin with a short description of the key milestones in the economic changes, as Andres Sanchez had write:

> With the adoption of the import replacement industrialization model, in the 1970s Medellín became Colombia's key industrial hub. This success was based on the rise of the textile cluster, which constituted the primary source of urban economic growth during most of the twentieth century. However, extreme specialization in textile production produced a sharp industrial crisis associated with economic liberalization of the 1990s. In addition, the emergence of organized drug trafficking significantly increased the levels of violence (Sanchez, 2013).

---

[1] Testimony of one of the researchers who worked in the Interchange project.

During the 1980s and 1990s most youngsters of the city accepted the *narco* (narcotics trafficking) culture as an imperative, the face of their future.

The role of drug lords and gangsters in politics, economics, and social fields was expansive, promoting a value scale based on wild competition and premises such as *money equals success, get rich or die trying it, you don't need to study to get rich*, and *women are objects that must be attractive, devoted and silent.* The combination of these two factors (a textile industry broken and the rise of narcotics trafficking) activated a deep urban crisis characterized by institutional instability, corruption, impunity and fear.

However, in a parallel process a group of institutions became the base of a new pathway. Proantioquia (2016) is an NGO that has been one of this institutions that from 1975 invested resources from the private sector to change the cultural mindset of the people in the state of Antioquia and, particularly in Medellín by helping the different education bureaus in Antioquia, to find new ways of teaching and thinking. Also this NGO has been a lighthouse for the city because it's been making an independent annual report of well-being in Medellín, a document that has served as a blueprint for identifying when and where the city needs to invest its own resources.

In 1989 the Antioquia's Centro de Ciencia y Tecnología de Antioquia (2018) (CTA) was founded. CTA is a nonprofit private corporation that works as an articulator of the regional innovation system. Its goal is to link the universities, the public sector and the private sector in designing and executing plans to communicate the science and technology knowledge throughout the state of Antioquia. Its work in STEAM has had an important impact in public schools through K-12 event in universities, and also in improving the entrepreneurship ecosystem.

Those organizations and more cultural initiatives have been the key and the basis of the change in the city. But those organizations must work with the public sector and with the will of the policy makers in order to make the right and deep impact in the city.

For that reason, from 2002 to 2018, a group of policy makers, including mayors and the Town Council, have been continually investing in social and cultural transformations that are largely consistent with long-term plans that span administrations. Below are descriptions of a few of these initiatives.

- *A transportation system for low income people*. The Medellín Metro (Metro de Medellín 2019) has been a great example of a public institution with a big impact in the cultural mindset of the citizens. Clean areas, order, safety, and respect are values that people experience inside of the subway, cable cars, trolleys and integrated buses that are all part of the system.
- *A library network*. Since 2002 the Mayor's Office, through the bureau of citizenship culture, has been building and supporting a net of 35 libraries including 9 library-parks, where the whole community has, besides accessing books, the opportunity to connect with rich cultural offerings: movies, lectures, dance, music, writing programs, among many others (Sistema de Bibliotecas Públicas de Medellín n.d.).

- *Using water storage areas as public spaces.* Unidades de Vida Articulada (Units of Articulated Life) (UVAs) (Empresas Publicas Medellín 2019) is an abbreviation for a strategy to use eighteen water storage tanks in city neighborhoods as parks, libraries and kinder gardens, with activities for all the family members. These places foster local ownership and care for the public spaces, respecting diversity and promoting social inclusion within the framework of Corporate Social Responsibility.

These initiatives are part of the social inclusive architecture in the city that helps to connect low incomes families with new opportunities and most importantly with a new mindset in the city, with a feeling of hope. But added to the infrastructure to connect the students, different programs in STEAM have become a great tool to also inspire them, because if you want to foster an innovation culture, there is a need to inspire and connect the whole community.

## 7.3 About the STEAM Programs

### *7.3.1 STEAM-LABS and STEAMakers*

STEAM-LABS (Zintgraff et al. 2018; Zintgraff et al. 2015) was an early example of the kinds of programs described in this chapter. The program was later enhanced and renamed to STEAMakers. The program was a collaboration between Medellín's Secretary of Education, Sapiencia (the city's higher education agency), Parque Explora (the city's children's and family museum), and The University of Texas at Austin (UT Austin).

In 2014, the program brought together eleven schools in Medellín, certified 23 teachers, and trained an additional 142 teachers via activities by the certified teachers in their schools. The teacher training ultimately led to 1440 students between sixth and eleventh grade being served. Additional teachers and students were served in later years of the program.

UT Austin's efforts were led by Dr. Cliff Zintgraff from the university's $IC^2$ Institute, and by Dr. Carol Fletcher from the university's Center for STEM Education within its College of Education. The $IC^2$ Institute (Innovation, Creativity, and Capital) ($IC^2$ Institute 2019) had thirty-plus years of experience working with entrepreneurs, innovators and policymakers on the technology-based development of regions. During this program, the Institute was specifically studying the role of secondary education in development of a technopolis, applying prior experience from similar programs known to $IC^2$. The Center for STEM Education was a recognized leader in developing STEM capacity in Texas. For example, their Texas Regional Collaboratives program had trained 30,000 teachers in STEM methodologies.

The *first stone* of STEAM-LABS was a decision to create student projects based on Medellín's development plan, *Medellín 2021*. This decision was based on the theory that alignment with the city's development priorities would best connect the program to participant energy and potential resources. The fundamental instructional design decision of the program was to base the teacher training content on the project-based learning methodology of the Buck Institute for Education (Markham et al. 2003). That instructional methodology emphasized, among other complementary principles, the ideas of authenticity, real-world relevance, and adult mentor connections for students.

The 1440 students served far exceeded the program goal of 500. This expansion was a direct result of trained teachers taking the initiative to train more teachers on their own campuses, who in turn delivered the program to students. Program leaders attributed this "burst of energy" to the decision to align with Medellín's development plan, in concert with preparing the teachers using proven curriculum and methods. The following observation is not scientific, but it was obvious to the program leaders how real-world relevance energized all who participated in the program.

STEAM-LABS was also one early impetus for a focus on STEAM education, versus solely on STEM. This conclusion was reached the first morning of a December 2013 opening symposium attended by a wide cross-section of stakeholders including the Secretary of Education, government leaders, policymakers, university rectors and senior administrators, industry representatives, and K-11 principals, leaders and many teachers. There existed a strong consensus that Medellín's long history as a center of culture, art and political thought demanded a cross disciplinary approach that moved beyond the mainly technical elements of STEM.

STEAMakers added a heroic narrative into the projects performed by students. Via this narrative, students were inspired toward science careers. In STEAMakers, twice per year, the program conducted a large science celebration, with music and different artistic performances, connecting students not only with the science, but also with the ancient art of storytelling.

The PBL model at the core of the STEAMakers program facilitated teachers working together and thinking about real-world problems during performance of their jobs. Program leaders observed that working on projects helped teachers focus on designing a learning experience and on students' learning problems, rather than on teacher-centered problems. For many teachers, this user-centered focus was a new perspective. User-centered design is a different pathway to share learning experiences with the students, far from the idea of a classroom "dictator."

STEAM-LABS/STEAMakers was one of several programs that seeded STEAM education initiatives in Medellín, alongside the Horizons program in Ruta N, and STEM+H (humanities), an educational strategy from educational bureau of Medellín. Ruta N would also choose to add the *A* to STEM, including arts as a powerful way to create meaningful learning experiences and to connect technology with creative industries.

### *7.3.2 Horizons Program*

Ruta N Medellín (2011), a public corporation and the center of business and innovation for Medellín, created the Horizons program. Horizons is a STEAM education program designed to develop talent for the new economy, and also to help develop the culture of the new city. Several productive (private) sector companies were involving in the creation of the program. Those companies shared their ideas of the skill sets needed in the labor market. Their inputs were important inputs at the origin of Horizons program.

This program has inspired middle and high school students from public schools around the city. The program connected those students within the STEAM framework through three different projects: *Interchange, Engineering at N* and *Innobotica*. Each project had specific content as a means to develop interest in the students towards STEAM careers. The first slogan of the Horizons program was: Conquering borders is our heritage. Each project reflected this philosophy. The name of the program was based on the classic painting of Francisco Antonio Cano named Horizontes, where you see a traditional family on a hill, as they are seeing the big mountains that surround the Medellin's valley. The mountains are a metaphor for a big challenge, with the knowledge economy as the new challenge to embrace, the new hill to conquer.

#### 7.3.2.1 Interchange Project

In April 2013, a group of researchers from Purdue University came to the Spanish Library located in District 1 of Medellín, one of the districts with the lowest incomes in the city. The link between those researchers and the local library team began the exchange of ideas about approaching a group of high school students through a STEAM program that would become known as the Interchange Project. The role of Professor Carol Handwerker, a researcher in materials engineering at Purdue University (Purdue University Materials Engineering n.d.) was critical to create this project; her will to connect Ph.D. students from Purdue with Colombian high school students was the *first stone* in this building. During 2013, the Purdue University team and the library team developed a six-week course of nanotechnology for forty students from public schools in District 1, inspired by the Nanodays initiative in the USA (National Information STEM Education Network n.d.). During 2014, members of the Ruta N Corporation and a group of faculty members from Purdue, led by Professor Handwerker, developed a methodology to scale the projects from 40 students to 730 in 2015.

Also in 2014, an annual international exchange started with a first cohort of fourteen high school students. The students participated in the Gifted Education Research Institute (GERI) Summer Camp at Purdue. The objectives of this project were:

- Increase high school students' interest and knowledge about science.
- Encourage students to pursue university studies in science, technology, engineering, arts and mathematics (STEAM) disciplines.
- Inspire the next generation of leaders in science, for both the high school students and the graduate students participating in the program.
- Create strong, productive partnerships between Colombia and Purdue University, and other universities around the world.
- Develop a model for broader interactions in K-12 education.

The methodology of the Interchange Project was based on an interactive experience between Ph.D. students from Purdue University (later with Ph.D. students from other universities) and high school students from Medellín, through virtual platforms like Hangout or Skype. Participants shared science concepts related to nanotechnology, biotechnology, astrophysics, entomology and others, doing hands-on activities, and closing each session with a *questions and answers* moment. This last moment in each one of the sessions became the most attractive moment. The students asked more about the person behind the screen than the research itself, seeing in them a real role model.

One highlight from these experiences was the link between a Colombian Astrophysicist from Harvard University, Antonio Copete, and Johan Stiven Bedoya, a Medellín tenth grade high school student in 2015. After the video conference with Antonio Copete, Johan started thinking about studying astronomy. He had always loved to see the stars, but he had never thought of astronomy as a professional opportunity. He wrote to the Antonio via a message on Facebook, a message that he never thought would receive a response. But Antonio's response was almost immediate, full of pride for his new friend. Here a short part of the letter from Antonio to Johan:

> Hello Johan,
>
> Thank you very much for your message. The first thing I would say is that I'm glad you've taken advantage of the video conferences and that it is precisely because of students like you that we are making the effort.

That was one of many amazing links between students and researchers that happened during the Interchange Project. From 2013 to 2015, fifty international researchers connected with high school students in order to promote the development of scientific and global thinking skills. Those links became an opportunity to increase students' interest in higher education, and create long term relationships between the researchers and the students.

In 2018, the Interchange Project hosted the fifth cohort of campers to the GERI Summer Camp (Geri-Camp) at Purdue University (Purdue University College of Education 2019) (Fig. 7.1). Since 1974, this summer camp has connected the lives of many students around the world, identifying and boosting their talents. During five years of the Interchange project, 48 students across five cohorts lived for two weeks in the same building with students from France, Greece, China, Korea and Israel, and from different states of the USA.

Geri-Camp was a language-immersive experience for the participating Medellín public school students. The experience was especially challenging because the students were forced to increase their English skills to communicate, and also because they were forced to develop various social skills to interact and be a part of these multi-cultural teams. Program leaders have observed that Geri-Camp has been a life changing experience for many students, opening their minds to other cultures, and breaking biases about religion, economic level and nationality, and generally increasing students' self-confidence.

**Fig. 7.1** Colombian high school students at Geri-Camp

The company Arukay worked with the project, implementing the methodology, and scaled the project from near 40 students to more than 700 at the end of 2015. This was the first Horizons project and the basis of the Horizons Program. As a result of Horizons, led by the Ruta N Corporation, and in alliance with universities and educational technology companies, there is now a complete offering of contents and activities for students throughout the public schools of the city.

#### 7.3.2.2 Engineering N

In the Engineering N project (Universidad EAFIT n.d.), twenty schools from eleven different districts in the city were selected for an immersion process in EAFIT University, a well-recognized private university in the city. Students from 13 to 17 years old were taught by high-level EAFIT researchers, with workshops based on design thinking methodology.

Through this project, 300 students from Medellín public schools experienced how to think as engineers. Program leaders noted that many students viewed engineering as something distant and complicated, a subject just for geniuses or "boring researchers." The program presented engineering as something both familiar and fun. The methodology aimed to make students recognize and feel confident about their own skills. The program attempted to both share engineering knowledge and engender an engineering mindset.

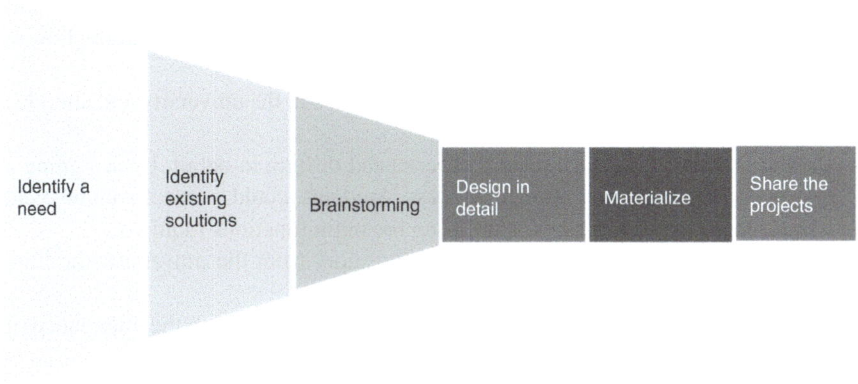


**Fig. 7.2** The Engineering N design process

**Fig. 7.3** Two medellín high school students congratulate each other during Engineering N

The six stages of the program introduced students to the problem-solving approach. Figure 7.2 illustrates the approach, which has the six steps listed below. Figure 7.3 shows students at the Engineering N event.

1. *Identify a need.* In the first stage, students investigated and drew a map of existing solutions in the world to solve the problems they previously identified in their communities. The students learned how to use advanced filters in Google for searching, and they learned to arrive at a design question.

2. *Identify existing solutions*. The aim of the second stage was to inspire the students with other solutions. Creative exercises, brainstorming and learning how to approach new material was especially important.
3. *Brainstorming*. Students then visited different labs in the university and attended some university lectures.
4. *Design in detail*. The participants experienced design, in detail, by designing a materials-based product. In the process, students could come to understand which materials were better for fulfilling the main functions required.
5. *Materialize*. With the support of their counselors from the university, the high school students materialized their ideas.
6. *Share the projects*. Students shared their projects in a fair with the attendance of their parents and teachers from their schools.

#### 7.3.2.3 Innobotica

Horizon's *Innobotica* program started in 2014. The program focuses on robotics, programming, and basic electronics, with content designed for 11 to 14-year-old students. In 2014, 1250 youth worked in teams to learn coding, basic electronics, communication skills, and home automation using Arduino software and hardware.

The education philosophy of the program is learning by doing. Home automation, also known as domotics, was chosen as the students' learn-and-do laboratory. In one story of the impact of Innobotica, selected students shared how they remember the first day they understood how a remote control works, and they told program leaders about trying to teach this content to their parents, sharing useful knowledge that they received in their class.

In 2015, over a 25-week period, intensive teacher professional development was performed, with 85 working groups participating in workshops of four hours per week. The workshops involved seventy teachers, and the 33 facilitators were current engineering students. At the end of that year, 85 projects were presented to the city in a major science fair with more than 2.500 students from all over the city. Those projects had something in common besides the robotics and domotics. Projects were rooted in an effort to change the participants' vision of the world, leading them to understand the needs of their own communities and to see their own environment as an opportunity to innovate.

The Innobotica project had students and teacher involved from 44 schools, but was it a meaningful experience for the students? Interviews were conducted at the time of the program to gain insights into the views of students. David Garcia, a student from San Juan Bautista de la Salle School, shared this thought: "I thought I was not capable of learning and doing all that I did." Comments like this one were common among the students participating in the program.

In fact, while program leaders valued the hard skills developed, they were more interested in students' development of confidence and self-esteem, both in Innobotica and in other Horizons programs. Silvio Marin, one of the facilitators of Innobotica, shared that idea as follows:

> When your robot does what you tell it to it's because you have solved a problem using your knowledge of mathematics and physics and that fills you with happiness.

The joy of knowledge was one of the most important expressions in the classes. Program leaders witnessed this with students, and also with the facilitators. The program leaders attribute the low dropout rate of 10% to the feeling of wonder experienced in a class delivering constant new learnings to facilitators and students.

Juan Camilo Hidalgo was one of the students who was selected to attend the GERI Summer Camp at Purdue University in the summer of 2015. He was selected because of his inquisitive spirit and his excellent performance in Innobotica. Today, he is an electromechanical engineering student in a public college in Medellín. He previously worked with one of the partner educational technology companies. He was teaching robotics to other students in public schools in the city. This quote from an email written by Juan shares part of his story, and it illustrates the awareness that has been generated in Juan's mind about STEAM education, skills and careers:

> At the beginning of the year 2014 when I was studying the eleventh grade, a proposal came to the classroom that would help us grow personally and see the world with eyes full of innovation. It was a project being carried out by the science and technology innovation center of Medellín, Ruta N, and the entity interested in empowering human talent in children and young people through STEAM skills...The proposal consisted in being in extracurricular classes learning to arm and disarm robots, learning different programming languages and creating an innovation project with which we would participate by a trip to the Geri camp at Purdue University (West Lafayette – Indiana).

Through the program, many students have made an impact in their own communities. Again from Juan Camilo Hidalgo:

> They put their trust in me, and I was able to take my great experience as a student, innovator and facilitator to the different schools in Medellín, which led many of them to be motivated to do a project, and we could participate in the competitions that Pygmalion had proposed in the UVA of the Imagination in Villa Hermosa. What for many is a stroke of luck, I see it as the result of excellent teamwork and the support of people who put their hearts to what they do.

The Innobotica program has impacted a significant number of students, teachers and facilitators. In 2015, more than 2000 students came to the city's botanic garden and enjoyed a science contest. Three levels were available for students with different experience. During 2015, more than 100 facilitators (engineering students) went to more than 120 public schools sharing their related science experiments. More importantly, they shared the narrative that people can become a change maker for their own neighborhoods.

Those programs are not active currently, but they were the seed for the office of the Secretary of Education in the city, who started to implement robotics programs, and coding programs, with methodologies like PBL (project-based learning), promoting problem-solving skills in students. See Table 7.1 for a summary of all the Horizon STEAM programs.

**Table 7.1** Horizon STEAM program summary

| Program | Method | Number of students | Year |
|---|---|---|---|
| Interchange | Explore some amazing experiments, based on the research of PhD students, with engineering challenges from the EPICs method (service learning design) and demonstrations with fun hands-on STEM activities | 730 | 2014–2015 |
| Innobotica | PBL method based on coding and robotics competitions framework | 4000 | 2014–2016 |
| Engineering N | PBL based on the design thinking steps | 300 | 2014 |

## 7.4 Policies That Support the Programs

As previously noted, Ruta N Medellín (2011) is the center of business and innovation in Medellín. It is a public corporation, founded by Empresas Publicas Medellín (EPM), a public utilities company serving Medellín and beyond; UNE, a telecommunication company; and the Mayor's office. Those three stakeholders have been key in the consolidation of the innovation ecosystem.

From 2009 to now, this corporation has the responsibility to implement the Science, Technology, and Innovation (ST + i) policy for the city. The biggest challenge, but also the biggest aim, is to help to transform the industrial economy to a knowledge economy by increasing investment in ST + i, increasing infrastructure to prototype new solutions for companies and government, and to connect citizens with cutting edge technologies that inspire new talent for a knowledge economy. The corporation has led implementation of STEAM education programs as part of its mission to implement ST + i public policy, while applying principles of equity and transparency in order to promote an innovative cultural mindset.

### *7.4.1 Overall Setting and Plans in Medellin*

The ST + i Plan was born as a result of a collective construction of the Regional Innovation System (SRI) and became a public policy of the city of Medellín in Agreement 024 of 2012 of its Municipal Council. The ST+i Plan was designed as a holistic coordination and integration system that meets the demand for robust economic development, and for environmentally sustainable and socially cohesive strategy. That strategy includes different public and private entities related to science, technology and innovation, strengthening the city and region's innovation system. The overall objective of the ST+i Plan is to encourage, promote and coordinate policies; to support R&D in science, technology and innovation; all with the goal of identifying and capitalizing on new knowledge businesses.

This plan has become part of the mindset of the entrepreneurs in the city, but also large companies like Haceb (domestic appliances), and Bancolombia (banking).

Bancolombia has created their own innovation hub. Medellín's footprint of innovation is seen in both new businesses and in the adoption of new technologies in large companies.

However, one of the biggest challenges of the ST + i plan, a challenge specifically addressed in the plan's design, is to show results in education by 2021. The talent development and STEAM orientation pursued for Medellín in 2021 is clearly envisioned:

> The entire population of children and young people are enrolled in primary and secondary schools. Eighty per cent of high school graduates continue to study at university, especially at the university level. In the new disciplines of engineering and biosciences. 50% of the graduates of these careers continue their studies in doctoral programs in Colombia and in global centers of excellence. The remaining 20% of high school graduates register in technical and technological training programs or are part of training programs in the humanities (Ruta N Medellín, 2011, p. 8).

The plan's STEAM framework has been an important driver in spreading the ideas and pillars of the ST + i Plan for public school students. The plan specifically shares this goal: "Intensify the academic programs in basic sciences in primary and secondary using new technologies" (Ruta N Medellín 2011, p. 30). The STEAM framework has been important to achieve the goals of the larger plan. It has stated in a public and high-profile document that a new approach to science in schools is important, and that new technologies are important to accelerate social appropriation of knowledge from primary schools to colleges.

The most recent development comes from the current city administration. Starting in 2016 and continuing into 2019, the city's Secretary of Education developed a new program, *STEAM + H Territory*. This program has further extended the consistent pursuit of the Medellín 2021 vision. The new program's goal is to achieve an even broader voluntary agreement to transform Medellín into a territory that engages students and teachers in sciences and investigation, while strengthening at the same time the human component that allows them to be knowledge citizens.

## 7.5 Education Philosophies and Methods

### *7.5.1 Key Education Principles*

As the flagship STEAM program of Ruta N, the Horizons program has been the program working consistently over several years to expand STEAM education consistent with the Medellín 2021 vision. The Horizon program has embraced the STEAM framework through the following principles and methodologies.

- Students always have the opportunity to understand an object or phenomenon.
- Students explain, discuss and argue their ideas about an object or phenomenon.
- Instructional content is shared progressively in order to enhance the ability of students to take ownership of concepts.

- Activities are planned to always allow the students to be autonomous—do-it-yourself approach.
- Failure is viewed as an important aspect of student learning.
- Learning experiences are based on a service learning design. All programs enhance the ability of the students to analyze their own context (school, neighborhood, city), identify problems, and think about solutions.

### *7.5.2 Deep Educational Philosophies*

This principles of the Horizons program are based on a philosophy strongly tied to the local needs of Medellín. The primary goal of the program is to promote self-esteem in every student. In Ruta N's view, and in the view of other stakeholders, the most important task is to give citizens, male and female, a strong sense of self-worth. It is also important to engender common sense about what kinds of activities in life are productive and which ones are not, given many citizens are subject to constant offers from the illegal market.

Meanwhile, having the right talent for the XXI century labor market is important, but it is secondary to the prior concerns. Citizens: (1) with a high level of critical thinking; and (2) focused on creating solutions to the problems that surround them; are less likely to be influenced by unproductive alternatives. Citizens with a global view will better understand where opportunities exist to bring in solutions from all over the world while making strong and positive bonds across cultures and countries.

A live illustration of these ideas can be heard in the voice of student Maria Daniela Gamba. In her closing speech of the ceremony of Engineering N, on October 24, 2014 at San Cristóbal School, in front of hundreds of students and their families, she said:

> We have been able to realize that science also represents something playful and attractive for young people, with it we can build our life project from now on.

In Horizon programs, Ruta N wishes to leave students with those kinds of ideas. First, it is intended that students close the last session feeling that humanity has barely explored its own small neighborhood. Second, it is intended that students see how life oscillates between endless possibilities. These are the most important learnings about innovation in education, because when students feel the abundance of ideas, they start to see their own world as a place full of opportunities. It is critical for Medellín that students see their city as a place full of opportunities.

Ruta N leaders believe these goals are achieved for many students. Abundance thinking is one of the most important achievements in this program. Abundance thinking is the basis for hope, the core educational philosophy of the Horizons program, and the basis for the educational methods of the program. Wanting to learn and having curiosity were the main keys for Horizons participants to overcome their fears, to leave their prejudices behind, and to understand that STEAM topics and the STEAM approach to life is not exclusively for experts.

## 7.6 Outcomes

The structure of Medellín's education system is based on the national structure of the education system. The Ministry of National Education (MNE) delegates to the Secretariats of Education responsibility for implementing the system and ensuring its proper functioning. The system is organized as follows: formal education institutions at the pre-school and basic levels, and higher education; as well as informal educational institutions and programs, strategies and activities. In informal education in Medellín, the public sector, the private sector and academia have articulated a commitment to education as a decisive factor in social transformation.

### *7.6.1 Quantitative Outcomes*

Since 2002, the city has been increasing the quantity and the quality of its educational offerings. Enrollment in Medellín in 2016 was 450,256 students in 631 private and public schools (Education Secretariat of Medellín 2017). The quality of Colombian education is measured with a national index issued by the Ministry of Education. In the report of the ranking for the year 2018, Medellín's quality results had increased 13% in middle school, 11% in high school, and 5% in primary. Notwithstanding these results, Medellín must bring more attention to primary students and teachers. Still, as measured by the index, the number of students taking a good class, in a good classroom, with an inspiring teacher is increasing year by year.

There has been a boost through use of technology and media in the number of educational institutions served: 135 in 2016, to 169 in 2018. Across 229 educational institutions, there have been 884 STEAM training activities working with sixteen partners and the participation of more than 500.

Transformation in education has mixed with other factors (e.g., transportation, culture) to drive a transformation in *security*. The rate of murders per 100,000 people has decreased from 368.7 in 1992 to 21.7 in 2017.

### *7.6.2 Qualitative Outcomes*

Program leaders have observed two important, if less measurable outcomes. First, STEAM programs come with opportunities to travel, to learn about other countries and cultures, and to learn that the outside world is big but not that far away. Such an awareness among staff, teachers and students has led to many positive developments, especially many science and robotics clubs, and the related opportunity to compete with students from all over the world.

Second, each month high level managers, CEOs, directors, entrepreneurs, policy makers, and innovators have a date in Ruta N for a regular strategy meeting. In that meeting, important projects are shared with ecosystem leaders. The meetings involve not only domestic organizations, but also leaders of global companies with operations in Medellín. Those global players are looking for the talents of Medellín students. This meeting is one indicator of a developing relationship between education and industry.

## 7.7 Lessons Learned

Many lessons have been learned over the years of these programs. Some lessons reinforced original beliefs, and others have required the program to adapt. Below is a summary of the main learnings.

### *7.7.1 Reinforcing Lessons*

- Government initiatives that expand access to science and arts have helped students see beyond their individual circumstances and to think about their lives more globally.
- The STEAM framework allows students and teachers think about how they can have a positive impact in their own communities.
- One of the next steps is to strengthen the link between companies and the talent of the students involved in STEAM programs. This is in order to increase students' understanding of the labor market, but also for students' self-esteem for the coming time when that person approaches a real situation in a science, technology or innovation job.
- Over time the students and teachers became more confident and globally connected with the STEAM framework. Many of them understood the link between that framework and the skills that industry and government need to confront XXI century challenges. *This is the start of a virtuous cycle that benefits our community.*

### *7.7.2 Unexpected Lessons and/or Those Requiring Adaptation*

- During the last five years, Ruta N Corporation has been working mostly with students in order to provide them with a STEAM orientation. Over time, it became more and more relevant to link the teacher with user-centered design experiences in order to change the "teacher as dictator" mindset to a "learning experiences designer" mindset.
- For Medellín students, due to the culture, being smart in school is not always popular and is subject to negative peer pressure. STEM content spreads two bet-

ter messages, that students can create real solutions for their own communities, and that it is relevant and even cool to be intelligent, inquisitive, creative, collaborative, and a team player.
- It is important to have gender-oriented STEAM programs because there is gender inequity in the IT industry, which is male centered.

## 7.8 Conclusion

It was noted that Medellín, Colombia has seen a drastic and wonderful improvement in the security situation of the city. The greater than 90% decrease in the murder rate helps us to understand several things: how dark things were in the past, how bright things are now, how even more reform can make things even better, and how a return to scarcity thinking and drug dealing could undermine the progress achieved.

The city and its people have come far, and these changes will only continue with sustained work such as described here. Without programs like the ones described in this chapter, Medellín could repeat its past stories, where students grow up thinking that to be a bad guy—a *badass* in the local lingo—is something to be proud of. This perception remains a shadow in the mindset of many students.

Talent development needs to be the core of a close connection between education and economy. In particular, the education and industry connection needs to continue to grow. The STEAM model is the right model to strengthen the bond. In addition, the global perspective of staff, students and teachers is promising. The way that students in STEAM programs and many more in the city are thinking about their own lives is more global, more connected with science and arts, and populated with greater hope for the future.

Declaring Medellín as a STEM + H Territory, a Medellín for science, innovation and humanities, will further radiate this educational focus into the secondary and the primary basic schools, and will provide a platform to focus training efforts on teachers. This community of teachers will share an abundance thinking mentality with students, using STEAM educational models, aligned with the goals of Medellín 2021, and enhancing the virtuous development cycle already underway in Medellín.

## References

Centro de Ciencia y Tecnología de Antioquia. (2018). CTA. Retrieved July 12, 2019 from https://cta.org.co/.

Education Secretariat of Medellín. (2017, May). Educación en cifras. Retrieved July 13, 2019 from https://medellin.edu.co/secretaria/educacion-en-cifras?showall=1.

Empresas Publicas Medellín. (2019, June). Unidades de Vida Articulada – UVA. Retrieved July 12, 2019 from https://www.epm.com.co/site/nuestros-proyectos/proyecto-uva/unidades-de-vida-articulada.

Markham, T., Larmer, J., & Ravitz, J. (2003). Project based learning handbook: A guide to standards-focused project based learning for middle and high school teachers. Buck Institute for Education.
Metro de Medellín. (2019, May). Metro de Medellín: Calidad de vida. Retrieved July 12, 2019 from https://www.metrodeMedellín.gov.co/.
National Information STEM Education Network. (n.d.). NanoDays. Retrieved July 12, 2019 from http://www.nisenet.org/nanodays.
Proantioqia. (2016). Proantioquia: Fundacíon para el desarollo. Retrieved July 12, 2019 from https://www.proantioquia.org.co/.
Purdue University Materials Engineering. (n.d.). Carol Handwerker. Retrieved July 12, 2019 from https://engineering.purdue.edu/MSE/people/ptProfile?resource_id=11509.
Ruta N Corporation. (n.d.). Ruta N Medellín: Centro de innovación y negocios. Retrieved July 12, 2019 from https://www.rutanMedellín.org.
Ruta N Medellín. (2011, July). Plan de Ciencia, Technología e Innovación de Medellín – 2010. Retrieved July 13, 2019 from https://www.rutanmedellin.org/images/programas/plan_cti/Documentos/Plan-de-CTi-de-Medellin.pdf.
Sanchez, A. (2013). The reinvention of Medellín. *Lecturas de Economía, 78*, 185–227.
Sistema de Bibliotecas Públicas de Medellín. (n.d.). Sistema de Bibliotecas Públicas de Medellín: Un programa de la Secretaría de Cultura Ciudadana. Retrieved July 12, 2019 from https://bibliotecasmedellin.gov.co/cms/.
Universidad EAFIT. (n.d.). Ingeniería N. Retrieved July 12, 2019 from http://www.eafit.edu.co/ninos/otros-proyectos/a-la-medida/Paginas/ingenieria_n.aspx.
University of Texas at Austin $IC^2$ Institute. (2019). $IC^2$ Institute. Retrieved July 12, 2019 from www.ic2.utexas.edu.
Zintgraff, C., Fletcher, C., Jordan-Kaszuba, J., & Webb, J. (2015). StemDev Medellín and an instrument proposal to assess regional STEM-economic development alignment. In J. Slovak & D. Gibson (Eds.), *Building sustainable R&D centers in emerging technology regions* (pp. 209–242). Brno: Masaryk University.
Zintgraff, C., Daza, M. F., Rodriguez Vides, A., Fletcher, C., Kazsuba, J. J., & Webb, J. M. (2018). Cultural and historical influences on a project-based learning training program in Medellín, Colombia. In P. Young (Section Ed.) and J. M. Spector, B. B. Lockee, & M. Childress (Volume Eds.), *Learning, design, and technology: An international compendium of theory, research, practice, and policy*. New York: Springer.

# Chapter 8
# San Antonio's Cybersecurity Cluster and CyberPatriot

**Joe Sánchez and Cliff Zintgraff**

**Abstract** San Antonio, Texas, U.S. has a robust cybersecurity cluster and the second highest number of information security professionals in the United States. In that setting, the CyberTexas Foundation runs a program called CyberPatriot, a local instance of the national program by the same name. In the local program, over 300 middle and high school student teams, led by teachers and supported by cyber security professionals/mentors, compete in a schoolyear-long mainly after-school program where students learn cybersecurity knowledge, skills and best practices. The project-based learning and competitive approach draws the support of industry, colleges and universities, government, non-profits, and of numerous professionals in the region. Explicit inclusion of cybersecurity education in various policy documents encourages long-term development, and the resulting talent pipeline has played a role in attracting new business. This chapter describes the industry history, program, industry cluster support, pedagogy, partnerships, and city and associated policies that drive the program forward. The chapter presents a version of the *STEM in the Technopolis* virtuous cycle described in Chap. 1 of the current volume, one made specific to CyberPatriot in San Antonio.

## 8.1 Introduction

San Antonio, Texas, USA has a robust cybersecurity cluster and the second highest number of information security professionals in the nation, second only to the Washington D.C. area. In that setting, an area non-profit, the CyberTexas Foundation, runs a program called CyberPatriot, a local instance of the national program by the

J. Sánchez (✉)
CyberTexas Foundation, San Antonio, TX, USA
e-mail: jsanchez@cybertexas.org

C. Zintgraff
The University of Texas at Austin, Austin, TX, USA
e-mail: cliff.zintgraff@utexas.edu



C. Zintgraff et al. (eds.), *STEM in the Technopolis: The Power of STEM Education in Regional Technology Policy*,
https://doi.org/10.1007/978-3-030-39851-4_8

same name. San Antonio's explicit inclusion of cybersecurity education in various policy documents has encouraged long-term development of the program. The resulting talent pipeline has played a role in attracting new business and in developing career professionals. Those businesses and professionals in turn support CyberPatriot.

The current chapter describes in greater detail San Antonio's CyberPatriot program. The program's roots in a sector of high importance to the local community form a strong foundation for content, vocal support, volunteers, mentors and funding. With the highest number of competing teams in CyberPatriot in the U.S., CyberPatriot is a rich source of quality STEM education experiences for area students. It is a strong example of: (1) a virtuous cycle: (2) supported by regional STEM education policy; and (3) that connects K-12 STEM education to the region's high priority economic sectors.

## 8.2 About the Case

### *8.2.1 About the Program*

CyberPatriot is a national program run by the Air Force Association (AFA) to inspire K-12 students toward careers in cybersecurity or other science, technology, engineering, and mathematics (STEM) disciplines critical to our nation's future. The Center for Infrastructure Assurance and Security (CIAS) at the University of Texas at San Antonio (UTSA) designed and developed CyberPatriot after it had developed a university level competition, the National Collegiate Cyber Defense Competition (NCCDC). The CIAS has been operating the CyberPatriot competition since the inception of the program.

CyberPatriot, a national program, began in 2008. In 2010, as AFA and UTSA emerged beyond their two pilot years, other San Antonio area leaders learned of the program and worked with high schools to form teams for CyberPatriot's first open competition. This first year San Antonio fielded 24 teams among the 674 teams nationally, with one San Antonio team placing third nationally in the Open Division.

Today, San Antonio's local CyberPatriot program is operated by the CyberTexas Foundation, an area educational non-profit focused on cybersecurity education from primary school to the university level. Main elements of the program include the following.

- CyberPatriot students volunteer to participate in the program.
- CyberPatriot has two competition divisions: Open Division and Junior ROTC (all services).
- CyberPatriot activities consist of team formation, student training, teacher and mentor training, sequential, virtual rounds of competitions from September to the national finals round in March/April.
- Teams are selected and determined by the team's coach, who must be a school staff member.

- Prior to each competition round, teams download prepared operating systems (Microsoft, Linux variants, etc.) which are corrupted with vulnerabilities.
- Learning occurs primarily in afterschool programs. Some schools may allocate elective class time, study periods, or other times for students during the school day.
- Students learn from training materials prepared and delivered by their coach (teacher), mentor or provided online by CyberPatriot. They also learn from training plans passed-on by previous team members who have graduated.
- In the competition, over the course of a six hour round, teams must "find and fix" as many vulnerabilities and misconfigurations as possible and as quickly as possible.
- Teams also compete in network configuration and digital forensics tests.
- Now in its twelfth year, the competition had over 6500 high school and middle school teams compete from all states and in many foreign countries.
- San Antonio had 317 teams registered in the 2018–2019 competition year.
- For the fourth year in a row, the total number was more than any other city in the nation.

San Antonio's 317 teams, with up to six members each, consisted of approximately 1900 students. There were 130 teachers and 50 mentors supporting the program's 2018–2019 efforts. The project-based learning and competitive approach draws the support of industry, non-profits, and of the numerous professionals in the region.

CyberTexas culminates CyberPatriot's competition year with a celebration. The San Antonio Mayor's Cyber Cup Luncheon and College Fair is held each March to thank the participants, coaches, mentors and sponsors for their wonderful work. The event includes a College Fair with professors from local centers of excellence sharing their respective programs with high school and middle school CyberPatriots. In addition, area military, Department of Defense and industry partners are in attendance to share the type of critical missions they have and the types of jobs available.

All area CyberPatriot participants are recognized, while also honoring the best among eight categories including all-girls teams, middle school teams, rookie teams, Junior ROTC teams and the best performing team. With this event at the center, CyberTexas helps facilitate $2 million in scholarships and internship opportunities at the event. Foundation partners and sponsors provide the internship funds and provide summer internships to graduating seniors, internships that often continue into the students' college careers.

### 8.2.2 History of Cybersecurity in San Antonio

San Antonio has been the home of significant technological and STEM-based advancements for over a century. It is home to the nation's first military flight, the creation of the first computer local area network (ARCnet), and the first video teleconferencing device (MINX). San Antonio is home to the first balloon-expandable heart stent, the first military Intrusion Detection System, and the first military Computer Emergency Response Team (CERT).

The seeds for cybersecurity development and the sector's economic strength in San Antonio were planted by the U. S. Air Force. The United States Air Force Security Service, known as USAFSS, located in San Antonio in 1949. It was the service's decision in 1985 to consolidate and physically move its mission of Communications Security (COMSEC), Computer Security (COMPUSEC) and TEMPEST to San Antonio that led to today's military missions and the resulting cybersecurity ecosystem. It was from the Air Force mission of Information Security that a small number of community leaders moved forward with a STEM-based idea to develop academic programs to educate its employees and develop an academic and workforce pipeline.

In 1999, the city's landscape of Information Security was predominately found behind highly classified walls of a cluster of buildings on Joint Base San Antonio (JBSA)-Lackland known as Security Hill. Most, if not all, information security companies in town supported Security Hill's numerous missions and programs. On the academic front, only one book chapter in one Computer Science Network course at the University of Texas at San Antonio touched on Information Security. Likely the same could be said for most American universities at the time. Air Force civilian leaders from Security Hill identified the need to collaborate with area colleges and universities to develop programs to educate their military and civilian workforce.

### *8.2.3 Development of Education in the Cybersecurity Technopolis*

In that timeframe, this chapter's lead author expressed a desire to the newly installed President of the University of Texas at San Antonio (UTSA), Dr. Ricardo Romo, to develop academic programs and research and development opportunities in cybersecurity, all in partnership with the U. S. Air Force's Air Intelligence Agency (AIA). At the same time, a young entrepreneur named David Spencer also recognized the need for information security educational programs. David Spencer helped recruit other local leaders, Marc Gravely and the lead author. The full team included area university representatives and professors, military and civilian information security specialists, business leaders and city and county staff members. From this initial partnership, one can track a twenty-year history of successful programs at primary, secondary and post-secondary educational levels, adult and youth training programs, cybersecurity R&D, and related economic development.

Interest grew between the University of Texas at San Antonio (UTSA) and the Air Force to collaborate on R&D and bachelor's degree development. This collaboration led to formation in 2001 of the UTSA Center for Infrastructure Assurance and Security (CIAS) by Dr. Glenn Dietrich and the lead author. In 2003, Dr. Dietrich and your lead author led the way for UTSA as it earned the designation of a Center of Academic Excellence in Information Assurance Education from the U. S. National Security Agency. The formation of the CIAS was pivotal in San Antonio's drive to develop our community's cybersecurity technopolis. A year later, a dual college credit program, the Information Technology and Security Academy (ITSA), for high school students was established. The authors were actively involved in ITSA's creation.

Two months after San Antonio's first CyberPatiot participation, when the ITSA team placed third in the Open Division, all members of that team were selected for internships with AIA's Air Force Computer Emergency Response Team (AFCERT) on Security Hill. With their gained network operational knowledge, Secret clearance and adult mentorship, the ITSA team of all seniors won the 2011–2012 CyberPatriot National Championship. Figures 8.1 and 8.2 are pictures of the airport welcome given to this national championship team.

In ensuing years, San Antonio's emerging CyberPatriot training model was driven by local leaders with strong Air Force, Security Hill and community roots. Local teams benefited from mentors from the military and then burgeoning cyber industries to prepare them for the evermore complex competition operated by the CIAS. Area cyber and IT business offered their experts, facilities and funding to deliver cyber training. Each of these components required detailed planning and resources. The current CyberPatriot training program involves 2–3 training sessions per year for coaches, weekly sessions for students in their respective schools, and an additional 4–6 sessions for teams competing at the national finals.

**Fig. 8.1** San Antonio Mayor Julian Castro's welcome home to the CyberPatriot National Championship team from ITSA

**Fig. 8.2** Airport welcome home to the CyberPatriot National Championship team from ITSA. Three students wearing their San Antonio Mayor's Cup bomber jackets, awarded for winning the San Antonio Mayor's Cyber Cup

In support, area cyber and IT companies formed coalitions and programs with the assistance of the San Antonio Chamber of Commerce, the San Antonio Hispanic Chamber of Commerce, the Alamo Chapter of the Armed Forces Communications and Electronic Association (AFCEA), and the local Information Systems Security Association (ISSA). Numbers of IT and cyber-focused companies in San Antonio grew from a few handfuls in the early 2000s to now over 1000 companies.

The CyberTexas Foundation led development of the model to sustain the city's Cyber Technopolis. The model has continuously been enhanced by CyberTexas, the greater cyber ecosystem, and by many sponsors and supporters. For many years the foundation connected cyber experts in industry and military with area CyberPatriot teams. In partnership with Rackspace, the foundation offered summer cyber and IT camps.

The CyberTexas Foundation's goal has been "all-in" community support. For the past four years, San Antonio has had the most CyberPatriot teams register to compete. The effort started with 24 teams in 2010–2011. In 2018–2019, there were 317 teams (among 6500+ nationally), with nearly 100 of the teams at the middle school level. With the exception of one year, San Antonio has placed at least one team in the National Finals, now totaling nearly 20 finals teams. In the 2016–2017 CyberPatriot year, the Holmes High School Air Force Junior ROTC Team won the Cisco Networking Challenge All-Service Division, defeating all other ROTC service teams at the National Championship.

The authors summarize the San Antonio Cyber Technopolis model as comprised of:

- Recognizing and utilizing our community's strengths
- Identifying and enabling leaders: main organization leaders, organizational collaborations, and sub-program leaders
- Urging academic institutions to establish and grow cyber programs

- Collaboration with area military organizations and leaders
- Collaboration with area cyber and IT businesses
- Collaboration with area school district superintendents and their respective school principals
- Support from local city and county leaders
- Recognition and celebration of student and organizational accomplishments

By helping to develop San Antonio's Cyber Technopolis, the CyberTexas Foundation helps meet the need for near-term and future cybersecurity experts to defend the U.S., its infrastructure and its citizens. San Antonio's Cyber Technopolis is strong, growing, and an example of a virtuous cycle that might be replicated in other communities, adapted to their local economic development priorities.

## 8.3 Industry Cluster Served

San Antonio has promoted itself, with significant outside acknowledgement, as Military City USA due to its long history of hosting military bases and its support for those missions. Since the April 1949, the organization known as 16th AF Air Force and its predecessor organizations have operated in San Antonio. That agency eventually became the owner of Air Force networks, and then of cybersecurity concerns, which led to development of a local cluster of military cyber missions, related defense contractors, cybersecurity startups, world-renowned education programs, five NSA- and DHS-designated programs at local colleges and universities, and the embrace of cybersecurity education by secondary educators. In addition, the National Security Agency (NSA) has operated a Cryptologic site in San Antonio known as NSA/CSS Texas since the 1960s.

In 2009, with the help of the lead author and CyberTexas Foundation Senior Advisor, Chris Cook, San Antonio was selected by the Air Force as the location for the new 24th Air Force, responsible for overseeing, monitoring and securing Air Force networks globally. The 25th Air Force at JBSA Lackland oversees, monitors and secures Top Secret Air Force networks and conducts intelligence, surveillance and reconnaissance missions. In the fall of 2019, 24th AF and 25th AF merged into the 16th AF.

The San Antonio cyber ecosystem has developed in many other ways. Other cyber-focused academic institutions were formed, such as the Institute for Cyber Security and the Cyber Center for Security and Analytics, both at UTSA, and the Center for Information Technology and Cybersecurity at Texas A&M University-San Antonio. Every college and university in San Antonio developed cybersecurity degrees and are members of the Cyber Innovation and Research Consortium (CIRC) formed in 2007. Many colleges and universities developed summer cyber camps and partnerships with area high schools and middle schools.

With the growth in area cybersecurity industry partners, since 2002, the city informally proclaimed itself CyberCity U.S.A. The informal proclamation reflected a speech delivered by Congressman Ciro Rodriguez (D-TX) in support of HR 3394, the Cyber Security Research and Development Act. He stated:

> San Antonio has been a leader in developing the type of technology and educational programs made possible under this bill. A growing partnership of educational, private enterprise and military expertise make San Antonio "Cyber City USA" (U.S. House of Representatives, 2002, p. H214).

In addition to the military and Department of Defense, the San Antonio area is fortunate to have many partners across industry, education and institutions. Lists of these partners and selected impacts are listed below:

- *Industry players include*: Abacus, Accenture, Akima, Air Force STEM, AT&T, Bank of America, Booz|Allen|Hamilton, CNF Technologies, Deloitte, Denim Group, Digital Defense, Digital Fire, Diligent, Fed ITC, Frost Bank, General Dynamics IT, HEB, Infocyte, Innove, IPSecure, Jefferson Bank, Jungle Disk, New Horizons Computer Learning Center, Noblis NSP, Parsons, Port San Antonio, Rackspace, SAIC, SecureLogix, Spurs Sports and Entertainment, Symantec, and USAA.
- *Local Government players include*: Bexar County, City of San Antonio, City Public Service
- *Federal Government players include*: 16th Air Force, 24th Air Force, 25th Air Force, Cryptologic and Cyber Systems Division, Federal Bureau of Investigation (F.B.I.), Joint Information Operations Warfare Center, National Security Agency/Central Security Service (NSA/CSS) Texas, and the United States Secret Service.
- *Educational players include*: Alamo Colleges (Northeast Lakeview, Northwest Vista, Palo Alto College, San Antonio College, St. Philip's College), Hallmark University, Incarnate Word University, Our Lady of the Lake University, Purdue University Global, St. Mary's University, Trinity University, Texas A&M – San Antonio, and the University of Texas at San Antonio.
- *Educational Center and Institution players include*: Center for Infrastructure Assurance and Security (UTSA), Institute for Cyber Security (UTSA), Cyber Center for Security and Analytics (UTSA), Open Cloud Institute (UTSA), Center for Information Assurance Management and Leadership (OLLU), Center for Information Technology and Cybersecurity (Texas A&M – S.A.), and the National Security Collaboration Center (UTSA)
- *Institution and Association players include*: Alamo Air Force Association (AFA), Alamo Chapter of Armed Forces Communications and Electronics Association (AFCEA), Alamo Information Systems Security Association (ISSA), BSides San Antonio, San Antonio Chamber of Commerce, and the San Antonio Hispanic Chamber of Commerce.
- *Economic impact of the Information Technology cluster totals*: $10 million (Butler and Stefl 2014).
- *Number of Information Technology jobs*: 34,000 (Butler and Stefl 2014).

Figure 8.3 share accomplishments of the CyberPatriot program since its inception. This illustration highlights the deep involvement of industry cluster partners. Figure 8.4 is an illustration of the sectors of the technopolis at work supporting the CyberPatriot program. Note the presence of: (1) San Antonio Mayor Julian Castro; (2) university staff; (3) industry supporters, both on stage and represented on the screen; (4) parents in the audience; and (5) the students. Off stage are non-profit supporters operating the event.

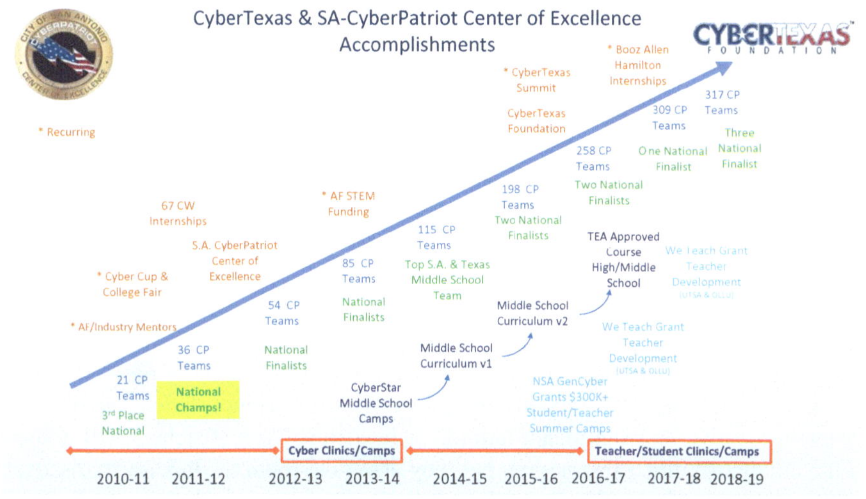


**Fig. 8.3** CyberPatriot accomplishments; Highlights industry cluster role

**Fig. 8.4** The 2011–2012 San Antonio Mayor's cup winners receive their awards

## 8.4 Education Philosophies and Methods

As a mainly out-of-school program, CyberPatriot applies two complementary educational methods, problem-based learning, and mentor-based learning. Mentor-based learning is not unique to San Antonio's CyberPatriot program, but its development is advanced relative to mentorship in most other K-12 STEM programs.

### 8.4.1 *Problem-Based Learning*

CyberPatriot's design is fundamentally based on the philosophy of problem-based learning (Prince and Felder, 2006). In problem-based learning, students are given a problem to solve and significant latitude in how to solve it. Teachers will deliver some amount of direct instruction, but they also give room to students to discover root causes and solve problems on their own. Guidance is provided via scaffolding, a process of giving students just enough support, including just-in-time support where needed (Hmelo-Silver et al. 2007).

As previously noted, industry professionals and others from outside schools find problem-based learning appealing because it reflects what happens in real-world scenarios. The job of the program designer and education professional is to make the proper adaptations, retaining as much real-world flavor as possible, while being realistic and supportive of the primary goals (Edelson et al. 1999). Those goals are student learning of technical content and learning of the so-called twenty-first century skills of communication, collaboration, critical thinking and creativity.

Selected teams are fortunate to have teachers who have degrees in the field or have past industry experience. This occurrence is more likely in San Antonio given the presence of the cybersecurity cluster, but still unusual. CyberPatriot program leaders have observed that teams receiving instruction during the school day perform much better than those only with after-school learning. Most CyberPatriot teachers are self-aware of subject matter limitations; therefore, many work to identify mentors who interact directly with their students, bringing them into the school day if possible (Zintgraff 2016). The mentoring role has proven quite important in the CyberPatriot program.

### 8.4.2 *Mentor-Based Learning*

Some of the most successful CyberPatriot teams have made mentoring the core of their instructional strategy. Mentoring is core to such an extent that the instructional strategy can fairly be called *mentor-based learning*. Mentoring is the core method through which students learn the lessons of program and become effective competitors. Zintgraff (2016), in his study of San Antonio's CyberPatriot program, noted that when positioned properly, students see mentors as role-model-worthy peers rather than as supervisors of their work, and that positioning is seen by CyberPatriot teachers as highly effective for student learning, and as a necessary complement to the teacher's most common role, that of facilitator of the overall learning experience.

In the local CyberPatriot program, the mentors' main role is directly delivering instruction to students. This delivery sometimes takes the form of direct instruction, but at least as often takes the form of interactive sessions, trial competition sessions, and projects that help students learn the technical and twenty-first-century-related lessons. Less frequently, mentors serve as resource providers and maintainers of equipment. In their full-time jobs, mentors are usually network administrators, cybersecurity professionals, systems administrators, web developers or program managers.

The effectiveness of mentors depends on their fit inside an overall structure, and especially on the relationship between coaches, mentors and industry partners. Table 8.1 lists related steps in rough order of execution that coaches, mentors and industry partners perform, demonstrating the relationship.

Over the years, innovative lead mentors, such as Capt. Justin Smith, Air Force, developed their own mentor programs by establishing a calendar depicting training topics; e.g., Microsoft Operating Systems, networks, script development, security, etc. Once a calendar was established, the lead mentor *volun-told* their employees and colleagues to lead classroom training for one to two weeks' worth of training activities, to share their respective expertise as a mentor. Local program leaders believe this method has resulted in back-to-back National Finals participation by the Alamo Heights Army Junior ROTC CyberPatriot team.

**Table 8.1** Tasks demonstrating coach, mentor, industry partner relationship

| Description | Performed by |
|---|---|
| Coach registers team at CyberPatriot website (www.uscyberpatriot.org) | Coach |
| CyberTexas places a call for mentors | CyberTexas, mentors |
| Coaches approach CyberTexas foundation for technical assistance | Coach |
| Vetted mentors are connected with coach at school | Coach, mentor |
| Mentors register as volunteer with CyberPatriot and school | Mentor |
| CyberTexas seeks industry partners under the "Adopt a School" program | CyberTexas, industry partner |
| CyberTexas conducts training for first year mentors and coaches | CyberTexas, mentors |
| Mentor determines the base knowledge of the team(s) | Mentor, coach, students |
| Mentor develops a "CyberPatriot season of training" to prepare team (windows, windows server, networks, security, digital forensics, etc.) | Mentor |
| Mentor brings in experts in the topics above to assist in training to spread the in-classroom mentoring to ease the burden on one individual | Mentor, coach, and students |
| Mentor and his/her support staff deliver training before each competition round | Mentor, coach, students |
| Industry partners ("Adopt a School") provide day-of support through food and beverages for the team | Industry partner |
| Teams compete | Coach, students |
| CyberTexas honors all teams at San Antonio Mayor's cyber cup and college fair | Celebration by all |

At the onset of the program in 2010–2011, current volunteer members of the CyberTexas Foundation sought and aligned mentors to the twenty four teams competing. It was not too difficult a task. As the number of San Antonio teams increased over the years, more mentors were required. The Foundation was able to keep up with mentors for only 2–3 years, at which point it was no longer possible to find mentors for all teams. With 317 teams in 2018–2019, approximately 50% of the teams had mentors. These mentors are identified in part by CyberPatriot program leaders, and in part through the recruiting efforts of teachers themselves.

This state of affairs highlights both the power of connecting STEM experiences to local priority industry clusters, and also its limits. San Antonio's CyberPatriot program has the highest participation in the nation in large measure due to the importance of the local industry cluster. The number of people available, expertise available, and enthusiasm for the work forms a strong foundation. Still, there are limits on creating the richest elements of the STEM experience. Markham et al. (2003), writing about the Buck Institute's project-based learning methodology, identified adult connections as one of six attributes important in their programs. Mentor availability and a program's ability to administer connections will meet natural limits, after which point this particular rich element of the STEM experience will wane for some teams.

The CyberTexas Foundation has been working to raise the ceiling on mentor availability and other participation by outside organizations. Their mechanism is called the Adopt a School Program. The program encourages area businesses to support CyberPatriot teams through funding for equipment, mentors, software, and perhaps most important to the participating students, pizza and drinks during their multiple six hour competitions throughout the competition year.

### 8.4.3 A Different Education Strategy: ITSA

In the San Antonio community, there is one very positive methodology exception, the Information Technology and Security Academy (ITSA). ITSA was identified earlier as the home school for the 2011–2012 national championship team. ITSA is a dual-credit high school program administered by San Antonio College, a local community (two-year) college in the Alamo Colleges District. ITSA is also supported by area school districts and industry partners.

ITSA is the second of five academies created as part of the Alamo Academies program. All five of those academies, which were founded over a fifteen-year period, either anticipated or reflect the *early college high school* model that has become popular in the U.S. In the Alamo Academies' particular model, high school juniors and seniors spend a half-day at their home campus and a half-day in IT and cybersecurity-specific instruction. Students study IT, networks, systems administration and cybersecurity while earning nearly 30 hours of college credit within the Alamo Colleges District. The program is free to students and parents.

Students must apply for entrance into ITSA during their sophomore year. They must take and pass the ACCUPLACER test, a commonly used test in the U.S. for program admissions. Once admitted, students are taught by San Antonio College IT instructors. CyberPatriot competition teams are formed, and the teams have adult professional mentors who are in active military service or in industry. The vast majority of students have received summer internships after their junior year, and tracking of future student accomplishment keeps the program well aligned to needs in the local workforce.

With the advanced knowledge shared by college instructors, ITSA students learn at a college level. Subsequently, ITSA CyberPatriot students have an advantage over typical high school CyberPatriot students. In after-school programs, students have teachers with varying cybersecurity skills, and mentors with varying skills, if they are fortunate to be among the fifty percent of students with mentors. Historically, ITSA CyberPatriot teams have performed better than San Antonio high school-based teams. In seven of the nine years CyberPatriot has been open for competition, at least one ITSA team has advanced to the National Finals.

Bringing programs like ITSA to greater scale, with the same depth of instruction and support as in the current program, is difficult, perhaps not possible. Still, ITSA serves an important education role in the San Antonio cybersecurity technopolis. It is a core program, one many students strive to enter. It sets the bar for knowledge, skill and competition success. Its integration with the college system and workforce sets the bar in a different way for local STEM programs. The college, school district and industry partnership that runs ITSA reflects the cybersecurity technopolis and principle of the virtuous cycle advanced in this volume.

## 8.5 Policies and Practices That Support the Program

A wide array of government, industry and other organizations have implemented a wide variety of policies in San Antonio in support of cybersecurity education. These policies move these organizations beyond vocal support. The act of creating the policies institutionalizes support, and it also makes a statement to the community about the importance of the supported programs.

SA2020, a city-initiated non-profit that advocates for and measures the city's development (SA2020 n.d.), has articulated a comprehensive plan and program for San Antonio's regional development. The plan addresses history, culture, jobs and workforce, and smart city concerns that face the city. Explicit in the SA2020 plan is the identification of the cybersecurity cluster as important to San Antonio's economic development. SA2020 specifically references the San Antonio Cyber Action Plan as a catalyst for economic competitiveness. SA2020's first yearly report in 2011 report cited the new Air Force mission at Joint Base San Antonio by noting "the 24th Air Force is a critical hub of the nation's cybersecurity" (SA2020 2011, p. 34).

The San Antonio Chamber pursues cybersecurity as a priority industry cluster (San Antonio Chamber of Commerce 2019; San Antonio Chamber of Commerce n.d.) The Chamber's Cybersecurity Council is composed of private-sector, public sector, federal government, academic, and entrepreneurial members dedicated to growing San Antonio's cybersecurity industry. The chamber renamed their Information Technology Committee to the Cybersecurity Council, reflecting the outsized role cybersecurity plays in the community.

The council serves as the driving force behind Cybersecurity San Antonio. The City of San Antonio and San Antonio Chamber of Commerce aligned strategy and funds to form the Cybersecurity San Antonio partnership, which seeks to "accelerate the growth and national reputation of San Antonio's cybersecurity sector by fostering a collaborative environment for innovation, job producing, investments and public-private partnerships" (San Antonio Chamber of Commerce n.d., Platform, para. 1).

The San Antonio Hispanic Chamber of Commerce (2017), a leading local promoter of STEM education, includes promotion of cybersecurity education and the importance of the cluster to the city's development. In particular, the Hispanic Chamber has integrated cybersecurity into their CORE4 STEM program. CORE4 STEM is a three-day celebration of education and career opportunities in STEM fields. It brings middle and high school students from underrepresented populations to a symposium at the city's major convention center. Additional activities occur at colleges around the city. The CORE4 Expo brings students, parents, and educators together with high-profile STEM professionals, corporations, government agencies, universities, and colleges.

Bexar County Commissioners Court acknowledged the need for future cyber experts by providing funds to assist students and parents with the cost of IT-related certification testing. The CyberTexas Foundation oversees and distributes the funding to Southwest High School and Business Careers/Holmes High School for their respective students as they test for industry standard certifications such as CompTIA's Network+ and Security+.

Port San Antonio is the redevelopment initiative for the former Kelly Air Force Base, closed during the 1990s. The Port has seen growth of more engineering, data analytics and cybersecurity high paying jobs at the Port. In the past 18 months (since spring 2018), more than 2000 new jobs have been created centered around STEM-based specialties, cybersecurity, aerospace and applied technology (robotics, AI and machine learning) (Port San Antonio 2019). Development of the cybersecurity cluster is a core part of Port San Antonio's strategy. Port San Antonio has engaged STEM-oriented education strategies through support of the San Antonio Museum of Science and Technology (SAMSAT), an upstart museum focused on delivering STEM experiences to students. SAMSAT is partnered with SASTEMIC, a STEM non-profit delivering STEM experiences to 14,000 students each year (SASTEMIC 2019). The policies of the Port San Antonio Board of Directors, who are appointed by the city's elected representatives, are contributing further to availability of cybersecurity education experiences.

CyberTexas worked with area Air Force organizations by seeking their support to have their experts serve as mentors. The 24th Air Force (Cyber) and 25th Air Force (Intelligence, Surveillance and Reconnaissance, and classified cyber missions) Commanders established written policy allowing their respective military and civilian workforce to serve as CyberPatriot mentors during duty hours as long as their respective missions were not impacted.

The Cyber Innovation and Research Consortium (CIRC) was established and managed by the lead author in 2007 as a means to bring together the cybersecurity and IT programs and professors across the community. The goal was to better understand the respective growing cybersecurity degrees and programs across the San Antonio metropolitan area. The CIRC fosters partnerships, enabling opportunities for colleges and universities to design, prepare and submit proposals and grants in the field. CyberTexas now administers the work of the CIRC (CyberTexas Foundation 2014).

The CIRC also help academic institutions to collaborate around work to become centers of excellence. Five area academic institutions earned have earned the National Security Agency/Department of Homeland Security designation as Centers of Academic Excellence (CAE). Beginning with UTSA in 2003, over the course of the next five years, San Antonio's designated institutions expanded to Our Lady of the Lake University, Texas A&M University – San Antonio, San Antonio College and St. Philip's College. No other city in the nation has five NSA Centers of Academic Excellence.

The University of Texas at San Antonio is one of approximately twenty institutions with three designations: CAE-Information Assurance Education, CAE-Information Assurance Research (CAE-R) and CAE-Cyber Operations. The most difficult to earn, the CAE-Cyber Operations designation will enable UTSA to "join the NSA with assistance in building a future workforce knowledgeable and trained in specialized intelligence, military and law-enforcement cyber operations (e.g., collection, exploitation and response) to enhance the national security of the United States" (Lutrell 2018, para. 4).

CyberTexas worked with area state legislators and local leaders to draft legislation to form the Texas Cybersecurity Council. This Council is comprised of state officials, representatives from state academic institutions and business leaders. This council established statewide cybersecurity operational and educational policies and goals.

## 8.6 Continued K-12 Talent Pipeline Development Outcomes and the Virtuous Cycle

In the most recent study of information technology's economic impact in San Antonio, Butler and Stefl (2014) concluded there were about 3500 federal and military cybersecurity professionals in San Antonio, and about 31,400 information technology professionals in federal, military and private sector positions, leading to $10 billion in yearly economic impact to the city. The authors estimate from these numbers that there are between 5000 and 10,000 cybersecurity professionals across all categories, and they lead to between $1.75 billion to $3.5 billion in yearly economic impact. While attributing benefits to any one effort is not possible, it is reasonable to believe that the city's cybersecurity education and talent pipeline development plays a large role is sustaining and growing this industry.

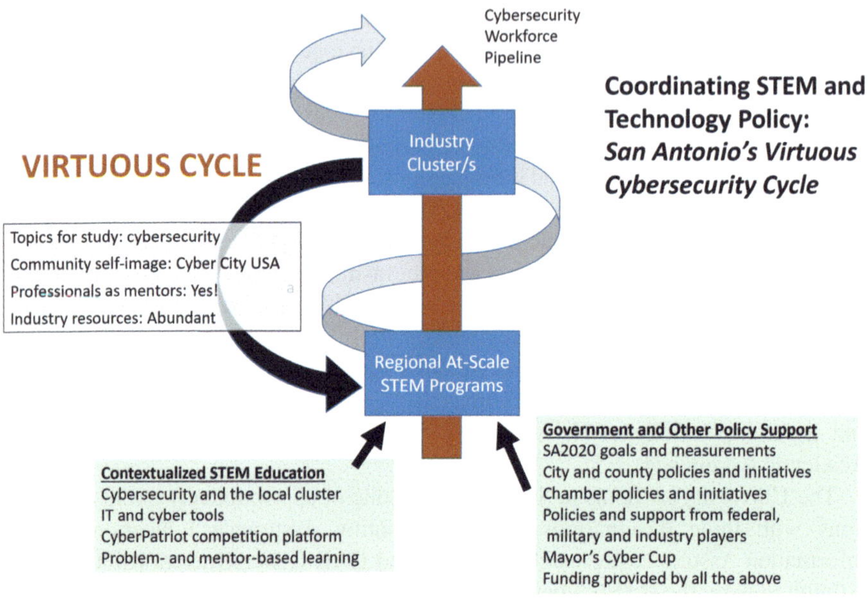


**Fig. 8.5** San Antonio cyber technopolis virtuous cycle

Figure 8.5 adapts the virtuous cycle illustration presented in the opening chapter of this volume. This illustration replaces the abstract ideas expressed in the opening chapter with details of the cybersecurity education virtuous cycle in San Antonio.

In 2015, the CyberTexas Foundation was recognized by FBI Director James Comey as one of the nation's winners of the FBI Director's Community Leadership Award. The award recognized CyberTexas for its educational cybersecurity community initiatives. These initiatives can be seen in the illustration of the virtuous cycle that has developed in San Antonio.

## 8.7 Conclusion

Air Force Brigadier General (Ret.) Bernie Skoch serves as the national commissioner of CyberPatriot. In 2016, Commissioner Skoch shared this thought during an event, commenting on San Antonio's role as the second Center of Academic Excellence (CyberPatriot 2013) in the program:

> San Antonio is doing what no one else has in promoting cyber education in youth. The partnerships you have formed among government, academe, the military, civic leaders, and your local sponsors are amazing.

This pronouncement exemplified nearly two decades of work by many passionate individuals in the field of cybersecurity in San Antonio. Their efforts exhibited vision, collaboration and the principle of acting on a community's strengths. The CyberPatriot program would not exist without the critical role UTSA's Center for Infrastructure Assurance played in establishment of the concept and development of the technical engine to operate and score the competition.

Cybersecurity is San Antonio's strength, one traced back at least decades, and perhaps even more to the original missions of the U.S. military in early twentieth century San Antonio. Other chapters in this volume track the history of other regions: the artistic, economic and cultural history of Medellín, Colombia; the development of modern China; the reaction of rural Fundão, Portugal to migration to big cities; the role of agriculture in São Carlos, Brazil. The reader should take away the lesson that local priority industry clusters are powerful platforms for student education. They are sources of inspiration, vocal support, volunteers, mentors and funding. They serve as a rallying point for joint effort in the community. They serve as a rich platform for teaching technical and other important knowledge and skills students need in the twenty-first century.

## References


Butler, R., & Stefl, M. (2014). The San Antonio IT economy: 2014 economic impact. Retrieved August 6, 2019 from https://drive.google.com/file/d/1_KLC63JGAC_iJjOH444Xb2ojl7ks-Da3B/view.

CyberPatriot. (2013). Centers of excellence. Retrieved August 6, 2019 from https://www.uscyberpatriot.org/Pages/About/Centers-of-Excellence.aspx.

CyberTexas Foundation. (2014). Higher education. Retrieved July 10, 2019 from https://www.cybertexas.org/collegeuniversity-cyber-degree-offerings/.

Edelson, D. C., Gordin, D. N., & Pea, R. D. (1999). Addressing the challenges of inquiry-based learning through technology and curriculum design. *The Journal of the Learning Sciences, 8*(3&4), 391–450.

Hmelo-Silver, C. E., Duncan, R. G., & Chinn, C. A. (2007). Scaffolding and achievement in problem-based and inquiry learning: A response to Kirschner, Sweller, and Clark (2006). *Educational Psychologist, 42*(2), 99–107.

Lutrell, P. (2018, June). NSA designates UTSA a national center of academic excellence in cyber operations. Retrieved July 10, 2019 from https://www.utsa.edu/today/2018/06/story/NSA-CyberDesignation.html.

Markham, T., Larmer, J., & Ravitz, J. (2003). Project based learning handbook: A guide to standards-focused project based learning for middle and high school teachers. Buck Institute for Education.

Port San Antonio. (2019, July). Progress report summer 2019. Retrieved August 6, 2019 from http://www.portsanantonio.us/progress-report-summer-2019.

Prince, M. & Felder, R. M. (2006). Inductive teaching and learning methods: Definitions, comparisons and research bases. *Journal of Engineering Education, 95*(2), 123–38.

SA2020. (2011). SA2020: Dream it. Map it. Do it. Retrieved August 6, 2019 from https://www.sa2020.org/wp-content/uploads/2017/10/SA2020_Final_Report.pdf.

SA2020 (n.d.). SA2020 impact report 2018. Retrieved August 6, 2019 from http://www.sa2020.org/wp-content/uploads/2019/04/2018ImpactReport_digital.pdf.

San Antonio Chamber of Commerce. (2019). Cybersecurity Council. Retrieved August 6, 2019 from https://www.sachamber.org/get-involved/cyber/.

San Antonio Chamber of Commerce. (n.d.). Cybersecurity San Antonio. Retrieved August 6, 2019 from http://cybersecuritysa.com/.

San Antonio Hispanic Chamber of Commerce. (2017). Programs: CORE4 STEM program. Retrieved July 10, 2019 from https://www.sahcc.org/programs/core-4-stem/.
SASTEMIC. (2019). SASTEMIC. Retrieved August 6, 2019 from http://sastemic.org.
U.S. House of Representatives. (2002, February). Congressional record: Proceedings and debates of the 107th Congress, second session. Retrieved August 10, 2019 from https://www.congress.gov/congressional-record/volume-148/house-section/page/H214.
Zintgraff Jr., A. C. (2016). STEM professional volunteers in K-12 competition programs: Educator practices and impact on pedagogy (Doctoral dissertation).

# Chapter 9
# Case Study: Taiwanese Government Policy, STEM Education, and Industrial Revolution 4.0

**Chao-Lung Yang, Yun-Chi Yang, Ting-An Chou, Hsiao-Yen Wei, Cheng-Yuan Chen, and Chung-Hsien Kuo**

**Abstract** The government of Taiwan is pursuing a national strategy focused on the Fourth Industrial Revolution and the associated developments in advanced manufacturing. As part of this strategy, the government is working with the secondary schools and with colleges to improve STEM education experiences and develop the country's talent pipeline. This case shares the experience of FIRST Robotics Team 6191 TFG from Taipei First Girls High School, highlighting the educational approach, rich outside-school collaborations, and the manner in which the national government has engaged the teachers and students as leaders for the national effort. The case also shares the experience of National Taiwan University of Science and Technology (NTUST), also known as Taiwan Tech, and how NTUST is using a hands-on STEM tools to teach college students to learn knowledge and skills needed to design and manage production processes. A Taiwan-specific model of the STEM Technopolis virtuous cycle is shared.

## 9.1 Introduction

*Industrial Revolution 4.0,* also called Industry 4.0, a trend referring to the fourth industrial revolution first initiated by Germany, takes the automation of manufacturing to a new level by introducing the customized and flexible mass production technologies. It recently has become a trend sweeping over the world (Kagermann et al. 2011). Taiwan, as 14th most competitive economy in the World Competitiveness Yearbook released in 2017 by Switzerland-based International Institute for

C.-L. Yang (✉) · C.-H. Kuo
National Taiwan University of Science and Technology, Taipei, Taiwan
e-mail: clyang@mail.ntust.edu.tw; chkuo@mail.ntust.edu.tw

Y.-C. Yang · T.-A. Chou · H.-Y. Wei · C.-Y. Chen
Taiwan First Girls High School, Taipei, Taiwan
e-mail: cychen2@gapps.fg.tp.edu.tw



C. Zintgraff et al. (eds.), *STEM in the Technopolis: The Power of STEM Education in Regional Technology Policy*,
https://doi.org/10.1007/978-3-030-39851-4_9

% of graduates in ICT, Engineering, Math & Natural Sciences

| | Nation | Percentage (%) | Number of people (million) | year | | Nation | Percentage (%) | Number of people (million) | year |
|---|---|---|---|---|---|---|---|---|---|
| 1 | Saud Arabia | 38.3 | 2.4 | 2013 | 12 | Finland | 27.0 | 1.5 | 2014 |
| 2 | Malaysia | 37.2 | 7.2 | 2015 | 13 | Thailand | 26.8 | 9.9 | 2015 |
| 3 | Germany | 34.8 | 18.4 | 2014 | 18 | England | 25.8 | 18.6 | 2014 |
| 4 | Qatar | 33.6 | 0.1 | 2012 | 20 | Sweden | 25.5 | 1.6 | 2014 |
| 5 | Korea | 31.0 | 13.4 | 2014 | 21 | France | 24.7 | 13.3 | 2014 |
| 6 | Austria | 30.2 | 1.5 | 2014 | 32 | Japan | 21.1 | 15.2 | 2014 |
| 7 | Greek | 29.6 | 2.0 | 2014 | 34 | Canada | 20.7 | 5.7 | 2013 |
| 8 | Taiwan | 29.5 | 10.3 | 2015 | 41 | New Zealand | 19.9 | 1.0 | 2014 |
| 9 | India | 29.2 | 257.5 | 2013 | 50 | Australia | 16.7 | 5.6 | 2014 |
| 10 | Mexico | 28.1 | 14.3 | 2014 | 52 | America | 15.3 | 48.4 | 2014 |

Resource：IMD World Competitiveness Yearbook 2017, OECD, and UNESCO Statistics database

**Fig. 9.1** Graduate in science index in 2017 (IMD 2017)

Management Development (IMD) (IMD 2017), also participates in this crusade to enhance its competitiveness in manufacturing sectors such as information and communications technology (ICT) and semiconductor industries.

The strong manufacturing competitiveness of Taiwan has relied on Taiwan's education system, especially on its emphases on science, technology, engineering and mathematics (STEM) education. How to cultivate talents in STEM fields is a main indicator of national competitiveness when facing Industry 4.0 challenges.

For decades, Taiwan's government has placed great emphasis on STEM-related topics in the existing education system, especially in strengthening the ties between the education system and economic development (DTVE 2019). According to the 2017 IMD World Competitiveness Yearbook, the percentage of graduates in STEM fields in Taiwan was 29.5% and ranked eighth in the world (IMD 2017) (see Fig. 9.1). According to a survey conducted by the Minster of Education (MOE) of Taiwan in 2017, the number of graduates in STEM-related fields in Taiwan's higher education system climbed to a peak of 135,000 in the year of 2004. Then, the number gradually has declined each year to 103,000 in 2015, a decline rate of 23.7% (see Fig. 9.2). The percentage of students who graduated from STEM-related fields has also declined from 40.2% to 33.3% between 2004 and 2015.

The declines might be due to the low national birth rate, shift of manufacturing sites to overseas, and/or rapid growth of the service industry. However, some indicators are encouraging: more students pursued graduate-level degrees in STEM-related fields since the year 2000. The number of graduates with master's and doctoral degrees in STEM has increased from 9051 in the year 1997 (9.3%) to 29,439 in 2013 (26.9%) as a peak, a threefold increase over 15 years. Recently, though, the number decreased since 2013 from 29,439 to 26,728. Under this trend, how to cultivate the next-generation of young STEM talents to embrace the Industry 4.0 era is a crucial challenge for all education systems.

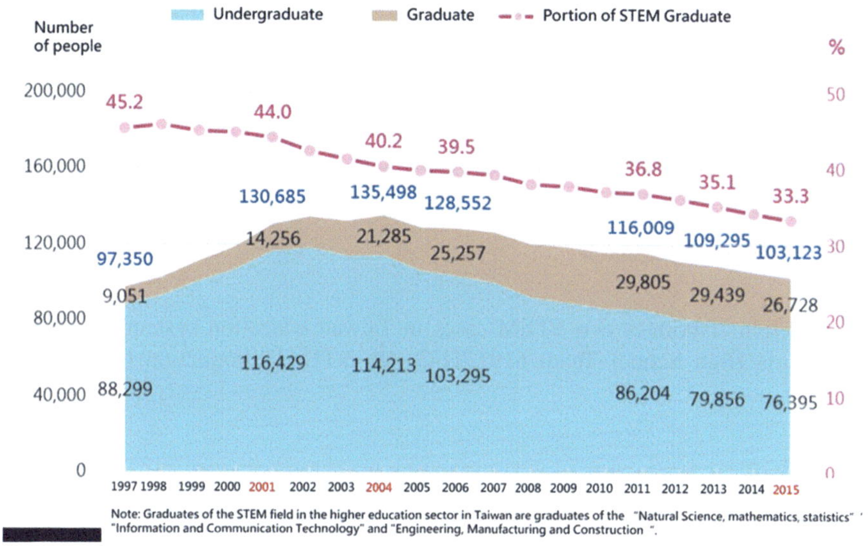


**Fig. 9.2** Number and percentage of graduates in STEM fields in higher education in Taiwan (MOE 2017)

In Taiwan, there are two major pathways for a student when he/she is graduated from junior high school around 15 years old (MOE 2018). The first pathway is known as the *general education system*, and it consists of senior high schools and universities. The students enter senior high school and university by taking entrance exams. The second pathway is Technological and Vocational Education (TVE), which consists of middle-level TVE and higher-level TVE. The middle-level TVE includes the Technical Skills programs at junior high schools and senior vocational schools, professional programs at general high schools, and vocational programs at comprehensive high schools. The higher-level TVE includes junior colleges, colleges of technology, and universities of science and technology (MOE 2014). As industry in Taiwan is facing a seemingly irreversible transition from Original Equipment Manufacturing (OEM) or Original Design Manufacturing (ODM) to high-tech industry, the curriculum in all levels of schools is gradually becoming disconnected, year by year, from the needs of employers. For example, the course guidelines will need to be changed from focusing on how to operate the machines/equipment to how to design and solve real-world problems. Also, the facility for educating students might need to be upgraded to meet the modern manufacturing environment. Taiwan has tried to strengthen STEM education within the existing education system by providing more extracurricular activities related to STEM and offering courses with hands-on STEM learning in the classroom.

This chapter presents two examples of promoting STEM education, one in senior high school in the general education system pathway and another one in TVE university in Taiwan. As practitioners of STEM education in Taiwan, the experiences of the First Girls High School and National Taiwan University of Science and

Technology (NTUST, also known as Taiwan Tech) are representing case studies of high school and Taiwan's technology university aligned with the government technology policy and the coming needs of the technology industry. Also, the collaborations among the government, academy and industry for providing students hands-on opportunities are addressed.

## 9.2 About the Cases

This section elaborates two STEM cases in Taiwan education system: (1) Taipei First Girls High School–Team 6191 TFG; (2) NTUST–Production Line STEM Practice.

### *9.2.1 Case Study #1: Taipei First Girls High School–Team 6191 TFG*

#### 9.2.1.1 Case Introduction

Team 6191 TFG, a robotics team at Taipei First Girls High School, offers an opportunity for high school students to step into the realm of robotics and *Industrial Revolution 4.0*. Taipei First Girls High School, founded in 1904, is one of the most prestigious girl's high schools in Taiwan. The FIRST Robotics Competition (FRC) is an international high school robotics competition that started in 1992. Each year, teams of high school students, coaches, and mentors work together for six weeks to build game-playing robots under certain rules and requirements. FIRST Robotics Competition Team 6191 TFG from Taipei First Girls High School is both the first all-girl robotics team and the first team from a public school in Taiwan. Due to its newly formed status, Team 6191 TFG initially lacked the important financial and technical support systems that are more readily available to the established teams. The team also faced the silent stereotype that girls in the realm of robotics would not be competitive.

However, these challenges did not deter the team. The lack of resources was made up by their effort. In 2017, Team 6191 TFG was a world finalist and was invited to join the world championship in Houston, Texas, USA in 2017. This achievement would not have come true without the help of others. The following subsections describe how the team obtained assistance from others outside their school, and from the government, to help them succeed. Team 6191 TFG achieved the goals of collaborating closely with the government and enterprises, encouraging females to join the robotics industry, and guiding youths step-by-step into Industrial Revolution 4.0 with STEM education. Figure 9.3 shows the photo taken of the team members in a STEM practice classroom.

**Fig. 9.3** Team 6191 TFG

#### 9.2.1.2 Transferring Knowledge to STEM Practice

Throughout the past two years, Team 6191 TFG concluded that the best way to teach students is to put the content learned into practice. For example, in class, students learn about torque, and how to calculate output force when using a gear, a crucial component for a robotics project. In practice, the output force is slightly smaller than the input force, and the lighter the gear, the smaller the difference. Considering the weight of the gear, therefore, is basic knowledge and important when performing a very practical step like purchasing equipment. But traditionally taught math and physics instruction in Taiwan idealizes phenomenon theory over practice. By joining Team 6191 TFG activities, the gap between theories and practical details on the robotics competition field is shortened. This example highlights a deficiency in Taiwanese education. When learning mathematics, students are asked to perform countless calculations, but they do not learn how the knowledge and skills they gain can be applied in real-world scenarios. This observation is consistent with critiques often made by companies who hire recent graduates.

Another indicator of the benefit of context-based learning is the team achievement in the Australia Region FRC. Team 6191 TFG has earned recognition from school faculty members, the principal, and even Taiwan's Ministry of Education. They believe the recognition is an indication of how rapidly they have grown in their knowledge and skills. Seeing how knowledge and skills are used in context has helped them learn the content better.

#### 9.2.1.3 Collaborations with Government, University, Industry, Non-Profit Organizations and Partners

NTUST and FRC team 3132 are the two biggest technical supporters for Team 6191 TFG. The collaboration with NTUST began in the first year. The team would probably not have completed the robot for the FIRST *2016 Stronghold competition* without their help. The professors and postgraduate students in NTUST assisted the team with coding, and gave advice on how to structure their robot.

There are multiple examples of Team 6191 receiving critical assistance from industry. In winter 2017, Omron Corporation taught team members about how to use sensors. The team also received help from the local manufacturers, blacksmiths, and one of the team member family's car repair shop. They supported the team in building the mechanical elements of the robots. Utilizing the help of community businesses, the team purchased motors, aluminum extrusions, carts, and other things they otherwise would have to buy overseas. This collaboration supported community small businesses, and it reduced the carbon footprint of their robots. FabLab (Fabrication Laboratory), a non-profit organization that provides makerspaces, technological courses, and tools, has built a lab on Taipei First Girls High School campus. The site stimulates STEM learning in the team, the school, and Taipei city.

In addition, the Ministry of Education, Taipei City Government, Quanta Computer, and Golf Gifts & Gallery sponsored Team 6191 TFG. Since the team established at the same time as the trend toward intelligence production, they became a part of an innovative secondary education effort supported by the government. Taiwan's Ministry of Education, along with the Taipei City Government, provided the team around $10,000 US for material purchases like sensors, controllers, chassis, and Mecanum wheels. The Ministry of Science and Technology (MOST) of Taiwan offered their school a major opportunity after learning the Team 6191 TFG story. MOST proposed a plan, with supporting resources, for technology-related firms, schools, research programs, and high-technology manufacturers to work together to cultivate maker education in secondary schools. As part of this effort, the team co-hosted summer camps in southern and northern Taiwan with the Southern Taiwan Science Park Administration, who supplied the camp with equipment and mentor training. In addition, MOST invited Team 6191 TFG to the inauguration of a robotics foundation in the Central Taiwan Science Park (CTSP).

In 2016, *FRC team 5987 Galaxia* in Israel brought to the team an idea called *Rookie Network*. With 15 veteran teams around the world, they created a series of robotics courses online in different languages. The course platform provides the organized content for rookie teams in this competition and a medium for exchanging questions and solutions without boundaries, which encourages more teachers and students to join the international "robotics party." Finally, Team 6191 TFG also invited government officials to participate and observe the FRC 2018 regionals in Australia. The CTSP Bureau of MOST was scheduled to host a future Taiwan regional tournament and an international conference.

#### 9.2.1.4 Aligning STEM Education to Industry Needs

Supporting young females to explore the realm of robotics is one of the missions of Taipei First Girls High School. The team planned to hold a summer robotics camp in eastern Taiwan, where the technology development is slower than other regions in Taiwan. Team 6191 TFG was actively in contact with an all-girl vocational school to help fund a team to participate in the FIRST Robotics Competition.

In the realm of Industrial Revolution 4.0, intelligent production has become a key concept to make manufacturing more effective by using the intelligence of machines. Taiwan has shown its desire to develop and apply the concept of intelligent production. President Tsai Ing-wen, the first female elected president of Taiwan, recently set up the Intelligent Machinery Office, and the Ministry of Economic Affairs followed up by launching Mix Taiwan 2.0, to focus policies and resources on intelligent production. These initiatives view intellectuals with creativity as key to the policy and as the infrastructure of Taiwan's economic growth. The team considers introducing robotics in secondary education as a major contributor to develop the workforce needed to support the Taiwan economy.

Meanwhile, Team 6191 TFG tried their best to promote STEM education across the country. Their mentors have been to many education exhibitions and have delivered speeches in high schools around Taiwan, sharing their stories, finding people who are interested in participating in FRC, and disseminating information about this new kind of hands-on teaching that integrates science, technology, engineering, mathematics, teamwork, and an enterprising spirit. Websites related to the team's projects are listed below. The timeline of outside assistance and the team's achievements are summarized in Table 9.1.

### *9.2.2 Case Study #2 NTUST–Production Line STEM Practice*

#### 9.2.2.1 Case Introduction

In order to reduce the gap between industry and university curricula, the Department of Industrial Management in NTUST transformed the existing Introduction to Electronic Business course offered to sophomores, by adding STEM practice. The concepts of Problem-Based Learning (PBL) were applied during the course development. PBL is promoted in Taiwan TVE universities and colleges to engage students with real-world problem-solving skills (Li and Tsai 2017). Essentially, under the design of PBL, the learners are expected to be responsible for their own learning. Students form small groups who determine what they need to learn, and they are given time to develop problem-solving and self-directed skills accessing the world's rich knowledge from many disciplines (Kek et al. 2017). Starting from 2018, the practice session of the Introduction to Electronic Business course, two hours per week, was developed to guide students to engage a problem that helped them learn what a production system is and how to build it.

**Table 9.1** Timeline of the team: Outside assistance and achievements

| Date | Description |
|---|---|
| Fall 2015 | Team 4253 from Taipei American School invited students of Taipei First Girls High School to visit their robotics laboratory and encouraged students to join FRC |
| 9 January 2016 | Team 6191 TFG was officially set up |
| 13–19 March 2016 | Team 6191 TFG participated in the 2016 FIRST Robotics Competition (FRC) (Australia Region) |
| 25 March 2016 | Dr. Genny Pang from Technische Universität München, the ex-mentor of Team 1967, visited Team 6191 TFG and shared her years of experience |
| 6 June 2016 | Robotics club was officially set up in school to create more opportunities for students to develop interest in robotics |
| 4 August 2016 | A team member became an exchange student and visited Dr. Genny Pang in Germany |
| 11 November 2016 | Students and teachers from Graf Eberhard Gymnasium in Germany visited Taipei First Girls High School, and Team 6191 TFG held a robotics lesson for them |
| November 2016 | The Rookie Network was launched |
| 16 December 2016 | Students and teachers from Urawadaiichi Girls' High School in Japan visited Taipei First Girls High School, and Team 6191 TFG held a robotics lesson for them |
| 20 January 2017 | Omron Corporation taught Team 6191 TFG about the knowledge and application of sensors |
| March 2017 | Ministry of Education, and Taipei City Government provided Team 6191 TFG financial aid |
| 13–15 March 2017 | Team 6191 TFG participated in 2017 FIRST Robotics Competition (FRC), (Australia Region), and was a finalist |
| 19–22 April 2017 | Team 6191 TFG participated in 2017 FIRST Championship. Golf Gifts & Gallery provided the team financial aid |
| May 2017 | Quanta computer provided the team financial aid |
| 6 July 2017 | Sir Alejandro Romero, mentor of Team 4262 ROBOHAWK visited Team 6191 TFG |
| 8 June 2017 | The mentor and captains from Team 3473 Sprocket visited Team 6191 TFG and shared their experience, strategies, and leadership style |
| 31 July 2017 4 August 2017 | High school freshmen robotics summer camp was held |
| 4–20 August 2017 | Southern Taiwan robotics summer camp was held |

In order to include PBL in the practice session of the course, the course designers start to introduce a STEM learning tool, *fischertechnik*, invented by Artur Fischer and produced by fischertechnik GmbH in Germany in the course curriculum in 2017. The aim of the tool is to teach automation and robotics theory through hands-on learning. The objective of introducing STEM practice is to guide students to understand how electronic devices, such as sensors, robots, and controllers, can function together as a production line under the control of computer programs. For undergraduate students in management schools, this learning experience is unique because, before they attend this course, they presume electronic devices are some kind of *mystery box* taught in engineering courses only. The following case study presents the development of this course and how the course is incorporating STEM tools.

#### 9.2.2.2 Transferring Knowledge to STEM Practice

Howard S. Barrow, one of the key pioneers of PBL, has written extensively about the essentials of PBL which was implemented in medical education in the 1960s. When the course's practice session was developed, the essentials of PBL were reviewed, and course tutors and teaching assistants were trained to make them familiar with PBL essentials. Table 9.2 lists the essentials of PBL and the corresponding implementation details for the mentioned Introduction to Electronic Business course.

Due to the lack of engineering background, college students in management school will need an easy-to-use device that can demonstrate a real-world industrial production line. Therefore, choosing the right STEM tool is very important. Two systems were considered during the early stage of course development: LEGO Mindstorms EV3 and fischertechnik. LEGO Mindstorms EV3 and its previous version NXT have gained wide acknowledgement in STEM education in Taiwan. In fact, a lot of high schools and universities have utilized LEGO Mindstorms in the extracurricular activities and regular courses (LEGO 2013). In comparison with LEGO, fischertechnik is less famous in Taiwan. Similar to LEGO EV3, the fischertechnik package also contains multiple sensors, building blocks, and a controller. Although fischertechnik has less presence and support in Taiwan, its learning package contains more versatile building blocks and components that can function more flexibly to construct a production line. Therefore, in order to meet the learning objective of building a production line system, fischertechnik was chosen for the practice session.

Once the STEM tool was determined, fischertechnik representatives in Taiwan helped train teaching assistants. Weekly student-centered teaching materials and learning milestones based on the PBL concept were also developed. The lessons encouraged students to learn about the subject through the experience of solving open-ended problems. Students were divided into groups and each student took on a role within the group. The instructor focuses on the students' reflection and reasoning to construct their own learning.

As the weeks progress, students must combine the multiple devices constructed every week into a bigger project. At the end of the semester, every group of students must design their own production system and be peer reviewed by other groups. The grades of STEM practice are determined by the completion level of production line, peer review results and instructor's comments. Figure 9.4 shows a production system consisting of warehouse, robot, and conveys built using fischertechnik building blocks, sensors, and controllers.

**Table 9.2** Essentials of PBL on practice session of the course based on Kek et al. (2017)

| PBL essentials | Concept | Implementation |
|---|---|---|
| Problems should present as they do in the real world | Present problems to learners that they will encounter in their work after graduation with only the information that would be initially available | The problem "how to build a modern production line with robotics technology" is the real-world case for students after graduation |
| Problem-solving skill development | With problems presented, learners practice and develop effective and efficient problem-solving skills guided by tutors | Students create a facility in which both the product and related information progress through the system effectively and efficiently |
| Student-centered | Learners recall and apply the unique knowledge and skills they already possess to understanding of the problem | Students will utilize knowledge they learned from first and second year of college course work |
| Self-directed learning skill development | Learners determine what to learn and how to obtain the knowledge | Students setup their own construction plan and determine what kind of production line will be built |
| Integrated knowledge | Learners integrate information to obtain an in-depth understanding of the problem and interrelation of information | Students will need to learn controller, sensor, and programming skills from STEM tools and know how to integrate them together |
| Small group collaborative learning | Learners develop skills through small group work with peer- and self-assessment | Students are clustered as small group to work together and assess each other |
| Reiterative | What was learned must be applied back to the problem at hand | The knowledge should be connected to "building a production line" |
| Reflective | Learners review what they have learned and reflect on what abstractions and generalization might be developed | Students should be able to develop the product flow and information flow based on the production line design |
| Self- and peer-assessment | Each learner assesses his/her own gain in knowledge. Others in the group assess each learner | Peer review and assessment will be performed inside and across groups |
| Skilled tutors | Tutors are skilled in facilitating learning | Senior-year students who took the course before and graduate students are deployed as tutors |

Practice sessions were held over 8 weeks in a semester. In order to complete the final project by combining the devices built each week, students needed to organize their teammates to learn different pieces, and host one or more brainstorming meetings to make sure the final project would contain all of the needed components. Extra-credit was given every week to students who accomplished a difficult design or functional element. The extra-credit system stimulated students to learn more knowledge on the internet and spread the active learning atmosphere to each group.

**Fig. 9.4** Illustration of a production line built by fischertechnik building blocks, sensors, and controllers (Fischertechnik GmbH 2019)

#### 9.2.2.3 Linkage Between Abstract and Physical Views of Production Lines

By their gradual learning, progressing from how to use electronic devices to how to construct a production system, students extend their knowledge from how to manage the production system to how that system is built. For example, the hands-on problem of detecting different items by their colors with color sensors requires students to learn how to operate the color sensor on their production line. Once the items can be color detected, students are challenged with the task of conveying items of different colors to different operations in the production line. This learning can be easily extended to similar problems, such as detecting different items with different available sensors—for example, shape or QR codes. Students must consider within the context of the system, thinking about how and when to implement.

This exercise encouraged students to apply hands-on experience in what they had learned from the courses related to production systems in the third year of college. Students can link their "abstract" view, which they usually learn from pictures, video on textbook, or website, to a "physical" view of a production line which they created and built using STEM tools. Due to the integration of learning, this course has been designated as a "Capstone" course in the Department of Industrial Management for undergraduate students. Figure 9.5 shows photos taken during the discussion on building a production line in class using the fischertechnik package.

Fig. 9.5 Discussion on building a production line in class using the fischertechnik package

#### 9.2.2.4 Linkage Between STEM Learning in High School and College

In Case Study #1, the story of Team 6191 TFG presents a group of female high school students collaborating with outside sources to engage in a robotics competition. In this case, students demonstrated the use of STEM tools, social networking during the FIRST Robotics Competition, and completion-based activity to gain knowledge of robotics, especially via extracurricular activities.

Due to the course requirements and the goal of learning for entering college, high school students in Taiwan have multiple predetermined subjects that will be tested via the university/college entrance exam. It also means students and their parents are very concerned about the learning progression on those subjects. It is important to know how to link the fundamental knowledge gained from those predetermined subjects, such as physics or chemistry (which might focus more on calculation and memorizing), with college-level courses, which might be more related to real-world problems. From Case Study #1, we learned that engaging students in hands-on projects can stimulate them to map knowledge from textbook calculations to practical usage on a real-world device design. Then learning can be more interesting, and the learning the skills for how to correlate, simulate, and extend textbook knowledge to real-world cases or other subjects is crucial for college learning. In fact, all these learning skills (correlation, simulation, and extension) are expected from college students.

In Case Study #2, a course in NTUST's management school that was originally developed for students without engineering backgrounds demonstrates the possibility of utilizing STEM tools to implement PBL in a course about production line design. As mentioned before, determining the problem that will be provided to students is very important for a PBL course. In this case, third-year college students were challenged to utilize what they had learned during two years in college in the development and design of a production line. Developing critical-thinking, problem-solving and collaborative skills are the main goals of this course. The learning of these skills can be pursued early, during high school. For example, the practice session in learning object detection using production line sensor technology, in the mentioned college course, can be coupled with high school robotics learning to correlate, simulate, and extend the knowledge of sensing technology from a single robot design to a combination of automation systems. To mimic the problem of the PBL course after the real-world problem, college students definitely need to oppose critical-thinking/problem-solving skills and collaborative skills. If those can be learned from high school, students might be in a better position to face future challenges.

## 9.3 Education Philosophies and Methods

### *9.3.1 Initiating NTUST Industry 4.0 STEM Education*

In 2016, NTUST established a school-level Industry 4.0 Implementation Center for the purposes of developing Industry 4.0-related technologies, cultivating talents in specialized fields, and achieving short-term and long-term strategic development goals. The main functions of this center are to consolidate resources, assist interdisciplinary programs, support multidisciplinary courses, and provide domestic industries with an enhanced ability to access and develop techniques and manpower training. Five missions of the NTUST Industry 4.0 Implementation Center are described in Table 9.3 below: lecturer training, interdisciplinary program coordination, introducing cutting-edge equipment and technologies, bridging Industry 4.0 enterprises and partner schools, and facilitating international cooperation.

### *9.3.2 Cultivating Industry 4.0 Talents to Meet Future Needs*

Facing the huge impact of the fourth Industrial Revolution, Taiwan's MOE initiated several projects to cultivate Industry 4.0 talents in Taiwan. Since 2015, NTUST has received annual grants to develop courses and upgrade facilities to cultivate Industry 4.0 talents, and especially to strengthen the mechanism for cultivating interdisciplinary skills. With grant support, the course curriculums of TVE universities/

**Table 9.3** Missions of the NTUST industry 4.0 implementation center

| Mission | Description |
|---|---|
| Lecturer training | The center provides cutting-edge professional STEM courses and workshops for teachers/lecturers to enhance their capability to offer industry 4.0 related STEM courses. For example, the information system integration workshop focusing on machine-to-machine (M2M) and the industrial internet of things (IIoT) workshops are offered to students and faculty members for learning industrial machine network development |
| Coordinating interdisciplinary programs | The center consolidates resources from partner schools and departments and provides the experimentation site to implement industry 4.0 dedicated courses. Furthermore, the center aims to coordinate industrial-level programs for students to participate in interdisciplinary hands-on projects |
| Introducing cutting-edge equipment and technologies | The center creates industry 4.0 level production lines with support from industry, to demonstrate future production management concepts and to provide industrial-level equipment and training to students. More and more new facilities will be provided to integrate M2M, IIoT, and 5G telecommunications into future manufacturing |
| Bridging industry 4.0 enterprises and partner schools | The center provides partner schools opportunities and resources to interact with industries. For example, students from partner universities can participate in workshops or courses hosted by the industrial companies. The center also fosters industry-university cooperation via student internships. Through the internship program, college students get the opportunity to work for a company on a project basis. For example, in 2019, around ten college students had summer internships at companies with long-term collaborations with the center. In order to transfer knowledge of industry 4.0, the center has built an industry 4.0 demonstration production line to assist small and medium enterprises in transforming and upgrading existing production models |
| Facilitating international cooperation | The center aims to collaborate with international industry 4.0 universities and enterprises by offering students exchange programs and international internships that follow standard international norms and practices |

colleges (DTVE 2019) were updated to include more hands-on STEM projects and practical training. The strategies of course development were: (1) decompose key technologies of the cyber-physical system and core of Industry 4.0 technology based on their characteristics; (2) develop interdisciplinary programs and learning environments; (3) build a demonstration site for learning; and (4) work closely with industry. From engineering schools to management schools, new introductory courses related to Industry 4.0 are offered to students.

Figure 9.6 shows an overview of Industry 4.0 talent cultivation in Taiwan. Essentially, the driving forces behind talent cultivation are industry acting through labor unions, and government counselors such as the counselor board of MOE or MOST. High schools and technological universities play important roles in meeting industry's needs by developing key technology as well as interdisciplinary programs and courses. The construction of an Industry 4.0 demonstration site on campus provides a learning environment to support STEM learning.

On the basis of the strategies mentioned above, the NTUST Industry 4.0 Implementation Center has developed a multidisciplinary program for

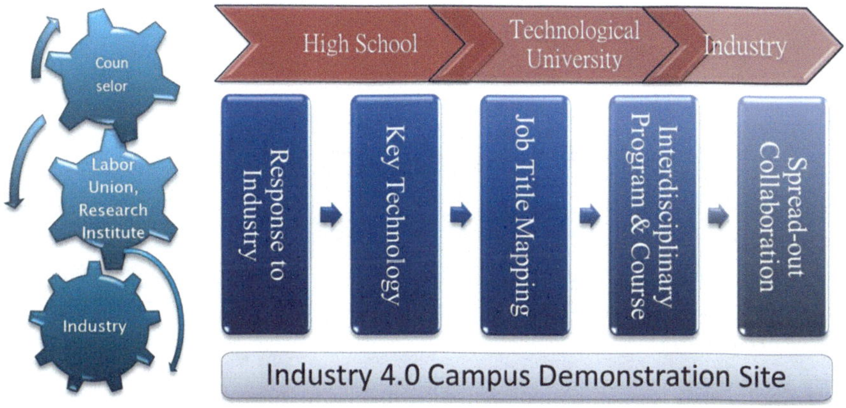


**Fig. 9.6** Overview of cultivation of industry 4.0 talents in Taiwan

undergraduate students. The core value of the program revolves around the combination of "mission-oriented project" and "practice-oriented education." This program aims to serve as the foundation for technological as well as scientific upgrading in domestic industries.

Moreover, NTUST aims to become a pioneer in assisting partner schools to establish local educational demonstration factories by "revising, improving, and updating teaching manuals," "cultivating seedling teachers," and "sharing and connecting demonstration factories" (MOE n.d., para. 4). NTUST develops industrial training courses to support industry–university collaborative projects and make up for the shortage of Industry 4.0 manpower in domestic enterprises, thereby enhancing their strategic competitiveness. The collaborations also offer students the benefits of taking internships to receive practical training in industries related to their core specializations.

In order to encourage more students to learn the most up-to-date technology and contribute their expertise to society, NTUST collaborates with global universities such as the Tokyo Institute of Technology (Japan) and Esslingen University of Applied Sciences (Germany). NTUST is dedicated to promote industry–university collaborations and foster a spirit of educational innovation. NTUST has established many successful graduate programs, one of which combines an international graduate program with industry personnel training. It not only helps students enrich their professional skills and broaden their global view, but also helps companies find outstanding employees.

## 9.4 Industry Cluster Served

The industries that support NTUST's Industry 4.0 Implementation Center consists of manufacturers of ICT, semiconductors, molds, assembly, and machinery. These companies sponsor facilities and donate funds to help the center cultivate STEM

**Fig. 9.7** The first Taiwan industry 4.0 competition

talents. In the winter of 2017, ten companies sponsored the center's organization of the first Taiwan Industry 4.0 student project competition. Figure 9.7 shows a photo taken during the competition on the NTUST campus, and Fig. 9.8 shows representatives of the ten sponsors (Fair Friend Enterprise Group, Hiwin Mikrosystem Corp., LNC Technology, CimForce, Delta Electronics, Garmin Ltd., Techman Ltd., Career Technology (Mfg.) Co., Ltd., Yatec Engineering Corp., and Sha Yang Ye Inc.).

In addition, other companies sponsored the remodeling of the factory and the acquisition of high-definition machines and hardware and software packages in the Industry 4.0 Implementation Center. Figure 9.9 shows the sponsors' company logos. Besides the facility sponsorship, the companies also help in developing course curriculum. For example, the engineers and managers of CimForce, one of the major software developers for Industry 4.0 applications in Taiwan, help develop practice sessions for multiple courses in the Mechanical Engineering Department of NTUST. CimForce employees also participate in the course as mentors and teaching assistants to help students learn to operate the system.

## 9.5 Policies That Support the Program

In order to cultivate STEM talents, MOST in Taiwan has set up the Artificial Intelligence (AI) Robotics Hub at CTSP to promote and accelerate the development of AI and the robotics industry. This hub cooperates with manufacturers, scientific research institutions, startup companies, makers, associations, colleges, and universities to create an innovation environment to develop intelligent robotics.

With the goal of cultivating talents needed by the industry in the future, this hub sets up hardware and software facilities and training and testing spaces for AI robotics. It also provides a variety of makerspaces, co-working spaces, and international

Fig. 9.8 Representatives of the Taiwan industry 4.0 competition sponsors

Fig. 9.9 Logos of NTUST industry 4.0 implementation center sponsors

competition sites. Moreover, this hub integrates resources from multiple disciplines to create a professional demonstration site where solutions are provided through AI robotics technologies. AI Robotics Hub at CTSP leads the industry in upgrading and establishing a new model for sustainable industrial competitiveness in Taiwan (Central Taiwan Science Park 2018a).

The AI Robotics Hub at CTSP provides four major services:

- *Maker Equipment*. The hub provides multiple equipment, including 3D printers, AI computing devices or computers, and robotics tools for the students.
- *AI Robot Contest Registration and Training*. The hub hosts various robotics competitions and training sessions. The competitions include the FIRST Robotics Competition (FRC) and Federation of International Robot-Sport Association (FIRA) and PICKATHON events.
- *IP Resource Sharing Service*. The hub provides online sharing of intellectual property resources for students to view existing methods or patents.
- *AI Robotics Course Promotion*. The hub offers multiple robotics courses such as NVIDIA image processing and practice, deep learning, collaborative robots, industrial robots, special cutting, 3D printing, circuit board engraving, and drilling and milling machine theory and implementation (Central Taiwan Science Park 2018b).

Figure 9.10 shows the opening ceremony of the CTSP 2019 FRC Taiwan Preliminary, part of the AI Robot Contest service. The AI Robotics Hub at CTSP not only helps teams to register for international contests, but also provides opportunities to communicate and study with other teams. Seventeen teams from senior high schools and senior vocational schools participated in this contest. As a result, STEM learning has received major attention in the senior high and vocational schools of Taiwan.

**Fig. 9.10** CTSP 2019 FRC Taiwan preliminary opening ceremony

Besides CTSP, the K–12 Education Administration of MOE Taiwan also started projects to create makerspaces in schools throughout the country (see Table 9.4) and to provide services to the primary and secondary schools in the surrounding areas. The main goal of these projects is to cultivate more talent for Taiwan's advanced technology industry (K–12 Education Administration 2017).

**Table 9.4** Makerspaces in Senior/Vocational High Schools in Taiwan

| City/County | School name | Laboratory name | Start date |
|---|---|---|---|
| Hualien City | National Hualien Industrial Vocational Senior High School | Fab Lab-Hualien | 2015.04.29 |
| Kaohsiung City | National Fongshan Senior Commercial & Industrial Vocational School | Kaohsiung and Pintong Self-Built Laboratory | 2015.05.05 |
| Taipei County | New Taipei Industrial Vocational High School | FabLab-NTVS | 2015.05.09 |
| Taichung City | Taichung Municipal Taichung Industrial High School | Central District Self-Building Laboratory | 2015.05.26 |
| Taipei City | Taipei First Girls High School | Fab Lab North Area | 2016.01.05 |
| Tainan City | National Tainan Second Senior High School | FABLAB TNSSH | 2016.03.02 |
| Taipei City | National Normal University High School | Fi-Lab | 2016.03.24 |
| Taichung City | Taichung Municipal Cingshuei Senior High School | i Do Fab Lab | 2016.04.07 |
| Tainan City | National Tainan Senior Marine Fishery Vocational School | 3D Creative Self-Made Laboratory | 2016.05.13 |
| Miaoli County | National Miao-Li Agricultural and Industrial Vocational High School | Big M fab lab | 2016.05.30 |
| Taoyuan City | Taoyuan Municipal Nei-Li Senior High School | NLHS MakerSpace | 2016.06.07 |
| Taipei County | National Overseas Chinese High School | FabLab-NOCSH | 2016.06.16 |
| Taichung City | Taichung Home Economics and Commercial High School | Fab Lab@Taichung Home Economics and Commercial High School | 2016.06.23 |
| Taichung City | Taichung Municipal Taichung Girls Senior High School | i make lab | 2016.06.27 |
| Yilan County | National Lotung Industrial Vocational High School | Fab Lab East Area | 2017.03.22 |
| Nantou City | National Nantou Senior High School | Fab Lab Central Area | 2017.04.20 |
| Kaohsiung City | The Affiliated Senior High School of National Kaohsiung Normal University | Fab Lab South Area | 2017.05.31 |
| Pingtung County | National Pingbei Senior High School | Fab Lab Southern Area | 2017.06.03 |

## 9.6 Outcomes/The Virtuous Cycle

Taiwanese companies in the semiconductor, machine tool, and electronic manufacturing services sectors have established long-term collaborations with TVE universities and colleges such as NTUST. Since 2015, more and more industry-academia projects have been launched to target the development of Industry 4.0 technology, such as Industry Internet of Things (IIoT), AI, and Industrial Big Data on Smart Manufacturing. In 2018, NTUST's Industry 4.0 Implementation Center received more than 50 industry-academia collaboration projects and technology transfers (total valued at US$1 million) to solve real industrial problems as well as increase the companies' competitiveness. Through collaborations supported by the industry-academia project, industry can (1) provide internship opportunities to students, (2) specify real-world problems and reframe them as research problems with university faculty, and (3) help upgrade/update the learning environment based on the needs of the project. In addition, the industry can supply industrial professionals as experts/mentors of STEM education. Therefore, industry-academia project support can be a key element of the virtuous cycle of cultivating Industry 4.0 talents in Taiwan. Figure 9.11 illustrates the virtuous cycle of coordinating STEM, industry, and technology policy to cultivate industry 4.0 talents in Taiwan.

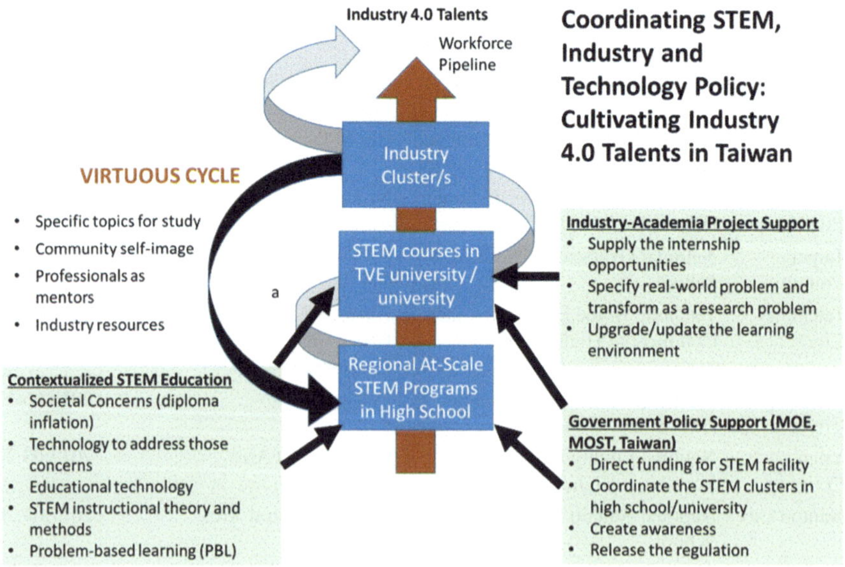


**Fig. 9.11** Virtuous cycle of STEM education in Taiwan

## 9.7 Lessons Learned

STEM plays an important role in Taiwan's education system. Although the total number of students is in decline due to the low birth rate, the number of master's and doctoral graduates in STEM is on the increase. STEM students supply the talents needed to develop Industry 4.0 technology. In this chapter, two case studies were introduced. One is the success story of an all-girl robotics team from Taipei First Girls High School. Team 6191 has participated in the FIRST Robotics Competition for two years. Another story described how a course was enhanced by adding STEM education for college students in a TVE university management school.

Team 6191 TFG has effectively demonstrated how robotics can be used to deliver STEM education to students in a senior high school, particularly an all-girl school. Through a variety of support measures from the government, university, industries, and nonprofit organizations, many activities were conducted to encourage the study of robotics among the students. In collaborating with those entities and with other robotics teams around the world, Team 6191 TFG has moved beyond immediate student colleagues and begun to contribute to the success of robotics throughout Taiwan, as well as globally. In doing so, Team 6191 TFG has embraced the FIRST organization's key principles of *coopertition* and *gracious professionalism* (Kamen 2019), which will benefit all who compete in the twenty-first century economy.

As one of the most privileged TVE universities in Taiwan, NTUST has invested a lot of resources in cultivating talents to face the challenge of Industry 4.0. The case study of adding STEM education to the typical management school course Introduction to Electronic Business demonstrates that STEM education can stimulate high-quality learning in a regular college course, even for management school students. Through step-by-step learning and support from teaching assistants, management school students can complete a hands-on project of building a production line system using a STEM tool package. This learning experience helps them link abstract knowledge and a physical system together.

Although the success of said case study can be foreseen, there are still some challenges of engaging more students in STEM learning in Taiwan. First of all, the mindset of parents and of the entire education system needs to be motivated. For a long time, most Taiwanese families have encouraged their children to pursue as much higher education as possible, such as a master's degree or even a doctoral degree. This mindset in fact creates severe "diploma inflation" in Taiwan, and also influences the possibility of promoting STEM education. For example, in high school, families or parents might opt for STEM education only if it can help their children succeed at entering a good university. Another issue related to STEM education, whether at the high school or the college level, is that it is time consuming. Under the PBL scheme, students will spend quite a lot of time determining, discussing, and solving problems by learning STEM tools. Compared with traditional courses, in addition to the time students actually need to spend in the classroom, more time must be spent on home study. Experiments indicate that highly self-motivated and self-directed students can learn much better than those who need tutors or lecturers to "push" them to learn. Therefore, how to engage students with weak self-motivation in STEM learning is still an open question. Last but not least, STEM education needs more resource support, especially from skilled tutors. The findings of our study show that experienced tutors can really create and motivate a good learning atmosphere for students. How to keep training and maintaining a reasonable number of tutors for a continued course is a key issue for both high school and college STEM course development.

## References


Central Taiwan Science Park (2018a). AI Robotics Hub at CTSP. Retrieved August 8, 2019 from https://ctsphub.tw/en/aboutbase/.

Central Taiwan Science Park (2018b). Main service of AI Robotics Hub at CTSP. Retrieved August 8, 2019 from https://ctsphub.tw/aboutus/service/.

DTVE (2019, August). An overview of technological and vocational education. Retrieved August 8, 2019 from https://english.moe.gov.tw/cpview-4-15156-E7588-1.html.

Fischertechnik GmbH (2019). The history of fischertechnik. Retrieved August 8, 2019 from https://www.fischertechnik.de/en/about-us

IMD (2017). IMD World Digital Competitiveness Ranking 2017 [Press release]. Retrieved August 8, 2019 from https://www.imd.org/globalassets/wcc/docs/release-2017/world_digital_competitiveness_yearbook_2017.pdf.

K–12 Education Administration, MOE (2017). 3D printing popularization programs for senior secondary schools. Retrieved August 8, 2019 from http://3d.tchcvs.tw/3dHistory/index.php?node=FabLab&w=school

Kagermann, V. H., Lukas, W. D., & Wahlster, W. (2011, April 1). Industrie 4.0: Mit dem Internet der Dinge auf dem Weg zur 4. industriellen Revolution. *VDI Nachrichten*. Retrieved August 8, 2019 from http://www.wolfgang-wahlster.de/wordpress/wpcontent/uploads/Industrie_4_0_Mit_dem_Internet_der_Dinge_auf_dem_Weg_zur_vierten_industriellen_Revolution_2.pdf.

Kamen, D. (2019). FIRST vision and mission. Retrieved August 8, 2019 from https://www.firstinspires.org/about/vision-and-mission.

Kek, M., Yih, C. A., & Huijser, H. (2017). *Problem-based learning into the future: Imagining an agile PBL ecology for learning*. Singapore: Springer Singapore.

LEGO (2013). The introduction of LEGO mindstorms EV3. Retrieved August 8, 2019 from https://www.lego.com/en-us/mindstorms/about-ev3.

Li, H. C., & Tsai, T. L. (2017). The implementation of problem-based learning in a Taiwanese primary mathematics classroom: Lessons learned from the students' side of the story. *Educational Studies, 43*(3), 354–369.

MOE (2014). *Technological & vocational education in Taiwan, ROC*. Retrieved August 8, 2019 from https://ws.moe.edu.tw/001/Upload/5/RelFile/7801/38355/2014_TVE-En.pdf.

MOE (2017). The graduate student in STEM fields higher education in Taiwan. *Education Statistic Indicator, 73*. Retrieved August 8, 2019 from http://stats.moe.gov.tw/files/brief/%E6%88%91%E5%9C%8B%E9%AB%98% E7%AD%89%E6%95%99%E8%82%B2STEM%E9%A0%98%E5%9F%9F%E7 %95%A2%E6%A5%AD%E7%94%9F%E6%A6%82%E6%B3%81.pdf.

MOE. (2018). *Education in Taiwan*. Taiwan: Ministry of Education.

MOE (n.d.). Talent development program (Ministry of Education) talents cultivation plan (Ministry of Education). Retrieved August 8, 2019 from https://www.industry4.ntust.edu.tw/files/11-1108-5926.php?Lang=zh-tw.

# Chapter 10
# Greater Austin STEM Ecosystem

**Tricia Berry**

**Abstract** The Greater Austin STEM Ecosystem strives to foster deeper collaboration across networks and systems to ensure STEM programming is learner-centered and accessible to all students throughout the Greater Austin. Historically, initiatives such as the STEM Pipeline Collaborative, the Central Texas Summer STEM Funders Collaborative, and the Semiconductor Workforce Development Collaboration fostered ecosystem-wide STEM collaborations and advanced STEM learning. Today, similar volunteer-led, industry-engaged collaborations advance the goals of the Greater Austin STEM Ecosystem through ecosystem-engaged networks, community-wide technology solutions that promote STEM learning opportunities and foster STEM volunteering, and community-wide networking forums, convenings and amplifications. While still in the early stages of defining and assessing ecosystem-wide common metrics, the collaborative efforts are driving increased participation and engagement throughout the STEM community.

## 10.1 Introduction

The Greater Austin area has a long history of rich STEM programming and STEM-focused non-profits, a robust STEM industry, strong higher education and school district systems, and a vibrant community of entrepreneurs, educators, funders, volunteers, and others invested in advancing STEM learning and opportunities throughout the community. Collaborations have identified gaps in STEM learning or workforce development, developed solutions to advance STEM programming, and enriched opportunities for learners throughout the community. Formalizing the collaboration efforts with the recognized designation of the Greater Austin STEM Ecosystem as a STEM Learning Ecosystem takes advantage of existing systems,

T. Berry (✉)
The University of Texas at Austin, Austin, TX, USA
e-mail: triciaberry@utexas.edu



C. Zintgraff et al. (eds.), *STEM in the Technopolis: The Power of STEM Education in Regional Technology Policy*,
https://doi.org/10.1007/978-3-030-39851-4_10

solutions, and networks and provides a strategic framework for advancing STEM learning throughout the Greater Austin community.

A STEM Learning Ecosystem encompasses schools, community settings such as after-school and summer programs, science centers and museums, and informal experiences at home and in a variety of environments that together constitute a rich array of learning opportunities for young people. A learning ecosystem harnesses the unique contributions of all these different settings in symbiosis to deliver STEM learning for all children. Designed pathways enable young people to become engaged, knowledgeable, and skilled in the STEM disciplines as they progress through childhood into adolescence and early adulthood (STEM Funders Network 2019).

In November 2015, the STEM Funders Network announced the Greater Austin STEM Ecosystem as one of 27 communities selected to pilot the national STEM Ecosystems Initiative.

The Greater Austin STEM Ecosystem strives to foster deeper collaboration across networks and systems to ensure STEM programming is accessible to all students throughout the Greater Austin (Greater Austin STEM Ecosystem, n.d.-c). The ecosystem has four goals: (1) develop and engage an effective and sustainable network of academic, business and community partners who are interested in providing STEM opportunities for pre-kindergarten through grade 16 learners throughout Central Texas; (2) support excellence in formal and informal STEM education throughout Central Texas; (3) incorporate an equity driven approach to promote awareness of and access to STEM pathways; and (4) ensure long-term sustainability of the Greater Austin STEM Ecosystem. The vision of the ecosystem is two-fold: (1) pre-kindergarten through college learners have the opportunity to obtain the requisite STEM skills to enter competitive twenty-first century STEM jobs in Central Texas and (2) all educators are provided the tools and support to ensure that students of all ages are STEM competent and STEM literate.

The ecosystem is learning-centered, advancing STEM education at all levels by connecting, supporting, and fostering collaborations between networks, organizations, and individuals. The ecosystem seeks to identify gaps and opportunities in STEM networks, systems, and learning throughout the community, reduce duplication, and amplify individual STEM organizations and their collective impact. Given the Greater Austin area's long history of STEM industries, educational programs and systems, and community engagement, the ecosystem is positioned to take advantage of historical collaborations and advancements along with existing systems and solutions to advance STEM learning for all.

## 10.2 History of Ecosystem-Related Networks, Projects, and Collaborations

The Greater Austin area has a rich history of STEM-related community collaborations that have advanced STEM opportunities and learning. A collaboration between schools, companies, and a data-driven education organization led to an increase in the number and diversity of students pursuing STEM classes through high school. A collaboration across funders led to streamlined grant processes and increased access to high-quality STEM programming by diverse and low-income students throughout the area. Additionally, a collaboration between semiconductor companies, the Austin Community College, and a local workforce development organization led to closing the gap in supply of semiconductor technicians in the community and support of precollege programs throughout the area. The Greater Austin STEM Ecosystem recognizes the power of these community and organization-led efforts and seeks to foster and support similar collaborative efforts in the future.

### *10.2.1 STEM Pipeline Collaborative*

The STEM Pipeline Collaborative was launched by $E^3$ Alliance in 2008 to help address a gap between the supply of STEM students and the workforce demand throughout Central Texas. The collaborative included 12 school districts, four higher education institutions, and 10 community organizations and industry partners such as 3 M and IBM who worked together to on efforts designed to expand and diversify the STEM pipeline ($E^3$ Alliance 2012). With the collaborative, $E^3$ Alliance secured a $150,000 National Science Foundation grant to provide “teacher professional development opportunities, business mentors for middle school robotics teams, and money to support engineering programs to 10 area school districts with high school-level engineering” ($E^3$ Alliance 2008, para. 2).

After 5 years, $E^3$ Alliance found the efforts of the STEM Pipeline Collaborative expanded the diversity of those participating in the STEM pipeline and increased the overall number of students participating in Project Lead The Way courses by 430% ($E^3$ Alliance 2012). Project Lead The Way is a non-profit organization that provides high school and middle school hands-on engineering, computer science and biomedical science curriculum and educator training. According to a 2012 $E^3$ Alliance report summarizing the impact of the initiative, four key outcomes were realized that continue to impact the STEM pipeline today:

1. The number of students participating in the national engineering curriculum Project Lead The Way quadrupled from 1493 students in 2007–2008 to 6373 in 2011–2012, a rate twice the state’s rate of growth;
2. Project Lead The Way students outperformed their non-Project Lead The Way peers with approximately 10% more Project Lead The Way students exceeding grade level expectations on the 11th grade standardized math test and approximately 10% more Project Lead The Way students enrolling in college; and
3. Participation by Hispanic students in Project Lead The Way courses increased at nearly 1.5 times the overall rate and both female students and low income students increased at approximately 2 times the overall rate ($E^3$ Alliance 2012).

Enrollment in Project Lead The Way courses continues to increase throughout the Greater Austin STEM Ecosystem and the STEM Pipeline Collaborative served 13,579 students during the 2017–2018 academic year ($E^3$ Alliance 2018).

### 10.2.2 Central Texas Summer STEM Funders Collaborative

The Central Texas Summer STEM Funders Collaborative, comprised of private, corporate, and community foundation funders, was founded in 2015 and collectively contributed $one million to Summer STEM programs its first year. With a goal to combat summer learning loss and increase access to high-quality STEM programming for low-income youth throughout the community, the collaborative established a common application and review process for grant proposals and implemented a common assessment methodology to allow for comparison across programs. The collaborative has provided professional development for grantees including topics of interest to funders such as strategies to engage diverse students and strategies to engage role models and volunteers in programming. The Summer STEM Funders Collaborative also initiated the proposal to bring the STEM Ecosystem effort to the Greater Austin area with representatives from the collaborative serving on the founding ecosystem steering committee. Companies engaged include many throughout the technology and semiconductor sectors including AMD, Applied Materials, Freescale Semiconductor, Google, IBM, Intel, and Samsung.

The collaborative continues today as the Summer Learning Investment Hub and is managed by Learn All The Time, a local non-profit that guides research-based approaches in out-of-school time systems and fosters a network of organizations and agencies providing out-of-school time services to youth and families. With the goal to combat summer learning loss central to its mission, the collaborative now expands its reach beyond STEM programming while maintaining common measurements of student outcomes and program quality, providing professional development for grantees, and providing a streamlined process for grant applications and reviews. In 2019, the Summer Learning Investment Hub awarded over $900,000 in grants to 15 high quality programs reaching over 3000 students with 90% of students qualifying for free or reduced lunch (Learn All The Time 2018).

### 10.2.3 Semiconductor Workforce Development Collaboration

The Skillpoint Alliance Semiconductor Executive Council was created in response to workforce challenges throughout the local semiconductor industry. Driven by a gap in the supply of semiconductor technicians, a collaboration between Austin Community College (ACC), the Capital Area Training Foundation (later known as Skillpoint Alliance), and Austin-area semiconductor companies developed and launched the ACC Semiconductor Manufacturing Technology degree program in 1995 to address the workforce needs within semiconductor manufacturing (Murphy 1997). Companies supported curriculum development, internships, and marketing efforts to promote the program to high school students and others in the community.

The Semiconductor Executive Council also supported precollege outreach and education in the community to raise awareness of semiconductor manufacturing career options and to excite the next generation of diverse STEM workers. The council provided funding and volunteers to engage in programs such as The University of Texas at Austin Society of Women Engineers student organization's Encounter with Engineering, a day-long program for high school girls designed to generate interest in engineering careers through role model interactions, hands-on activities, and interactive presentations. The council supported Introduce a Girl to Engineering Day and other precollege outreach programs reaching elementary and middle school students led by the Women in Engineering Program and the Equal Opportunity in Engineering Program at The University of Texas at Austin. Support of Central Texas Discover Engineering provided funding and volunteers from semiconductor companies for classroom visits throughout the area to expose students to semiconductor manufacturing and careers in STEM. The Skillpoint Alliance Semiconductor Executive Council later changed to the Technology and Engineering Executive Council as STEM workforce needs changed throughout the ecosystem and continued to support STEM outreach and education throughout the community.

## 10.3 Current Efforts Advancing the STEM Ecosystem

Led by a steering committee of volunteers from diverse STEM-related networks, organizations, and companies, the Greater Austin STEM Ecosystem recognizes and connects current efforts to advance STEM learning throughout the community while taking advantage of past community collaborations and learnings. Greater Austin STEM Networking Forums and Community Convenings support the ecosystem's goal to develop and engage an effective and sustainable network of academic, business, and community partners who are interested in providing STEM opportunities for pre-kindergarten through college learners throughout Central Texas (Greater Austin STEM Ecosystem, n.d.-c). Partnering networks such as Central Texas Discover Engineering, Learn All The Time, and Texas Girls Collaborative Project collaboratively lead STEM networking and learning events to advance the ecosystem's goal of supporting excellence in formal and informal STEM education throughout Central Texas. The ecosystem also engages STEM advocates and organizations across the community to collectively amplify STEM programs around themes such as the past 2017 Solar Eclipse and annual Computer Science Education Week and Engineers Week celebrations. With engaged networks and industry, collaborative events and amplifications, and shared structures and systems in place, the Greater Austin STEM Ecosystem continues to advance its goals.

### *10.3.1 Structures, Systems and Solutions That Support Ecosystem Advancement*

The Greater Austin STEM Ecosystem has embraced nimble structures, unique systems, and existing solutions that support the mission of fostering collaboration and advancing accessible STEM programming for all students. Many ecosystems throughout the national STEM Learning Ecosystem initiative have created non-profit organizations, hired staff, managed a membership structure for participants, or created STEM programming for students or educators. Given the robust STEM community and rich STEM programming in the Greater Austin area, this approach does not make sense for the community. The Greater Austin STEM Ecosystem takes advantage of the area's rich history of STEM programming, passion of STEM leaders, alignment in mission of STEM organizations and programs, and opportunities for collaboration while remaining actively nimble with an eye on the future.

### *10.3.2 Volunteer, Community-Based, Shared Network Structure*

The Greater Austin STEM Ecosystem is led by a volunteer steering committee of STEM educators, funders, community leaders, and company representatives. The original steering committee consisted of a small group of dedicated leaders who crafted the proposal to the STEM Funders Network to establish the ecosystem initiative in the Austin area. As the ecosystem has evolved, the steering committee has evolved. The steering committee now consists of partners from strategic organizations and networks throughout the community:

- Higher Education including Austin Community College, Huston-Tillotson University, and The University of Texas at Austin;
- School Districts including Austin Independent School District's Austin Partners in Education;
- Funders and companies such as the Andy Roddick Foundation and Cisco Systems;
- STEM networks such as Central Texas Discover Engineering, Learn All The Time, and Texas Girls Collaborative Project;
- Government-related or connected organizations such as the Austin Public Library, Early Head Start provider Child Inc., and STARBASE Austin with Camp Mabry; and
- Non-profits such as 4-H CAPITAL, Austin Science Education Foundation and Thinkery.

Steering committee volunteers share their time, expertise, meeting space, funding, networks, and other resources to advance ecosystem efforts. As gaps in knowledge, resources, or networks are identified, the steering committee adjusts its membership and will continue to evolve as the ecosystem grows, changes, and adapts to STEM learning needs throughout the community.

### *10.3.3 Ecosystem-Engaged Networks*

The Greater Austin STEM Ecosystem embraces the concept of collective impact where networks and organizations from various parts of the STEM ecosystem come together around a common agenda to solve a shared community challenge. Central Texas Discover Engineering, Learn All The Time Network, and Texas Girls Collaborative Project are all network-based organizations with goals that align with the Greater Austin STEM Ecosystem to foster collaborations and advance learning throughout the community. Each network contributes connections, systems, and organizational strengths to the ecosystem while gaining access to a broader community network and expanded visibility through amplified ecosystem efforts.

#### 10.3.3.1 Central Texas Discover Engineering

Central Texas Discover Engineering (CTDE) is a non-profit with a mission "to excite Central Texas K-12 students to pursue careers in math, science, and engineering through in-class, hands-on engineering activities and volunteer-driven initiatives of the engineering and education communities" (Central Texas Discover Engineering 2019). The organization has connected STEM volunteers and companies with Central Texas educators for over 20 years to bring STEM role models and hands-on STEM activities into classrooms. The organization is led by a volunteer steering committee with members representing local companies, professional engineering societies, non-profits, school districts, and The University of Texas at Austin who connect their organizations and broader networks into the efforts of CTDE. CTDE is funded by grants from companies such as 3 M, Emerson, IBM and National Instruments that contribute volunteer support to the steering committee and classroom visits throughout the community.

CTDE also collaborates with non-profits and organizations in the community to connect STEM volunteers with students and educators or families in public STEM events. In collaboration with the Bullock Texas State History Museum, CTDE leads Science Thursdays for students on field trips where STEM volunteers lead hands-on activities and demonstrations connected to exhibits and films at the museum. CTDE also collaborates with the Bullock Texas State History Museum to bring a museum-related hands-on activity to kindergarten through eighth grade participants in Girl Day at UT Austin, an annual STEM festival on The University of Texas at Austin campus hosted by the Women in Engineering Program. In collaboration with the American Society of Civil Engineers and Thinkery, the local STEM-focused children's museum, CTDE supports "Engineering Day at the Museum", a day where STEM volunteers lead hands-on activities and demonstrations for museum visitors in celebration of Engineers Week. CTDE also supports company and school STEM nights and family events to further connect STEM volunteers with students.

CTDE plays an important role in the Greater Austin STEM Ecosystem with its long history of STEM engagement, established systems and collaborations, and vast network of educators, organizations, and STEM professionals. Through its efforts, CTDE reaches over 14,000 K-12 students and engages with hundreds of volunteers and educators annually (Central Texas Discover Engineering 2019). CTDE contributes its STEM volunteer and educator network and system of connecting STEM volunteers with requests across the community to the ecosystem efforts while gaining access to the broader ecosystem network and increased visibility through ecosystem events.

#### 10.3.3.2 Learn All The Time

Learn All The Time (LATT) is a network of organizations, non-profits, school districts, universities, and public agencies that focus on quality, outcomes, and the capacity of out of school time programs (Learn All The Time 2016a, b). While not solely focused on STEM, LATT's mission overlaps with that of the Greater Austin STEM Ecosystem to reach all students throughout the area and provide educators with the tools they need to be successful. LATT has had representation on the Greater Austin STEM Ecosystem steering committee since its founding, providing critical connections between out of school time programs and the STEM community. LATT assumed management of the Central Texas Summer STEM Funders Collaborative, rebranding it as the Summer Learning Investment Hub while continuing to seek funding from technology and semiconductor sector companies to support STEM programming. The Greater Austin STEM Ecosystem contributes to the STEM learning within LATT and connects LATT members to a broader network of organizations and individuals engaged in informal out of school time programming such as that of Central Texas Discover Engineering.

#### 10.3.3.3 Texas Girls Collaborative Project

The Texas Girls Collaborative Project (TxGCP) is a collaborative, state-wide network of organizations and individuals advancing girls' interest and access to STEM programs and career pathways. Founded in 2007, TxGCP "provides forums, curriculum, best practices, and resources to foster collaborations, build capacity of participating organizations, and create a state-wide network of informed and connected informal and formal STEM educators and advocates" (Texas Girls Collaborative Project 2018a). TxGCP is one of over 30 Collaboratives making up the National Girls Collaborative Project network which serves more than 36,400 organizations across over 40 states (National Girls Collaborative Project 2018). TxGCP is led by the Women in Engineering Program at The University of Texas at Austin with regional volunteer leadership teams across Texas made up of STEM advocates from schools and school districts, universities and community colleges, companies, government agencies, and non-profits. The Austin-based regional leadership team drives TxGCP strategy and programming throughout community and contributes network resources, technology solutions, volunteers, and industry connections to the Greater Austin STEM Ecosystem.

The Greater Austin STEM Ecosystem has adopted TxGCP's model of regional STEM networking forums and collaborates on a regional STEM email listserv, the Central Texas K-12 STEM Outreach Google Group. Both have fostered connections with hundreds of STEM-related organizations, educators, and professionals across the Greater Austin area. The ecosystem has also adopted two additional TxGCP-supported systems for fostering STEM connections between programs, families, students, and industry. The Connectory connects educators, families, and students to STEM learning opportunities and events throughout the region. Texas STEM Connections, an online STEM volunteer matching system, provides connections between STEM industry volunteers and STEM volunteer opportunities throughout area classrooms and programs.

### *10.3.4 Community Engagement System*

The Greater Austin STEM Ecosystem has devised a system to allow community members to easily jump in and out of ecosystem events, programs and initiatives. There is no membership structure to join – anyone engaged in STEM education or outreach or programming in the community is part of the ecosystem. The ecosystem is open to all who want to join in, present, share, lead, amplify efforts, and be engaged.

The Greater Austin STEM Networking Forums and Community Convenings are examples of the community engagement system in action. Organizations across the ecosystem volunteer space and resources to host events. Individuals throughout the area contribute by presenting about their programs or organizations, sharing announcements in their networks, and leading or supporting parts of the events such as driving the agenda, hosting the check-in area, and taking notes of the discussion. The engaged community generates the content and direction for ecosystem events.

### *10.3.5 Technology Solutions*

Through the enlightened self-interest of organizations and networks throughout the Greater Austin STEM Ecosystem, technology solutions developed or supported by individual organizations are shared broadly throughout the community. Three technology solutions focused on making connections, fostering connections, and advancing STEM learning have been adopted by the ecosystem and are amplified through ecosystem communication channels and events. All are shared resources that contribute to the mission and goals of the ecosystem as well as those of collaborating organizations and networks.

#### 10.3.5.1 The Connectory

The Connectory is a searchable, online portal of STEM opportunities and events for families to explore within their communities. Developed and managed globally by the National Girls Collaborative Project, The Connectory is a collaboration with organizations such as the Afterschool Alliance, Association of Science-Technology Centers, and Maker Ed.

The Connectory makes the connections that inspire young people to explore, discover, and create. By offering the most comprehensive collection of STEM opportunities and programs, The Connectory is the go-to place for families to discover local STEM opportunities for the children in their lives and for program providers to find partners with which to collaborate. (The Connectory 2018).

As the regional hub for the National Girls Collaborative Project, the Texas Girls Collaborative Project supports The Connectory within the Greater Austin STEM Ecosystem.

For the Greater Austin area, The Connectory experienced a 30 percent growth in STEM opportunities posted in 2018 compared to 2017 and over 3000 unique visitors explored the online portal during that time. With the ecosystem embracing The Connectory as the central portal for promoting STEM opportunities and events, a wider network of program providers utilize the tool and more families and educators are aware of its existence, thus extending STEM learning further throughout the community.

#### 10.3.5.2 Texas STEM Connections

Texas STEM Connections is an online portal for connecting STEM volunteers to in-person and virtual STEM volunteer opportunities across Texas. The system is designed to make it easy for STEM professionals to connect with educators, classrooms, summer camps, afterschool and weekend programs, and others seeking STEM volunteers, role models, and mentors. Requests for STEM volunteers are searchable by zip code and are curated so volunteers in the system easily find opportunities that align with their skillset, interests, or location.

"Texas STEM Connections is a collaboration led by the Texas Girls Collaborative Project and supported by many organizations interested in creating meaningful STEM connections and role model interactions for students including: Texas Partnership for Out of School Time, Informal Science Education Association of Texas, Greater Austin STEM Ecosystem, Central Texas Discover Engineering, The University of Texas at Austin K12 STEM Collaborative, and The University of Texas at Austin Women in Engineering Program" (Texas Girls Collaborative Project 2018a, b para. 2). Its development and launch was funded by Intel as the company saw the community need to better connect STEM professionals and employee with students and classrooms. The Greater Austin STEM Ecosystem promotes Texas STEM Connections through various communication channels in addition to showcasing the system in community events. With over 1000 educators and volunteers in the system, over 46,000 students were impacted through virtual and in-person connections from 2016 through 2018.

#### 10.3.5.3 Central Texas STEM Google Group

The Central Texas K-12 STEM Google Group was created in 2010 by the Texas Girls Collaborative Project as a way to create an open communication and collaboration channel for anyone engaged in STEM throughout the community. The Google Group is open to anyone engaged in STEM outreach and creates connections, fosters collaborations, shares opportunities for funding and professional development, disseminates resources, and creates a sense of community across the region. It's an easy way to share information across all engaged in STEM outreach and to get the latest information on grants, research, best practices, and events. Organizations can use it to market their own events or seek resources to do what they are doing even better. Adding the Google Group to the suite of technology solutions supported and amplified by the ecosystem contributes to the Greater Austin STEM Ecosystem's common goal to connect and support those engaged in the STEM ecosystem throughout the community. The group reaches around 1000 STEM educators and advocates throughout the community, an increase of about 60% since 2015 when the Greater Austin STEM Ecosystem was founded.

## 10.4 Collaborative Events and Amplifications

The Greater Austin STEM Ecosystem has taken advantage of established models for community-wide STEM networking events and expanded visibility and reach of STEM organizations and programs through community-wide amplification efforts. Organizations such as Central Texas Discover Engineering, Learn All The Time, and Texas Girls Collaborative Project have a history of hosting community-wide events and amplifying efforts throughout the area. When STEM opportunities and community-wide events arise, the Greater Austin STEM Ecosystem has the opportunity to amplify programs and events as a collection and collaboration in order to amplify reach and impact.

### *10.4.1 STEM Networking Forums and Community Convenings*

Greater Austin STEM Networking Forums and Community Convenings are modeled off of similar regional networking events led by the Texas Girls Collaborative Project and community events led by Central Texas Discover Engineering and Learn All The Time. These events provide an opportunity for individuals from a variety of organizations throughout the community to connect, share resources and information, and foster relationships and collaborations. They are hosted at varied locations such as iFLY Austin, Cisco, STARBASE Austin at Camp Mabry, and the Girl Scouts of Central Texas to showcase STEM organizations in the Greater Austin area. Each event includes facilitated networking, community announcements, and

STEM program presentations or facilitated discussion focused on topics such as the Central Texas STEM Workforce, girls in STEM, STEM professional development providers, corporations, summer camp providers, or early STEM learners. STEM Networking Forums reach an average of 45 individuals at each event with approximately 100 unique individuals participating in at least one Networking Forum annually.

By bringing together those interested in particular topics, communities of practice are developing with the opportunity for increased learning and collaboration to benefit the overall ecosystem. For example, an initial Greater Austin STEM Ecosystem Corporate Convening focused on companies throughout the area engaged in STEM outreach and education led to a brainstorm of topics of interest to the corporate community such as best practices in STEM outreach and a showcase of corporate outreach programs. The corporate community's interest in continued and sustained engagement has led to quarterly Corporate Convenings. A Community Convening focused on STEM professional development providers has led to another community of practice where STEM professional development offerings are being mapped, opportunities for collaboration are being explored, and best practices in areas such as marketing and recruiting are topics for future convenings. With the ecosystem network and volunteer infrastructure supporting the Community Convenings, communities of practice can meet, develop, and explore opportunities to learn and collaborate to advance STEM learning throughout the community.

### *10.4.2 2017 Solar Eclipse Amplification*

On Monday, August 21, 2017, the moon obscured about 65% of the sun in Central Texas. Around Austin in days leading up to the solar eclipse and throughout the day of the eclipse, museums, universities, non-profits, public libraries, state parks, and other organizations hosted lectures, hands-on activities, and watching parties. The Greater Austin STEM Ecosystem compiled a list of events and shared widely through email newsletters, community listservs, and social media using the hashtag #AustinSTEM in an effort to showcase the breadth of STEM activity throughout community and to amplify the individual events (Greater Austin STEM Ecosystem. n.d.-b). Over 18 Greater Austin area events were shared throughout the community along with additional resources from NASA and other organizations providing additional STEM learning and engagement.

### *10.4.3 Engineers Week Amplification*

Celebrated annually the week of Presidents Day, Engineers Week is an international celebration of how engineers make a difference in our world. Engineers Week activities throughout Central Texas engage students and learners of all ages from

Engineering Days at area museums to engineering competitions and celebrations on university campuses to hands-on activities with engineering role models in local classrooms. The Greater Austin STEM Ecosystem compiles a list of Engineers Week events and shares widely through newsletters, community listservs, and social media using the hashtag #AustinSTEM in an effort to showcase the breadth of STEM activity throughout community and to amplify the individual events (Greater Austin STEM Ecosystem. n.d.-a). The list is also compiled into a flyer and press release that is easily shared throughout the community and printed to hand out at Engineers Week events such as Engineering Day at Thinkery and Girl Day at UT Austin. Nine community events including Engineering Day at Thinkery and Girl Day at UT Austin hosted by area museums, university programs, and non-profits were amplified in 2019.

#### 10.4.3.1 Engineering Day at Thinkery

Engineering Day at Thinkery is an example of collaborations throughout the Greater Austin STEM Ecosystem that connect students and families with STEM hands-on activities and demonstrations. Engineering Day was founded in 2002 at the Austin Children's Museum, the precursor to Thinkery, as a collaboration with the Austin branch of the American Society of Civil Engineers with support from the Central Texas Discover Engineering volunteer network. Throughout the years, volunteers from the American Society of Civil Engineers, companies such as Big Red Dog and National Instruments, university programs such as the Women in Engineering Program at The University of Texas at Austin, and university student organizations such as The University of Texas at Austin's Student Engineers Educating Kids and Student Engineering Council have led engineering-related hands-on activities and demonstrations for kids and families visiting the museum the first weekend of Engineers Week. With the Greater Austin STEM Ecosystem amplifications, Engineering Day at Thinkery participants are connected to upcoming Engineers Week activities to continue their engineering exploration throughout the week and beyond.

#### 10.4.3.2 Girl Day at UT Austin

Girl Day at UT Austin was founded in 2002 as Introduce a Girl to Engineering Day, an international celebration held annually during Engineers Week to help "focus a growing movement to inspire girls' futures so they learn they have a place in engineering a better world" (DiscoverE 2019). The Girl Day STEM Festival was added in 2016 to increase the number of computer science and other STEM discipline activities at the event. Girl Day at UT Austin has grown to a community-wide STEM festival engaging over 150 companies, non-profits, government agencies, high school programs, and university organizations, research groups, and programs. Over 8000 kindergarten through eighth grade students register annually with over 1500 volunteers leading hands-on activities and demonstrations throughout the

Cockrell School of Engineering complex on The University of Texas at Austin campus. The community-wide effort held the last Saturday of Engineers Week is proclaimed Introduce a Girl to Engineering Day by the mayor of the City of Austin and has been acknowledged as the largest Engineers Week Girl Day event by DiscoverE. Girl Day at UT Austin benefits from all the Greater Austin STEM Ecosystem amplifications leading into the last weekend of Engineers Week. Girl Day also has gained community partners after the Women in Engineering Program at UT Austin shares information about the event at Greater Austin STEM Networking Forums and Community Convenings.

## 10.5 Industry Engagement

Industry has been engaged in STEM education and workforce development throughout the Greater Austin area for a long time. The Greater Austin STEM Ecosystem has engaged industry from its founding with IBM and Samsung contributing letters of collaboration and support in the ecosystem application process. Industry throughout the ecosystem support STEM programs, encourage employees to volunteer, and identify gaps and opportunities in STEM workforce development to drive change. Examples of industry engagement include funding of the Central Texas STEM Funders Collaborative, volunteer engagement through Central Texas Discover Engineering, and hands-on activity engagement through Girl Day at UT Austin.

### *10.5.1 Central Texas STEM Funders Collaborative*

The Central Texas STEM Funders Collaborative is an example of a community-wide collaboration of companies, foundations, and organizations that invest in STEM programs throughout the Greater Austin STEM Ecosystem. Founded in 2015, the collaborative's goal is to combat summer learning loss and increase access to high-quality STEM programming for low-income youth throughout the community. Founding collaborative supporters included Andy Roddick Foundation, Austin Community Foundation, Central Texas Discover Engineering, Entrepreneur's Foundation, IBM, Intel, KDK-Harman Foundation, KLE Foundation, RGK Foundation, Samsung, Texas Girls Collaborative and Webber Family Foundation. The collaborative created a single grant application and a process for review by multiple funders that streamlined the process for both collaborative funders and grant seekers. The process fostered collaboration and sharing of best practices by funders, and increased overall impact of STEM programming throughout the ecosystem with shared metrics, common assessments, and increased opportunities for low-income youth. The Summer Learning Investment Hub annually awards over $900,000 in grants reaching over 3000 students with 90% of students qualifying for free or reduced lunch (Learn All The Time 2018).

### 10.5.2 Central Texas Discover Engineering

Central Texas Discover Engineering has a long history of engaging companies and corporate volunteers to engage in classrooms and excite students about STEM careers and pathways. Companies such as 3 M, Applied Materials, AT&T, Emerson, IBM, National Instruments, and Pape-Dawson Engineers contribute volunteers to serve on the organization's steering committee, meeting space, funding, materials and supplies, and more to ensure organization success. Companies invite the organization in for lunch and learns with employees to share best practices in engaging students in hands-on activities during classroom visits. Companies volunteer to take the lead for one or monthly Science Thursdays facilitated by Central Texas Discover Engineering at the Bullock Texas State History Museum. Company volunteers also adopt schools, visit classrooms, and share their career experiences with students throughout the community through connections made through the organization.

### 10.5.3 Girl Day at UT Austin

Girl Day at UT Austin is an example of an event where companies throughout the ecosystem join in to support a common goal of engaging more girls in STEM. Girl Day at UT Austin is a one-day event hosted by the Women in Engineering Program on The University of Texas at Austin campus where kindergarten through eighth grade girls and their families, teachers, Girl Scout troop leaders, and other adults explore STEM through hands-on activities, demonstrations, and shows led by STEM volunteers and mentors from industry, academia and the community. The program was founded in 2002 as Introduce a Girl to Engineering Day with 92 participating girls. Driven by interest from Google, the Girl Day STEM Festival was added to Girl Day at UT Austin in 2016 "to broaden participation across STEM disciplines, increase the number of computer science activities in the programming, and eliminate the cap on participant registrations to allow all interested students the opportunity to participate" (Women in Engineering Program at The University of Texas at Austin 2019). In 2019, 25 companies from across the area representing computing and technology, construction, manufacturing, semiconductors, and more participated in Girl Day. Girl Day at UT Austin has engaged over 41,000 elementary and middle school girls, over 11,000 volunteers and mentors contributing over 48,000 volunteer hours by leading hands-on activities and demonstrations, and over 50 local companies of all sizes since its founding in 2002.

## 10.6 Outcomes

The Greater Austin STEM Ecosystem has four goals: (1) develop and engage an effective and sustainable network of academic, business and community partners who are interested in providing STEM opportunities for pre-kindergarten through grade 16 learners throughout Central Texas; (2) support excellence in formal and informal STEM education throughout Central Texas; (3) incorporate an equity driven approach to promote awareness of, and access to, STEM pathways; and (4) ensure long-term sustainability of the Greater Austin STEM Ecosystem (Greater Austin STEM Ecosystem, n.d.-d). Preliminary results show the increase in engagement of STEM educators, organizations and advocates since the founding of the ecosystem. While still in the early stages of defining and assessing ecosystem-wide common metrics, the collaborative efforts are driving increased participation and engagement throughout the STEM community.

Outcomes related to the ecosystem's goal of increasing access to high-quality STEM programming throughout the ecosystem include the following:

- The Summer Learning Investment Hub is combatting summer learning loss and increasing access to high-quality STEM programming for low-income youth throughout the community with over $900,000 in grants to 15 high quality programs reaching over 3000 students with 90% of students qualifying for free or reduced lunch in 2019.
- Central Texas Discover Engineering reaches over 14,000 K-12 students and engages with hundreds of volunteers and educators annually using the ecosystem supported Texas STEM Connections volunteer matching online portal.
- Nine Engineers Week 2019 community events including Engineering Day at Thinkery and Girl Day at UT Austin hosted by museums, university programs and non-profits were amplified through collaborative, community-wide promotions to educators and families.
- The Connectory experienced a 30 percent growth in posts of Central Texas STEM opportunities and events in 2018 compared to 2017 and over 3000 unique Central Texas visitors explored the online portal during that time.
- In 2019, 25 companies from across the area representing computing and technology, construction, manufacturing, semiconductors, and more participated in Girl Day at UT Austin reaching over 8600 kindergarten through eighth grade girls and their families.

Outcomes from ecosystems efforts to develop and engage an effective and sustainable network of academic, business and community partners include the following:

- STEM Networking Forums reach an average of 45 individuals with approximately 100 unique individuals participating in at least one Networking Forum annually.
- STEM Community Convenings reach an average of 25 individuals at each event.
- Texas STEM Connections engages over 1000 educators and volunteers in the system to connect STEM volunteers with STEM volunteer opportunities in the community.
- The Central Texas K12 STEM Google Group reaches around 1000 STEM educators and advocates throughout the community, an increase of about 60% since 2015 when the Greater Austin STEM Ecosystem was founded.

- The Summer Learning Investment Hub created a single grant application and process for review by multiple funders that streamlined the process for both funders and grant seekers. The process fostered collaboration and sharing of best practices by funders, and increased overall impact of STEM programming throughout the ecosystem with shared metrics, common assessments, and increased opportunities for low-income youth.

## 10.7 Lessons Learned

The greatest lesson learned from the early years of the Greater Austin STEM Ecosystem is that the structure, goals, vision and initiatives of the ecosystem must take advantage of the particular strengths and opportunities of STEM organizations, industries, and individuals that exist in our community. The ecosystem structure must be flexible and adaptable to meet the STEM needs of the community and engage interests of various audiences and organizations involved. It takes time to understand the networks and organizations engaged in or supportive of STEM education, outreach, and workforce development. It takes time to build trust and explore the interests, opportunities, needs, and gaps in STEM learning throughout the community. It also takes time to develop a nimble infrastructure and to understand existing technology and network solutions that can enable quicker progress on ecosystem goals. With a dedicated and strategic ecosystem-wide effort, a vibrant, motivated, and passionate community of STEM programs, advocates, organizations, and industries will improve and advance STEM opportunities for all learners.

## References

Central Texas Discover Engineering (2019). Central texas discover engineering. Retrieved August 3, 2019 from http://www.centexeweek.org.

The Connectory (2018). The connectory. Retrieved August 3, 2019 from https://www.theconnectory.org.

DiscoverE (2019). Girl day. Retrieved August 3, 2019 from www.discovere.org/our-programs/girl-day.

E3 Alliance (2008, September 05). Central texas awarded $150,000 National science foundation grant [Press release]. Retrieved August 3, 2019 from https://e3alliance.org/2008/09/05/central-texas-awarded-150000-national-science-foundation-grant.

E3 Alliance (2012). $E^3$ Spotlight: Galvanizing the STEM pipeline. Retrieved August 3, 2019 from http://e3alliance.org/wp-content/uploads/2015/09/E3-Spotlight-Galvanizing-the-STEM-Pipeline-2.pdf.

E3 Alliance (2018). Students served. Retrieved August 3, 2019 from www.e3alliance.org/students-served.

Greater Austin STEM Ecosystem (n.d.-a). Engineers week 2019: Inspiring wonder. Retrieved August 3, 2019 from https://www.greateraustinstemecosystem.org/community-stem-events-opportunities/engineersweek.

Greater Austin STEM Ecosystem (n.d.-b). Greater Austin 2017 eclipse activities. Retrieved August 3, 2019 from https://www.greateraustinstemecosystem.org/community-stem-events-opportunities/2017eclipseactivities.
Greater Austin STEM Ecosystem (n.d.-c). Greater Austin STEM Ecosystem. Retrieved August 3, 2019 from https://greateraustinstemecosystem.org.
Greater Austin STEM Ecosystem (n.d.-d). Vision, Goals, Key Initiatives. Retrieved August 3, 2019 from https://greateraustinstemecosystem.org/about/goals.
Learn All The Time (2016a). Leadership series. Retrieved August 3, 2019 from https://learnallthetime.org/leadership-series-2.
Learn All The Time (2016b). Learn all the time. Retrieved August 3, 2019 from http://learnallthetime.org.
Learn All The Time (2018). Summer learning investment hub. Retrieved August 3, 2019 from http://www.learnallthetime.org/summerlearninginvestmenthub.
Murphy, S.L. (1997, July 20). ACC semiconductor training classes fall short of goals. Austin Business Journal.
National Girls Collaborative Project (2018). About NGCP. Retrieved August 3, 2019 from https://www.ngcproject.org/about-ngcp.
STEM Funders Network (2019). What are STEM Ecosystems? Retrieved August 3, 2019 from https://stemecosystems.org/what-are-stem-ecosystems/.
Texas Girls Collaborative Project (2018a). Texas girls collaborative project. Retrieved August 3, 2019 from https://www.txgcp.org.
Texas Girls Collaborative Project (2018b). Texas STEM connections. Retrieved August 3, 2019 from https://www.txgcp.org/texas-stem-connections.
Women in Engineering Program at The University of Texas at Austin (2019). Girl Day at UT Austin 2019 final report. Retrieved August 3, 2019 from https://utexas.app.box.com/v/UTGirlDayFinalReport2019.

# Chapter 11
# Fundão, Portugal: Using STEM Education to Help Build a New ICT Technopolis

**Ademar Aguiar and Sara Pereira**

**Abstract** In 2015, to reverse a rise in unemployment, a decline in economic growth and the population aging, the Municipality of Fundão designed the Strategic Plan for Innovation. The plan was thought to attract companies and people to Fundão and to encourage families and younger generations already in Fundão to live and work in their home territory. This would be accomplished by attracting investments and new businesses based on new technologies. The strategy had a significant positive impact on the number of jobs created in the city, its economic growth, and attraction of businesses and population. In effect, the municipality started from scratch and built a new ICT industry cluster to transform the economy for a globalized, digital age, addressing needs in the areas of software development, robotics, and technology-based solutions for traditional sectors. The consortium supporting the plan includes governmental organizations, universities, schools, civic associations, businesses, financial institutions, and innovation centers. After four years, Fundão hosts 14 new companies, including four multinationals. Those companies have created over 500 highly qualified jobs. The municipality also has 70 new startups and over 200 privately funded innovative projects. The training of young students on digital technologies, namely on programming, as well as the reskilling of adults, was an essential first step in the plan's implementation. This success was recognized with an award from the European Community–and it motivated the extension of the initiative into elementary schools, with the goal of covering all students of the municipality. This chapter examines the case within the lens of technopolis development and it includes interview insights from those involved in Fundão's sustainability plan.

A. Aguiar (✉)
University of Porto, Porto, Portugal
e-mail: ademar.aguiar@fe.up.pt

S. Pereira
University of Minho, Braga, Portugal
e-mail: sarapereira@ics.uminho.pt



C. Zintgraff et al. (eds.), *STEM in the Technopolis: The Power of STEM Education in Regional Technology Policy*,
https://doi.org/10.1007/978-3-030-39851-4_11

## 11.1 About the Case

Fundão is a small city, population of approximately 30,000, in the central region of Portugal, near the mountains of Serra da Estrela, 250 km far from the main cities of Lisbon and Porto (Fig. 11.1). Well-known by its colored fertile lands and genuine products with a reputation of unrivaled quality, the city has the largest cluster of cherry orchards in Portugal. Fundão also produces other high-

**Fig. 11.1** Location of Fundão, Portugal (Source: Adapted from image created by Rei-artur, January 2005, from the map Image:Mapa de Portugal.svg. Available at: https://en.wikipedia.org/wiki/Fund%C3%A3o,_Portugal#/media/File:LocalFundao.svg)

quality local products reflecting history and tradition: chestnuts, honey, wine, olive oil, cheese, wild mushrooms, and charcuterie (OSIRIS–Interreg Europe 2016).

Like many other small towns in Europe, Fundão started suffering unemployment rise, one of the typical problems of low density areas (under 10,000 inhabitants), especially in the younger generation under 30 years old, who despite being highly educated, could not find a job in the region.

In 2015, The Municipality of Fundão established a local development strategy (Projeto de Regulamento 2015), the Strategic Plan for Innovation, to quickly reverse unemployment rise, decline in economic growth, population aging and the exodus of young people from the region. Problems were aggravated by the 2011 economic crisis. That crisis was caused by the global recession, lack of competitiveness and limitations of being in the Eurozone (European Commission n.d.-a). The Strategic Plan was based on social innovation and cooperation between the municipality, businesses, schools, and universities. It was specifically designed to attract investments and enable new families and indigenous younger generations to live and work in their home territory.

The plan has been implemented through the attraction of companies and the incubation of new businesses, thus increasing the importance of innovation and technology in the region. Priority has been placed on offering excellent conditions to new businesses based on new technologies. In more detail, the municipality started providing infrastructure and service to foster entrepreneurship and new businesses, and it promoted an economic growth model based on applied knowledge and highly educated human resources.

The city created a key platform, the Living Lab Cova da Beira (LLCB). LLCB aimed to create regional cooperation networks, to leverage and disseminate innovation, and to integrate different stakeholders of the ecosystem. LLCB also aimed to create a culture of openness, collaboration, and community development; in other words, it aimed to be an urban incubator of companies and businesses.

The training of young students on programming and digital technologies, as well as the re-skilling of adults to satisfy the demand of IT companies, was an essential step in the implementation of the strategic plan. These efforts had a significant positive impact, namely, the creation of 500 new jobs in the city, attraction of businesses and population, and economic growth, as will be described throughout this case.

In this chapter, the authors describe the motivations, design, specific actions and outcomes related to the development of a new Fundão ICT technopolis that addressed desertification (the loss of population in a region, with all its negative effects). This information is both factual/statistical, and also qualitative and in the voice of those who developed the program. The narrative is about the development of this technopolis, its basis in the needs of the community, and its relationship to established industries and the importance of positioning this new development relative to the established economic activities of Fundão.

## 11.2 Building a Technopolis

The Municipality of Fundão is a good example of how and why a technology-based economic region, a technopolis (Smilor et al. 1989) starts. Fundão started a technopolis by targeting local economic development through the implementation of public policies and education programs targeting Science, Technology, Engineering and Mathematics (STEM).

The strategy defined was based on a model of open innovation (Projeto de Regulamento 2015), bringing together representatives of the quadruple helix (government, industry, education system, and civil participants) working towards the overall improvement of the local society. In Fundão's case, the education programs were not only universities, but also included schools.

### *11.2.1 Strategic Plan*

The vision of Paulo Fernandes, the mayor of Fundão, was embedded in the Strategic Innovation Plan. He stated that "a region is created by the people who live in it" (OSIRIS–Interreg Europe n.d.-a). By driving creation of this plan, Mayor Fernandes expected to mobilize the potential of the local community, and also to attract national and international entrepreneurs, promoting the development of the region and fighting against the tendencies of unemployment and desertification that were being felt in Portugal, and especially in the interior areas, caused by the economic crisis. This plan's purpose was, in short, to create jobs, fix people and retain talent.

The municipality designed the Strategic Innovation Plan in 2012, including actions at different levels under the common concept of Living Lab of Cova da Beira (LLCB). This consortium promoted by the Municipality of Fundão is a partnership with companies, universities, banks, public and private institutions, aiming to build a creative open ecosystem. The articulation between these valences is supported by the Urban Incubator for Enterprises and Businesses (UIEB).

UIEB includes: (a) incubation and co-working spaces for companies and new entrepreneurship projects; (b) the Fab Lab Aldeias do Xisto, prototyping laboratories, workshop-houses in the historic center of Fundão and nearby historical villages; (c) specific training centers and schools; and (d) a research and development center of health products and services to help the internationalization of products and entrepreneurs (Fig. 11.2). The Social and Business Incubator, the co-working spaces and the Fab Lab Aldeias do Xisto operate in the building 'A Praça', a building of the old Fundão market that was rehabilitated and adapted for this purpose.

The IT cluster is anchored on Centro de Negócios e Serviços Partilhados (CNSP) project (i.e. a business and shared services centre), started in 2013, which aims to transform the economy to a globalized digital age, addressing needs in areas such as software development, robotics, business services and tech-based solutions for traditional sectors (European Commission, 2018). The municipality invested around

**Fig. 11.2** Fundão case (Source: Projeto de Regulamento (2015)–Plano Estratégico para a Inovação do Município do Fundão)

2.4 million Euros with the support of European funds. After its first four years, CNSP now hosts 14 companies, including 4 multinationals, which created more than 500 high-qualified jobs, led to around 70 startups and over 200 privately-funded innovative projects (OSIRIS–Interreg Europe n.d.-b; Dinheiro Vivo/Lusa, 2018).

### *11.2.2 Integrating Primary and Post Secondary STEM Education in the Technopolis*

The Altran company, one of the world leaders in Engineering and R&D services, with offices in more than 30 countries, was one of the first big IT companies that decided to start activity in Fundão, creating 100 new jobs in 2014, and having grown to more than 400 professionals in 2019.

This success forced Fundão to face a good problem: a high demand from the new labor market in the IT sector and, on the other hand, not enough qualified human resources for the profiles demanded. Between 2017 and 2018, IT companies that have invested in Fundão created more than 500 new jobs and started hiring candidates with high technological skills. It is expected that Fundão would need, by the end of 2019, at least additional 300 professionals in the IT sector.

Since few young students of Fundão choose the IT academic curriculum, the city is providing incentives and benefits to attract professionals from other nearby towns and regions. Still, more qualified participants in the workforce are needed, and the local university and polytechnic (Universidade da Beira Interior and Instituto Politécnico de Castelo Branco) did not have the critical mass to respond to this appeal from the labor market.

### 11.2.3 Coding Training Programs

To immediately face the new demand, Fundão established a partnership with Academia de Código (Coding Academy) to offer intensive training programs on coding skills. Academia de Código was supported by the organization Portugal Social Innovation, financed by Social Impact Bonds[1]. The initial objective was to retrain people, most unemployed, 50% without a university education, in IT and coding programs, qualifying them to work as software developers in a short period of time. The training was intensive: it consisted of 650 hours of class time, over 14 weeks, to (re)qualify people in coding skills, turning them in software developers.

The result was a success. All forty participants of the first two cohorts found a job within two months. As of February 2019, nine cohorts were concluded, with 96% of the 180 participants employed (Brito, 2019). João Magalhães, CEO of Academia de Código, pointed out that "one of the challenges was to attract people not only from Fundão but also people who came from outside the region and wanted to come to Fundão to learn how to program and then work as software developers" (Personal interview, April 30, 2019).

Later, Academia de Código started to offer a curriculum option for students of elementary schools to develop their cognitive and non-cognitive skills and their digital literacy. The curriculum was designed to better prepare these students for future jobs and reduce the risk of unemployment in the medium term. This curriculum started with two pilot classes. Today, it is in all primary schools of Fundão (OSIRIS–Interreg Europe 2016). The original plan did not include secondary schools, but it is highly expected that additional curricula will be developed to support secondary students.

[1]A Social Impact Title is a mechanism designed to support innovative projects that respond to priority social problems within the competence of public policy, using pay-as-you-go logic. The project is implemented by one or more private entities and financed by one or several social investors, aiming for certain measurable social results using indicators and metrics previously validated by the Public Entity responsible for the sector policy. If the required results are achieved, the social investors are fully reimbursed. The Calouste Gulbenkian Foundation and the Shared Services & Outsourcing Platform Association were the social investors who financed the intervention of Academia de Código.

## 11.3 Education Philosophies and Methods

### *11.3.1 Pedagogy With Adults*

Academia de Código played an important role in adult training in Fundão. João Magalhães explains Academia de Código training in this way:

> We have developed a pedagogy, we have developed a curriculum, we have a team of more than twenty teachers that we have hired, that we have trained, because our pedagogy is very different from that of universities. Our programs are short, immersive and totally geared towards the needs of the market, because our goal is employment. They are very practical courses… our curriculum is 80% Java and 20% Javascript. The Java version is the most sought-after technology on the market right now by programmers and Javascript is highly valued and growing and is very complementary to Java. Moreover they are technologies that are very structured, especially the Java, and that in a logic of learning works very well. That is why we have developed a curriculum with this methodology.

Lucas et al. (2012) wrote about the pedagogy used in technical/vocational education. They described vocational education as "broadly hands-on, practical, experiential, real-world as well as, and often at the same time as, something which involves feedback, questioning, application and reflection and, when required, theoretical models and explanations" (p. 9). They wrote that vocational education pursues the following outcomes: (a) routine mastery of everyday procedures; (b) resourcefulness, the "aptitude to stop and think effectively" (p. 9); (c) verbal, mathematical, and digital literacy; (d) craftsmanship; (e) business awareness of the economic and social aspects of work; and (f) an inquisitive and resilient attitude.

João Magalhães also talked about the learning students need beyond technical mastery:

> Then it is not only the question of teaching the technical aspects, it is all the motivational part of soft skills, communication, team work, project work, … all those skills that a company needs when hiring a person in this area, all those tools that are necessary to technology and that we incorporate into our curriculum.

Effective vocational pedagogy cannot stop at mastery skills. In this volume's chapter on regional industry clusters, the authors include an argument for technical education as STEM education based on Richard Butler's decades of experience building such programs in his community (Zintgraff, Han, & Butler, this volume). While the fundamental goal is to prepare students for employment, the underlying pedagogy of technical education overlaps strongly with the pedagogy of the best STEM experiences, and the same core twenty-first century skills are learned.

Finally, vocational curriculum must be amenable to rapid adaptation, following closely behind the needs of employers. João Magalhães shared:

> The curriculum is also not watertight, it has been reviewed I don't know how many times since we started in 2015, because the market is always changing, technologies change at incredible speed, new versions always appear … and so our program does not quite the traditional of a university.

### 11.3.2 Pedagogy With Primary School Students

Academia de Código has also developed a curriculum and contents for teaching coding to children aged 6–10 years old. This project is not exclusive to Fundão. Its development began in Lisbon and today involves about 50,000 students throughout the country, in public and private schools (in the case of the latter, training is paid).

In Fundão the primary school curriculum also lets students experience coding as an extracurricular activity. The curriculum exhibits elements of maker education and design thinking, and also of constructionist methods (Papert, 1991) that require students to create artifacts, share them, and receive critique. João Magalhães, CEO of Academia de Código, shared these observations with the authors:

> What we perceive is that teaching programming to children has the most obvious advantages: thinking structure, problem solving, creativity… the biggest difference I see with the students that are with us is that they stop being users to become creators of technology, and that is a great transformation that I think we can do.

However, the overriding philosophy for the primary school curriculum is its alignment with the needs of the school system, maximizing the ease of adoption by teachers. Much effort is placed on teacher professional development and alignment to standards regarding the so-called twenty-first century skills. Given the scale of deployment—50,000 students using this solution—the training of teachers becomes the essential issue, along with developing a curriculum that works within the structure of schools. Furthermore, coding is optional in the school curriculum, which increases the importance of developing content teachers will choose to use. Schools that wish to integrate this project into their activities must apply. João Gonçalves described the 2015 process of development:

> Our pilot project in 2015 served to develop a curriculum and content to teach. Based on this experience, we have developed a tool to reach all children in Portugal and, in the future, outside Portugal. So, we developed a curriculum, developed the contents and developed the software. Today what we have is a solution for when a teacher, a school, or even a government want to introduce this new competence into the curriculum, we give a training to teachers, which is a training accredited by the Ministry of Education, so teachers earn credits for career progression, and then we have curricula and content for the first to sixth school year.

Individual and collective empowerment, through the development of the digital, creative and innovative potential of individuals, and its application in the region, is the broader educational philosophy for both the adult training and the primary school education programs. Both are in alignment with and supported by the Strategic Plan. Although the conception and initial development of the program was top-down, the teaching and learning philosophy is based on a bottom-up approach—each one must learn by doing for himself (Projeto de Regulamento 2015).

Based on the success of the plan, on attracting IT companies and creating demand for IT professionals, a new phase of educational program development is expected, extending in the future to secondary school students, 14–17 years old, completing the talent development pipeline for the Fundão ICT Technopolis.

## 11.4 Policies That Support the Program

The policy behind Fundão's Strategic Plan for Innovation is to promote a culture open to the outside world, to value applied knowledge, and to leverage the profitability of human capital to benefit of the citizens of Fundão. This policy was articulated in the Strategic Plan for Innovation. The policies were defined by the municipality, in 2013, when it decided to "launch a strong strategy, integrated and assertive, to try to diversify the economic activity in Fundão region, trying to promote the more traditional sectors, to attract investment, to create new jobs, and retain population", as stated by Ricardo Gonçalves, Coordinator of the Innovation and Investment Office of the Municipality of Fundão (Personal interview, May 21, 2019).

For new investment, the key principle was to target new technologies, because exponential growth was expected in the near future, and the approach could attract investment, with markets throughout the country and value created specifically in the region. Special attention was given to the attraction of investment and the creation of an environment conducive to the development of companies and their potential to create wealth and employment (Projeto de Regulamento, 2015). The plan called for providing the required base infrastructure, good connectivity, nurturing an open ecosystem, and promoting open participation, collaboration and empowerment of all involved: governmental organizations, industry, education system, and the community. Living Lab Cova da Beira was responsible for developing this collaborative work.

A specific decision was made to not restrict innovation to technological factors, but to also promote social entrepreneurship. This decision enabled diverse strengths. A more diverse group of agents were mobilized, and with them distinct areas of knowledge and services.

## 11.5 The Virtuous Cycle

The convergent approach of the Fundão's Strategic Plan for Innovation illustrates the virtuous cycle (see Chap. 1) created in this region, in a combination of complementary actions and synergies between the municipality, university and polytechnic institute, schools, civic associations and businesses (Fig. 11.3). One of the reasons for the success of Fundão's Strategic Plan lies precisely in the fact that it is based on a policy that focused on STEM Education to strengthen the industry in the region, knowing that the industry would benefit from professionals with STEM skills.

With STEM Education being contextualized in local development challenges, Fundão's Strategic Plan has driven a virtuous cycle that has contributed to increase the well-being, self-image and the engagement of the local community and to enhance economic growth. In the opening chapter of this volume, Zintgraff noted that the STEM Technopolis virtuous cycle is both practical and psychological. The

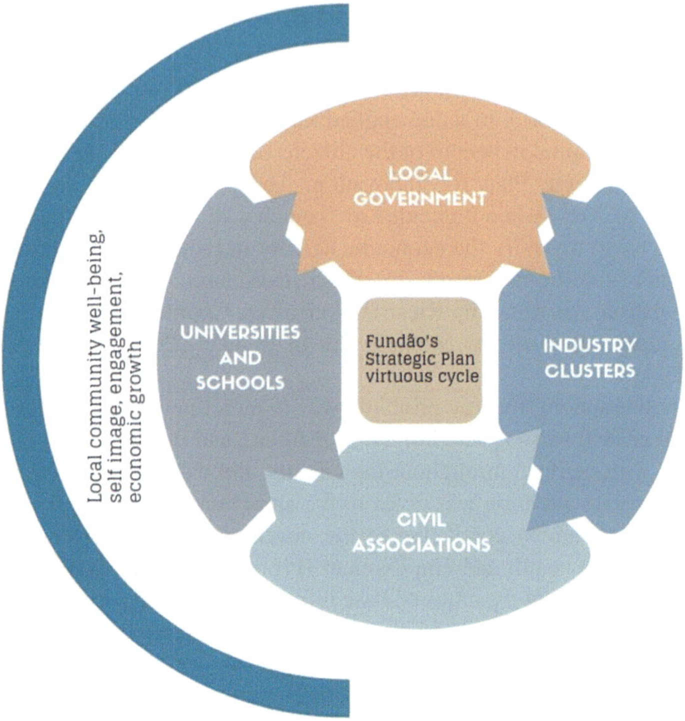


**Fig. 11.3** Fundão's STEM technopolis virtuous cycle

increase in hope, self-esteem, well-being and the quality of life of people living in the Fundão territory are undoubtedly one of the greatest gains of this plan.

In the interview with the President of the Coordination and Development Commission of the Central region, this idea of the virtuous cycle applied to the Fundão case is evident. Stating that Fundão's Strategic Plan has proven that" there are no condemned territories." Ana Abrunhosa stressed the importance of this plan to value endogenous resources, deepening and modernizing sectors that were more traditional, but also betting on new industries and diversifying the economic base. Ana Abrunhosa stated:

> There is a new life in this territory. The impact has to do with hope and self-esteem, this is fundamental to believe in these territories. It also has to do with the fact that in this project the Universidade da Beira Interior (UBI), the Polytechnic Institute and other companies have also been involved. It has made more dense the ecosystem that exists in the territory, attracted more people and so today we have a new hope in these territories and this strategic plan is certainly a reference for other territories that also suffer from the desertification of economic activity and people, and who believe that with a strategy, an action plan and with work, can reverse this trajectory (Personal interview, May 02, 2019).

Ricardo Gonçalves, Coordinator of the Innovation and Investment Office, of Fundão City Hall, also highlighted in his interview the creation of an ecosystem that involved the municipality, companies, the University of Beira Interior and the Institute of

Vocational Training, with whom they established a partnership to create an advanced training center for specific topics (for example, in advanced programming), which also involved the university. Throughout the ecosystem, a key principle was establishing win-win arrangements. All those involved invested, and each received a respective return for the investment made.

## 11.6 Outcomes

### *11.6.1 Awards*

The Centro de Negócios e Serviços Partilhados (CNSP) (i.e., Business and Shared Services Centre) received the RegioStars 2018 prize, an award from the European Commission that distinguishes innovative projects and good practices of regional development (European Commission n.d.-b). This European award, in addition to recognizing its contribution of 'Innovation for rural revival,' confirms the project's growth potential and shows one of the main lessons learned: its potential to be replicated in other municipalities. Back in 2015, Fundão had already won the European Enterprise Promotion Awards, and in 2016 also won the Municipality of the Year award in a Portugal-wide competition.

### *11.6.2 The Greatest Indicator: Reversal in Net Migration*

One very special result highlights outcomes of the work: a reverse in the direction of net migration. Paulo Fernandes, Fundão Council President, shared:

> For the first time in decades, Fundão has a positive migratory balance and is on the radar of national and international investors. This is thanks to this Center [CNSP].

### *11.6.3 Views of Key Actors*

The three key actors that were interviewed for this work, and who were directly or indirectly involved in the implementation of the Strategic Plan of Fundão, evaluate it in a very positive way. Their views reflect not only the results obtained but also the design and implementation process. In fact, their positive opinions about long-term sustainability lie precisely in the fact that it is a project designed with and for the local community, having as support the technological development.

Ricardo Gonçalves, Coordinator of the Innovation and Investment Office, Municipality of Fundão, focused on community participation and the importance of including traditional businesses and new technologies in the plan:

> It is an integrated and therefore participated strategy. This was done with a listening and permanent contact with the community. After a few months of the design of the Strategic Plan for Innovation, it was finally launched and it was very important all that process [of auscultation] because in fact the community felt that we were talking about a strategy that it was theirs. People felt involved, the community was involved and in fact this was one of the first conditions for their success. The strategy was based on the most traditional sectors of agri-food, so the first things we did was to create strategies to promote this sector, promote internationalization and new markets, but then we effectively focused on new technologies. (Personal interview, May 21, 2019).

Ana Abrunhosa, President of the Regional Coordination and Development Committee of the Center Region, focused especially on the need to involve "endogenous" resources, and the connection between that strategy and sustainability:

> There is one thing that is very important, is that this strategy is appropriated by the people of the territory itself. It's important that people feel that this is done with them and for them, and also to involve the actors in this territory - that suppliers and service providers are from this territory, betting on the companies that are already there. Because there cannot be that feeling that incentives and support exist only for those who come…It has to be a coordinated strategy. To support endogenous resources and traditional enterprises and to diversify the economic base. The development strategy is also limited if it is conditioned to the resources and actors of the territory, but this has to be done first with who is in the territory. This is what in first guarantees the sustainability of the projects. (Personal interview, May 02, 2019).

João Magalhães, CEO of Academia de Código, focused on the role of education in the emerged Fundão technopolis, the role education plays in the ecosystem.

> It is not just mobilizing companies, it is not just trying to bring new skills to education, it is creating an ecosystem that has been a bit of a secret. It is to be able to join education, with companies and with requalification. Building the ecosystem is what makes the project sustainable in the long run. That's why I'm a big fan of the project and that's why we wanted to be involved from the beginning. (…) In the case of Fundão, our role was to contribute to a project to create a technological pole. It is not only the requalification of people, it is also this support in the creation of a technological pole. (…) In these regions, such as Fundão or Terceira [in Azores], we contribute to something bigger, after all we contribute to the development of a technological pole and to the total change of a region or a population. So they are projects with much more impact. These are projects in which we also had some role, we are playing a role in the development of a region (João Magalhães, CEO of Academia de Código, personal interview, April 30, 2019).

Table 11.1 summarizes the essential strategies highlighted.

**Table 11.1** Essential strategies highlighted by key actors

| Actor | Role | Strategies highlighted |
|---|---|---|
| Ricardo Gonçalves | Coordinator of the innovation and investment office, municipality of Fundão | Community participation including both traditional businesses and new technologies in the plan |
| Ana Abrunhosa | President of the regional coordination and development committee of the Center Region | Involve "endogenous" resources connection between that strategy and sustainability |
| João Magalhães | CEO of academia de Código | The role of education in the ecosystem |

## 11.7 Conclusion

One of the best indicators of the success of the plan is the housing issues now arising in Fundão. Housing has become a priority for the Mayor Paulo Fernandes (Andrade and Botelho 2018). The municipality is promoting the rehabilitation of old houses and the construction of new ones, both in Fundão and nearby villages, bringing even more economic growth to the region.

Altogether, the new inhabitants of Fundão recognize that working and living in the region is not a real issue but rather an urban myth. Marta Caldeira, a staff member at the company Altran and responsible for the onboarding and housing of new employees, noted that people have a better quality of life, with more time, and are central, not only close to the main towns of Portugal, but also to Madrid and Salamanca, Spain.

The authors, who are from outside the area, conclude that the case of Fundão, Portugal is a strong example of a STEM Technopolis, and is especially compelling coming from a rural region. The municipality addressed its challenges, leveraged its strengths, ran a participatory process, involved all players from the technopolis, and solidly integrated vocational and STEM education philosophies, methods and actions. Fundão, in Portugal, has moved from the risk of desertification to being a full participant at the STEM Technopolis virtuous cycle.

## References

Andrade, V., & Botelho, N. (2018, August). Fundão atraiu empresas e pessoas, agora faltam casas. Jornal Expresso, 5 de Agosto de 2018. Retrieved April 26, 2019 from https://expresso.pt/economia/2018-08-05-Fundao-atraiu-empresas-e-pessoas-agora-faltam-casas#gs.7e8uw8.

Brito, P. (2019, February). Academia de Código já converteu 180. Rádio Cova da Beira. Retrieved April 26, 2019 from http://www.rcb-radiocovadabeira.pt/pag/51223.

Dinheiro Vivo/Lusa (2018, October). Fundão apostou nas novas tecnologias e criou 500 postos de trabalho qualificados. Retrieved April 26, 2019 from https://www.dinheirovivo.pt/empresas/fundao-apostou-nas-novas-tecnologias-e-criou-500-postos-de-trabalho-qualificados/.

European Commission (2018, September). Business and shared services centre supports smart growth in Portugal's centro region. European Union (2019). Retrieved April 26, 2019 from https://ec.europa.eu/regional_policy/en/projects/portugal/business-and-shared-services-centre-supports-smart-growth-in-portugals-centro-region.

European Commission (n.d.-a). What is the euro area? Retrieved April 26, 2019 from https://ec.europa.eu/info/business-economy-euro/euro-area/what-euro-area_en.

European Commission. (n.d.-b) REGIOSTARS Awards. Retrieved April 26, 2019 from https://ec.europa.eu/regional_policy/en/regio-stars-awards?Allnews=true#2.

Lucas, B., Spencer, E., & Claxton, G. (2012). *How to teach vocational education: A theory of vocational pedagogy*. London: City & Guilds Centre for Skills Development.

OSIRIS–Interreg Europe (2016, October). Partner presentation–Fundão. Open Social Innovation policies driven by co-creative Regional Innovation ecosystems (OSIRIS). European Union (2016–2020). Retrieved April 26, 2019 from https://www.interregeurope.eu/osiris/news/news-article/254/partner-presentation-fundao.

OSIRIS–Interreg Europe (n.d.-a). Mobilizing creative potential: In presov self governing region through open social innovations. European Union (2016–2020). Retrieved April 26, 2019 from https://www.interregeurope.eu/fileadmin/user_upload/tx_tevprojects/library/file_1545226236.pdf.

OSIRIS–Interreg Europe (n.d.-b) e-Catalogue of good practices. Open Social Innovation policies driven by co-creative Regional Innovation eco-systemS (OSIRIS). European Union (2016–2020). Retrieved April 26, 2019 from https://www.interregeurope.eu/fileadmin/user_upload/tx_tevprojects/library/file_1548405507.pdf.

Papert (1991) https://psycnet.apa.org/record/1991-99006-000.

Projeto de Regulamento (2015). Plano Estratégico para a Inovação do Município do Fundão. *Diário da República, 2*(14). Retrieved April 26, 2019 from https://www.cm-fundao.pt/ sites/default/files/CMF_DOCS/0213402137.pdf.

Smilor, R. W., Gibson, D. V., & Kozmetsky, G. (1989). Creating the technopolis: High-technology development in Austin, Texas. *Journal of Business Venturing, 4*(1), 49–67.

# Chapter 12
# Mexico's Movimiento STEM and Related Developments in the State of Querétaro

**Graciela Rojas and Laura Segura**

**Abstract** In Mexico, the organization *Movimiento STEM* has created Ecosistema STEM, an initiative to aggregate independent STEM education institutions, build partnerships, and build dynamic programs that leverage government STEM education initiatives. This chapter contains the story of Movimiento STEM's work, and focuses on efforts made in high schools in Querétaro, Mexico. After presenting the national context, the authors share how the United States-Mexico Foundation for Science (FUMEC) brought the Engineering Bases Program (EBP) to the highly industrial state of Querétaro. The goal of EBP is to increase theoretical and technical knowledge and skills of students, starting in high school, and moving them into technical vocations. EBP increases the chances of students joining the STEM labor force or continuing their education in STEM areas. EBP helps complete the Virtuous STEM and Economic Development Cycle described in this volume: (a) local government encourages collaborative work between industry and education; (b) extra-curricular education in STEM adds to official programs, and (c) local industry recruits prepared young people, some of whom give back by helping more youth enter STEM fields. Overall, by fostering scientific vocations through multiple levels of Mexico's talent pipeline, Movimiento STEM is helping drive hard-to-achieve rapid educational reform in Mexico's STEM education system.

## 12.1 Introduction

Mexico is facing the Fourth Industrial Revolution, like the rest of the world. And in a special way we know that the countries of Latin America have not yet effectively integrated STEM Education, which is the differentiating factor for countries to have possibilities to compete for new opportunities within the knowledge economy and at the same time reach the progress that the planet and humanity can still sustain,

G. Rojas · L. Segura (✉)
Movimiento STEM, Mexico City, Mexico
e-mail: presidencia@movimientostem.org; desarrollo@movimientostem.org



C. Zintgraff et al. (eds.), *STEM in the Technopolis: The Power of STEM Education in Regional Technology Policy*,
https://doi.org/10.1007/978-3-030-39851-4_12

meeting the needs presented by the achievement of the Sustainable Development Goals (United Nations 2015).

Although STEM competencies go beyond training hard science specialists, an adequate measurement to know how much progress is being made in preparing countries for the Fourth Revolution is the number of students in the areas of science, technology, engineering and mathematics, since this will give us clues about the possible quantity and quality of innovation.

Each country and region determine the strategic focus that they will give to their innovation efforts, according to their strengths, geographical, social and infrastructure conditions, as well as the needs and opportunities of the local, national and international market. This implies that the chain that includes innovation, research and development, industry, higher education and professional preparation, and basic and secondary education, must be closely linked in a consistent sequence (Smilor et al. 1989). That is, basic and secondary education feed the interest in a professional career in STEM, which promotes employability in industries that develop research and development (R&D) and therefore innovation that meets market needs.

Contextualizing the Virtuous Cycle diagram presented in Chap. 1 of this volume, coordinating STEM and technology policy in the Mexican context would have the characteristics displayed in Fig. 12.1.

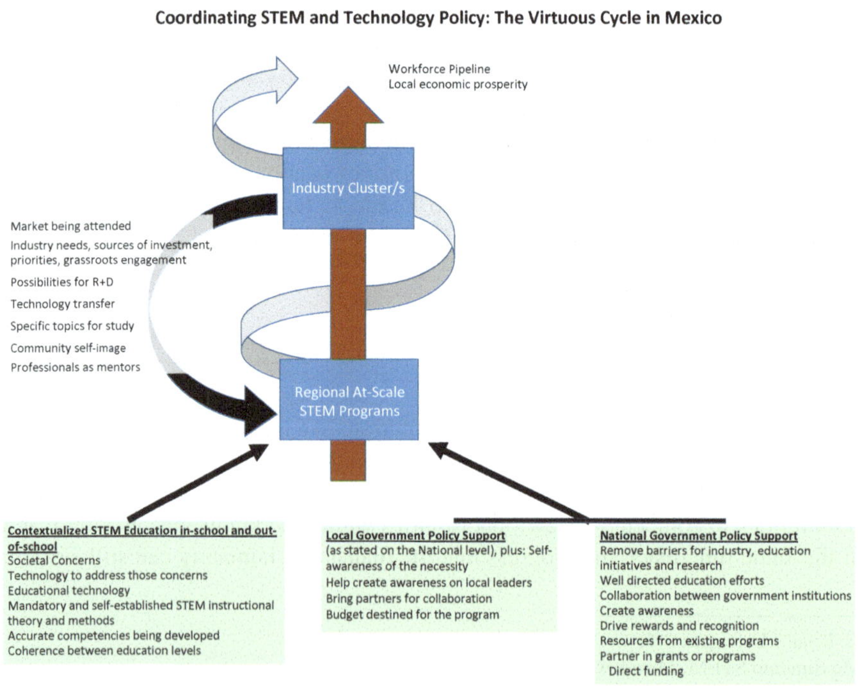


**Fig. 12.1** Virtuous STEM and economic development cycle, contextualized for Mexico. Adapted from Zintgraff (this volume)

The best way to achieve this sequence is the support that surrounds the process; that is, having the right conditions in government, investment, intersectoral alliances, and importantly, education in official curricula as well as extracurricular and informal education initiatives at the local and national levels, as the extracurricular spaces bring realistic and practical learning that increases or complements the competencies developed in the classroom of curricular education.

## 12.2 Mexico's Ecosistema STEM and Movimiento STEM

To emphasize the importance of Education in STEM, in Mexico there is a binding effort called Ecosistema STEM (STEM Ecosystem), from which the institutions that work Education in STEM with out-of-school programs (Friedman and James 2007) from different perspectives (which have also been named as EduSTEM institutions) are promoted, integrated and organized in the form of an economic sector. In this way, through the various programs implemented throughout the country, Mexico can grow the STEM ecosystem, and with it, have a greater impact on society.

The Ecosystem was founded and is led by Movimiento STEM, a non-profit founded in 2017 by the Mexican entrepreneur Graciela Rojas, who also founded Profesor Chiflado (translated as Wacky Professor) in 2004, a company with a methodology to educate and inspire children through entertainment (Rapeepisarn et al. 2006). Informing the Movimiento STEM action plan, research showed that there was no national showcase for initiatives which, like Profesor Chiflado, portrayed science and technology as interesting and desirable career paths for younger generations, leading to an increase in the number of scientists, engineers, mathematicians and related professionals, so Mexico can become more competitive in the twenty-first century.

This gap of information on the sector and the need for a structured and visionary plan for collective growth resulted in the creation of the Ecosistema STEM and its lines of action as defined by Movimiento STEM. These lines of action, which continue to evolve, are defined as follows:

- STEM Culture: To promote the importance of STEM Education and EduSTEM providers and their actions, among teachers, researchers, parents or guardians, young people and anyone interested in generating an exponential growth of STEM and having a greater impact on society. The strategy includes actions to: disseminate and position, influence public policy, conduct events such as the National Teacher Prize México and the National STEM Education Congress.
- Professionalization: Accreditation process for Movimiento STEM to guarantee that the programs of EduSTEM institutions are implemented with quality and optimal results, as they must demonstrate having accurate knowledge about STEM Education, carrying out educational actions in STEM areas, driving STEM transversely and generating STEM Competencies. The process consists

of a diagnosis, training and certification of facilitators, impact measurement and STEM specialized consulting.
- Community of learning and practice: Create workgroups for specific topics or tasks to benefit the progress and recognition of the whole ecosystem. The activities are linked to the international community and gather best practices in STEM Education, which encourages interaction and detonates the exponential growth of STEM to achieve a greater impact.
- Research: Movimiento STEM carries out projects to generate indicators, relevant research on STEM and STEM-related topics, as well as a shared vision on STEM and its impact on competitiveness facing the Fourth Industrial Revolution and on compliance with the UN 2030 Agenda. These projects bring a frame of reference for Ecosistema STEM members to direct their actions and portrays them as relevant actors for STEM advances in our country. The research projects are widely disseminated through different platforms and events.

This initiative makes it clear that the key aspect in STEM is education, which can be provided from many fronts, but with the common goal of increasing the competitiveness of the country and its different regions. This will be achieved by augmenting the critical mass of children, adolescents and young people with STEM skills and with a greater interest in pursuing a professional career in STEM. The work will be aligned with the characteristics of the jobs of the future and promoting innovation with a social and inclusive vision.

The STEM Ecosystem currently has nearly 100 members, most of them in Mexico City, with its network extending to 14 States in the country. All of these institutions run valuable and interesting programs, including Programa Bases de Ingeniería (the Engineering Bases Program, or EBP), in the State of Querétaro, highlighted below, chosen to be a main topic of this chapter due to its reflection of the Virtuous Cycle diagram in Chap. 1 for coordinating STEM and Technology Policy.

## 12.3 Program Example: The Engineering Bases Program

The Engineering Bases Program (EBP) (The United States-Mexico Foundation for Science, 2019) is carried out based on the methodology of Project Lead The Way (PLTW) and it creates a strong bond among upper secondary education, universities, and the labor market in the State. This methodology provides transformative learning experiences for PreK-12 students and teachers, creating an engaging, hands-on classroom environment, and it empowers students to develop in-demand knowledge and skills they need to thrive. It also provides teachers with the training, resources, and support they need to engage students in real-world learning (Project Lead The Way 2019).

## 12.4 Education and STEM Education in Mexico

The history of education in Mexico has gone through several stages of progress and setbacks in trying to establish a viable, egalitarian education system that meets the needs and characteristics of the national moment. Mexico's current General Education Law was published in 1993 and its last reform was made in September 2019 (Congreso General de los Estados Unidos Mexicanos 2019). Over the years it has undergone several modifications to ensure a current education that extends to everyone and that reaches its own quality standards and those of consultative bodies. Consistency with the Organisation for Economic Co-operation and Development (OECD) educational goals has been the guiding factor in recent years.

According to the report "Strong Foundations for Quality and Equity in Mexican Schools" (OECD 2018), the national education system and its current educational model (whose development began in 2012 and 2013 and its implementation in 2018) will have a positive effect on the search for a quality education with equity, which are the two central concerns, focusing education on student learning, and on the professionalization of teachers and the improvement of schools. This focus will occur as opposed to one exclusively on models of governance. However, governance models also remain a concern as they have not been sufficiently agile to cover one of the largest and most complex education systems in the OECD, with almost 31 million students enrolled in public and private institutions at compulsory education level in 2016. Basic education alone accounted for close to 26 million students, 1.2 million teachers and more than 225,000 institutions. Around five million additional students were enrolled at upper secondary level in 2016–2017 (OECD 2018).

The levels of education, according to information from the Ministry of Public Education, are distributed as shown in Fig. 12.2, and its basic demographics are stated in Table 12.1.

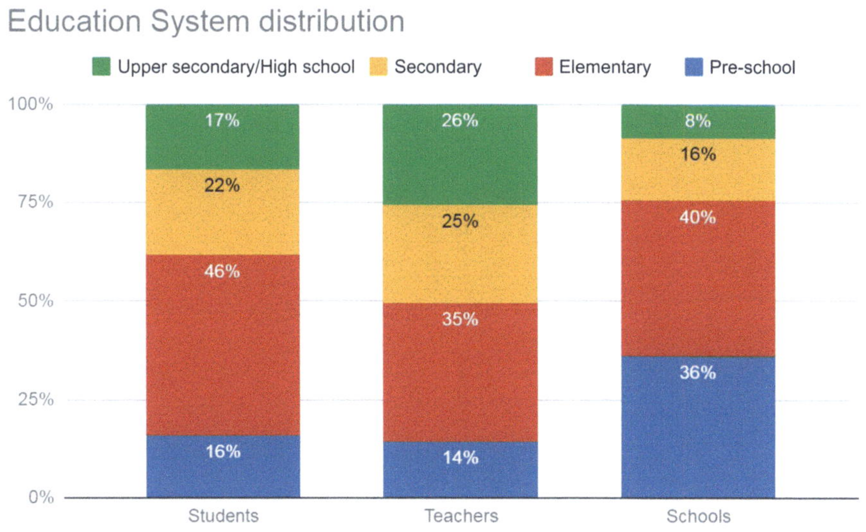


**Fig. 12.2** Elaborated based on information from the ministry of public education in the national and international statistics and educational indicators document, 2017–2018

**Table 12.1** Mexico PreK-12 education demographics (Ministry of Public Education in the National and International Statistics and Educational Indicators document: 2017–2018, 2018)

| Reach | Level | Ages | Students | | | Teachers | Schools |
|---|---|---|---|---|---|---|---|
| | | | Total | Women | Men | | |
| Basic, Compulsory | Pre-school | 3–5 | 4.891.002 | 49.55% | 50.45% | 238.133 | 89.579 |
| | Primary | 6–11 | 14.020.204 | 49.08% | 50.92% | 571.520 | 96.920 |
| | Lower secondary | 12–14 | 6.537.003 | 49.41% | 50.59% | 410.189 | 39.689 |
| Compulsory | Upper secondary | 15–18 | 5.237.003 | 50.63% | 49.37% | 423.754 | 20.852 |
| Total (millions) | | | 30.684.470 | 15.185.222 | 15.499.248 | 1.643.616 | 247.040 |

*Corrected education levels for Table : Upper secondary/High school Secondary Elementary Pre-schoolUpper secondary/High school Secondary Elementary Pre-school

*corrected education levels for graph above and table below: Upper secondary/ High school Secondary Elementary Pre-school

It is evident at first glance that there is a significant dropout rate from elementary school and up to the next two levels of education. Many children and young teenagers are missing out on opportunities, so it becomes a priority for education public policy to address the way teaching is being perceived and applied. In the new educational model, it is proposed that all these levels have a curricular progression in which the exit profile of a school grade is logically and consistently linked to the next grade, from pre-school to high school, creating strategic training paths aligned to the reality of the world of work. This educational model has turned to look closely at competencies development, including those related to STEM education.

The early development of these competencies and what can be accomplished in childhood and youth will determine not only personal success, but also national adequacy to the challenges to come. Mexico has a population marked by a high proportion of young people, one of the highest in the OECD, but the country has not turned this situation into a real advantage, as young generations are not engaged via new education trends, or even via older trends started twenty years ago. It is necessary to strengthen education in STEM where it already exists and start promoting it wherever it needs to be created.

Students begin to lose interest in science in the last years of elementary school or in the first years of secondary school, so when they reach high school, they choose non-STEM-related studies, but rather the social sciences (Valero et al. 2017). This highlights the importance of promoting curricular, non-curricular and informal education based on STEM, with initiatives aimed at students in elementary and early secondary grades, their parents, teachers and other social actors.

Currently within the Ecosistema STEM, a group of experts in STEM Education developed a rubric that links our education model to the characteristics of STEM competencies that are internationally agreed upon. The rubric begins with the features of the graduation profile of each educational level (Secretariat of Public Education 2017a, b). It further looks at the characteristics that the Global STEM Alliance stated in its STEM Education Framework (2016). Those characteristics reflect current education research, innovative and effective practices worldwide, and the understanding that the learning achieved by a student at one level will be the basis of the learning achieved in the next. The rubric found in Table 12.2 finally assesses to what extent a student has achieved "desirable traits" in the scale: Expert, Advanced, Good and Basic.

## 12.5 STEM Education and Careers in Mexico

This indicates that there is an intentional effort to bring education in Mexico toward global educational trends, committing technical knowledge with socio-emotional skills, to promote new generations that are critical, curious, creative, respectful, and inclusive and that are put to work to solve real problems of people and the environment. This path is the one that provides tools to cover the urgent demand of professionals in STEM careers.

**Table 12.2** Ecosistema STEM rubric (Alianza para la Promoción de STEM 2019, Visión STEM para México, pp. 32–33)

| Category | Description |
|---|---|
| 1. Critical thinking/Creativity/ Problem solving: Develops critical thinking and solves problems with creativity | Asks questions to solve problems of various kinds. Informs, analyzes and argues the solutions proposed and presents evidence for conclusions. Reflects on own thinking processes, relies on graphic organizers to represent them and evaluates their effectiveness. |
| 2. Problem solving: Strengthens mathematical thinking | Broadens own knowledge of mathematical techniques and concepts to pose and solve problems with different degrees of complexity, as well as to model and analyze situations. Values the qualities of mathematical thinking. |
| 3. Data literacy: Likes to explore and understand the natural and social world/Shows responsibility for the environment/Takes care of your body and avoids risky behaviors | a) Likes to explore and understand the natural and social world.<br>Identifies a variety of phenomena of the natural and social world, reads about them, reports in different sources, investigates applying principles of informed skepticism, formulates questions of increasing complexity, performs analysis and experiments. Systematizes own findings, builds answers to own questions and uses models to represent the phenomena. Understands the relevance of the natural and social sciences. |
| | b) Shows responsibility for the environment.<br>Promotes the care of the environment actively. Identifies problems related to the care of ecosystems and solutions that involve the use of natural resources with responsibility and rationality. Is committed to the application of sustainable actions in own environment. |
| | c) Takes care of own body and avoids risky behavior.<br>Activates bodily skills and adapts them to different situations that are faced in the game and school sports. Adopts a preventive approach by identifying the advantages of caring for own body, having a balanced diet and practicing physical activity regularly. |
| 4. Communication: Communicates with confidence and efficiency | Uses mother tongue to communicate effectively, respectfully and safely in different contexts with multiple purposes and interlocutors. If he/she speaks an indigenous language, also speaks in Spanish. Describes experiences, events, desires, aspirations, opinions and plans in English. |
| 5. Collaboration: Has initiative and favors collaboration | Recognizes, respects and appreciates the diversity of abilities and visions when working collaboratively. Has initiative, undertakes and strives to achieve personal and collective projects. |

(continued)

**Table 12.2** (continued)

| Category | Description |
|---|---|
| 6. Digital literacy and computer science | a) Compares and chooses the technological resources at own disposal and takes advantage of them with a multiplicity of purposes. Learns different ways to communicate and obtain information, selects it, analyzes it, evaluates it, discriminates it and builds knowledge. |
| | b) Compares and chooses the technological resources available and takes advantage of them with a multiplicity of purposes. Searches, selects, evaluates, classifies and interprets information, presents multimedia information, communicates, interacts with others, represents information, explores information, explores and experiments, manipulates dynamic representations of concepts and phenomena and creates products. Favors the development of critical, creative thinking, information management, communication, collaboration in the use of technology, digital citizenship, computational thinking. |

According to the Talent Sufficiency Survey (ManpowerGroup 2018), on average 45% of global employers are not able to cover their vacancies (highest level in the last 12 years); in Mexico the rate is 50% of employers. Job seekers are not fully prepared or do not have enough experience in their fields to do the work demanded. Also, there is an important lag in development of professional skills such as communication, collaboration, relationship building, creativity, empathy and desire of learning, important skills that will increase technological capabilities and will reduce the risk of replacement by automation.

With the recent arrival in Mexico of new companies and increased foreign direct investment, new production processes and new jobs were created for which there are almost no qualified personnel. In fact, the position of operator of machinery is the most difficult to cover precisely because of the arrival of investment in highly productive sectors such as automotive, general manufacturing, aerospace and aeronautics. This is particularly the case for the Mexican State of Querétaro.

This panorama puts in evidence that the current educational offer must be aligned to the labor requirements. Therefore, the education system must adopt the characteristics of training required by the industrial sector, which has been growing rapidly in recent years in Querétaro.

## 12.6 Querétaro: Society and Economy

### *12.6.1 Geography, Society*

Querétaro is a State located in the central zone of Mexico. It has 18 municipalities and is in a privileged location because it connects two zones with large industrial centers of the north-center and east-west of the country (see Fig. 12.3). There are three internal corridors of industrial and commercial vocation: San Juan del Rio-

**Fig. 12.3** Map of Mexico with location of Querétaro

Cadereyta de Montes, San Juan del Rio-Querétaro and the incipient Querétaro-Norte corridor.

Querétaro's population in 2020 is estimated at 2.279.637 (Consejo Nacional de Población CONAPO 2019). As of 2015, 61.6% of the population lived in the Metropolitan Area of Santiago de Querétaro (Metropolitan Areas of Mexico 2015). It is estimated that out of every 100 Queretans residing in the state territory, 28 are under 15 years old, 6 are over 65 years old and 66 people are aged between 15 and 64 years (Gobierno del Estado de Querétaro 2016). Stated differently, there is a demographic bonus ready to be exploited.

The state is poised for economic growth. It grew on average 5.8% per year between 2014 and 2017. This was mainly due to the strong development of the manufacturing industry, which is 28.2% of the state's GDP (El Economista 2019).

This growth has meant, among other social effects:

- By the first trimester 2019, according to the Strategic Indicators of Occupation and Employment, the manufacturing industry employed 23.6% of economically active Queretans. (INEGI, 2019).
- Increase in the average daily salary, which reached 424.68 pesos as of May 2019, third highest in Mexico. (Secretaría del trabajo y Previsión Social 2019)
- Overall reduction of poverty between 2016 and 2018 by 8.9% per year and reduction of extreme poverty by 31% in the same period (CONEVAL 2018).

**Fig. 12.4** Scheme of strategic sectors of Querétaro (Federal Government, Conacyt. Innovation Agenda in Querétaro, 2016)

- National immigration in the period 2010–2015 for people with higher education: 4th highest in Mexico with 5.200 aged 18–24, and 2nd in the country with 11.000 aged 25–49. (Almejo, Hernández, 2016).

## *12.6.2 Economic and Development Activity*

The accelerated economic growth is primarily due to the establishment and consolidation of clusters of the automotive, aeronautical and aerospace industries, information technologies, medical biotechnology and health (Fig. 12.4).

In addition to the establishment of the clusters, a factor for growth has been the integration of a chain of suppliers and direct efforts to articulate the academic and training programs with the productive sector, attending to the lines of action established in the State Development Plan 2016–2021. The State is also working to propel research centers, which works towards completing the virtuous cycle of innovation.

In general, the picture of Querétaro's infrastructure for innovation and development is comprised of the following elements:

- 1st national position for innovation of the economic sectors in 2014, according to the State Competitiveness Index (IMCO 2018).
- 45 industrial parks and zones (SEDESU 2018)
- 24 geographic clusters: pinpoints for economic and social development. (SEDESU 2018) (Fig. 12.5).

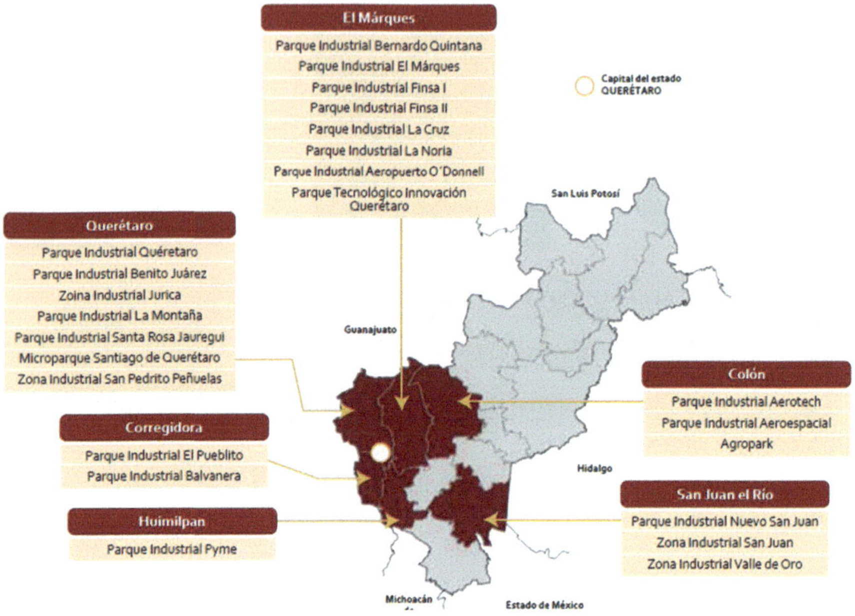


**Fig. 12.5** Distribution of the 24 geographic clusters

- 3rd place in the "National Ranking of Science, Technology and Innovation," for its training for science and technology
- 83 academic programs registered in the National Program of Quality Postgraduates, 66% of them in STEM-related areas, out of which 7 have international quality (CONACYT 2019)
- 1504 students on postgraduate studies within the state, through a CONACYT scholarship (CONCYTEQ 2017)
- 142 students have a CONACYT scholarship on national or foreign postgraduate courses (CONCYTEQ 2017)
- 53 research and technological development centers: 3678 researchers (SEDESU 2018)
- 719 researchers registered on the National System of Researchers (SNI); third place nationally for the highest ratio of researchers to citizens (CONACYT 2017). Eighty percent of these researchers are in STEM areas.

Regarding one of the main elements of this outlook, the establishment and growth of clusters, it has become a focal point to draw the attention of companies and groups of investors to explore possibilities of economic growth in Querétaro in areas and industries of interest. This becomes crucial as the National Expenditure on Research and Experimental Development (GIDE) is only 0.48% of GDP (CONACYT 2017), and 70.3% of this expenditure by 2015 came from government investment (OMI 2017), which provides a glance at how endangered this sectors' development is and the urgent need to boost its growth.

**Fig. 12.6** Position of Querétaro in the national index of science, technology and innovation

The extent and variety of industrial activity in Querétaro opens the opportunity to promote a greater participation of the productive private sector in the structuring of study plans that attend their special hiring needs, as well as to create the conditions for students to stay in companies and research centers, public and private within the State. In this sense, it is essential that research, technological development and innovation be strengthened, so that Querétaro students graduate from the undergraduate level with the necessary competencies for competitive professional performance.

In general and as a way to highlight the growth of innovation in Querétaro, the authors include a visual summary of a report from the Analysis Center for Research in Innovation, A.C. on various measures that reflect innovation (Fig. 12.6) (CAIINNO 2018).

### *12.6.3 The Government as an Ally of Development*

In its State Development Plan 2016-2021, the Querétaro State Government identified lines of action focused directly on industrial progress and its connection with the educational sphere, for strengthening access and quality of educational services in the State (Gobierno del Estado de Querétaro 2016). The main lines of action are:

- Promoting the link between educational institutions and the productive sector.

- Promoting scientific and technological education at all educational levels and in the municipalities of the entity.

For prosperity through the creation of favorable working conditions that favor the insertion of people in the labor market of the State, the lines of action are:

- Link people of economically active age with the productive sectors of the State.
- Articulate the academic and training programs with the productive sector.
- Promote employment options for vulnerable groups in society.
- Encourage the development of skills and technical skills of the economically active population.

Within the same document, the relationship among scientific, technological and innovation development is established in the "Prosperous Querétaro" guiding axis. The objective defined is "promoting the virtuous cycle of investment, employment and satisfaction of needs of consumption and savings of the population of Querétaro through the sustainable management of vocations and regional economic needs" (Gobierno del Estado de Querétaro 2016, p. 64).

With these goals and actions established, the State Government has participated in specific agreements with local and federal government offices, national and international organizations, businesses, and academic institutions, to create paths that facilitate growth and to foster programs that directly work towards sustainable development.

In pursuit of the objectives linked to education and prosperity, the United States-Mexico Foundation for Science (FUMEC), together with the School of Scientific and Technological Studies of the State of Querétaro (CECyTEQ), developed a business-education linkage program, called the Engineering Bases Program (in Spanish, *Bases de Ingeniería*), which is described next.

## 12.7 The Engineering Bases Program

### *12.7.1 Why It Is Necessary*

A determining factor to trigger the growth of STEM areas, and thus productivity, is the congruence among the components of the chain that start with education and reach the positioning of innovations in the market. However, in OECD PISA results (OECD 2016), 15-year-old students in Mexico are below the OECD average in science performance (416 points), reading (423 points) and mathematics (408 points). In these three areas, less than 1% of students in Mexico achieve excellence levels of competence (levels 5 and 6).

According to the same study, students in Mexico declare high levels of interest in science compared to their peers in other OECD countries, whether measured by their expectations of having a professional career related to science, or their beliefs in the importance of scientific research, or their motivation to learn science. However, these positive attitudes are weakly associated with the performance of students in mathematics.

These facts and the high risk of dropping out of school suggest that the Engineering Bases Program should be obligatory. It provides a fresh perspective and a motivation for young people to count on more and better tools with which they successfully insert themselves into a technically and technologically specialized work environment, as well as the possibility to follow an engineering career. The program also helps create the supply that meets the demand for a Querétaro-specific STEM workforce.

### 12.7.2 How EBP Became Reality

The Engineering Bases Program is driven by the United States-Mexico Foundation for Science (FUMEC) and the School of Scientific and Technological Studies of the State of Querétaro (CECyTEQ).

The first breakthrough took place in November 2008, during a national meeting for the implementation of the Comprehensive Reform of Higher Secondary Education (RIEMS). In this meeting FUMEC presented the possibility to replicate the Project Lead The Way (PLTW) program in higher secondary schools in Mexico. The General Director of CECyTEQ, Mtro. Carlos Ignacio Luhrs Eijkelboom, took this initiative as an opportunity to improve the profile of CECyTEQ students.

As part of the adoption of the program, CECyTEQ authorities visited US schools to review the performance of students immersed in Project Lead The Way. One of these schools was the The Science Academy of South Texas SAST in Mercedes, Texas, which is known for its excellent results in applying the PLTW program and is ranked as the Best Public School in the Brownsville Area and the 13th Best Public High School in Texas, with 85% proficiency in Math (Niche 2019).

From seeing the achievements of the successful school in Texas with PLTW, CECyTEQ authorities decided to take and contextualize the program for their institution and add it to their value chain offer, which comprises 8 programs for students to follow a successful path towards higher education or employability.

As the first step to implement the Engineering Bases Program, teachers with subjects related mainly to engineering and math are trained to operate it. As a next step, students are required to add one hour to their regular school day during the 3 years (6 semesters) of their high school preparation. Each semester the program emphasizes one step of the engineering design process, as listed in Table 12.3:

The Engineering Bases Program attends to a growing demand for technical and scientific talent capable of generating new ideas and of preparing for future work trends, and also encourages students to pursue careers in the areas of STEM. It offers a curriculum based on real projects within the partner companies, allowing students to perceive what their future work looks like.

The profile that results from this training is that of students self-directed to innovation and entrepreneurship, that are persistent and forward-learning. These characteristics become traits in students' personalities and help them be more confident and committed.

**Table 12.3** Main content of engineering bases program per semester

| Semester # | Content |
|---|---|
| 1 | Development of logical mathematical thinking |
| 2 | Introduction to engineering design I (application of design thinking and ideas visualization) |
| 3 | Introduction to engineering design II (transforming a creative idea into a prototype or physical model) |
| 4 | Design innovation and creativity (designing projects with a high level of feasibility, through engineering principles) |
| 5 | Applying engineering into products (applying engineering principles and methods to bring designs into reality, considering viability: manufacture) |
| 6 | Development of innovation process (integrating previous knowledge in the development of an engineering project of their choice) |

The first generation started the program in 2009, with 141 students on two campuses of CECyTEQ, Querétaro and Corregidora, which were chosen for the initial implementation because of their strategic location, high demand with students and with staff, and abundant career options in those locations related to engineering. This generation exited high school and the program in 2012.

From the beginning, it was well recognized that as students in the program worked in small groups and close to their professors, they received more personalized treatment, as well as instruction in additional subjects with an innovative approach for their activities.

### *12.7.3 Operation of the Program in CECyTEQ*

The School of Scientific and Technological Studies of the State of Querétaro (CECyTEQ) is a decentralized public body of the State Government. It was created with the purpose of expanding the educational offering at the technological baccalaureate level. This expansion was important because a growing number of high school graduates could not continue their studies because of problems of quota in the existing schools at the next educational level.

The technological nature of the studies, on the one hand, is aimed toward the training of technicians who join the labor market with the skills required for the strengthening and development of the productive sector of the entity. The same training may, on the other hand, increase the proportion of students entering university careers in the areas of engineering, technology, sciences and natural sciences, so necessary for the industry development.

CECyTEQ has 12 campuses (7 of those are part of the EBP program) and covers 14 technical careers (Table 12.4). Table 12.3 lists the number of students studying different topics.

**Table 12.4** Number of CECyTEQ students and their technical careers choices (CECYTEQ n.d.)

| Career choice | # Students |
|---|---|
| Administrative and management processes | 2058 |
| Programming | 1550 |
| Mechatronics | 1382 |
| Industrial maintenance | 1102 |
| Industrial production | 676 |
| Logistics | 466 |
| Electronics | 287 |
| Biotechnology | 270 |
| Electromechanics | 220 |
| Electricity | 176 |
| Transformation of plastics | 118 |
| Automotive maintenance | 92 |
| Machines and tools | 81 |
| Computer equipment support and maintenance | 32 |
| Total: | 8510 |

The 12 campuses are geographically distributed in correspondence to the population density of the areas and also to the location of the industry clusters. Each campus focuses on different careers depending on the needs of the surrounding area. The one covering the highest number of careers has 10 and the one with the fewest has only two. Most of the campuses are located in or near the State capital, with three more at the southeast and two in the north. Implementation at each campus is as follows:

Teachers of each campus are trained to implement the program, and their knowledge on the subjects is reaffirmed and / or updated.

For students, the first step is the dissemination of the program that is made in secondary schools so they can apply to the program.

The applicants go through a general exam, psychometric tests, selection and interviews, in order to get the best candidates.

The program requires one extra hour every weekday, during the 3 years of high school, and in that time they will be able to create their own engineering project.

Students can participate in programs from allied universities and with partner enterprises. They also have access to national and international fairs and competitions.

Around the learning experience, it has proven important to:

- Get parents as involved as their other activities allow them, so they accompany their children at times and be available for meetings.
- Have teachers join the program willingly and with the aim of becoming relevant role models.
- Gather data of the experience, so it becomes easier to revise or replicate.
- Identify success stories, whether to publicize them or to create a network of team leaders or support groups for following generations.
- Fortify a gender strategy, as girls comprise only 36.2% of beneficiaries.
- Diversify sources of funding to reduce dependency and risk.

Engineering Bases is compatible with other area programs, and partnering with those programs can be an effective strategy. For example, EBP can partner with initiatives working on other education levels, such as the System of Experiential and Indagatory Teaching of Science (SEVIC), imparted by the nonprofit Innovation in the Teaching of Science (INNOVEC 2019). This program brings to schools a different and interesting way for children to commit to their learning experiences. In Querétaro, the program attends to approximately 20,000 children from elementary to secondary school.

### 12.7.4 Relation with Industry

CECyTEQ is an active member of clusters in Automotive, Aeronautics, Information and Communication Technologies, and Biotechnology. The Querétaro campus is part of the state Mechatronics and Robotics Association. Students in the Engineering Bases Program visit companies such as Nestlé (food industry), Zafran Group (aviation), Construlita (lighting), DeAcero (steel) and Senoplast (plastic). They also visit research and development centers.

Engineering Bases Program students in the last semesters can also access internships and university stays. Those university stays are conducted in several higher education institutions of the State, most of them with a technological focus: Universidad Nacional Aeronáutica en Querétaro (UNAQ), Universidad Autónoma de Querétaro (UAQ), Instituto Tecnológico de San Juan del Río (ITSJR), Universidad Tecnológica de Querétaro (UTEQ), Universidad Politécnica de Querétaro (UPQ), Universidad Tecnológica de San Juan del Río (UTSJR), Universidad Tecnológica de Corregidora and Centro de Física Aplicada y Tecnología Avanzada (CFATA) of the Universidad Nacional Autónoma de México (UNAM).

Engineering Bases also has a specific program for alternative energies. It is held in the Pedro Escobedo campus. Students do some of their work at the El Sauz Combined Cycle Thermoelectric Power Plant, in the same municipality and only a 20-minute drive away.

As appointed on the State Development Plan 2016-2021, Querétaro must articulate its academic and training programs with the productive sector, and the current blossoming of industrial activity opens opportunities to work together and create logical and well-thought-out life and career paths for the next generations.

### 12.7.5 Program Implementation Results

As indicated by FUMEC by personal communication during May 2019, the program has been applied in Querétaro, Table 12.5 with results such as the following:

FUMEC has taken the success of the program in Querétaro to be extended to Estado de México with the help of the local government. There, three high school systems will be advised on how to implement the Engineering Bases Program.

**Table 12.5** Internal statistics about the engineering bases program

| Statistic | Measure or Comment |
|---|---|
| 7 program generations | 1594 students |
| Once they finish high school education: | 96.1% wants to keep studying in the field of interest. |
| Career choices: Aspirations | 77.5% Engineering<br>32.1% Masters<br>21.1% Doctorate<br>19.6% Universitary superior technician<br>16.7% Bachelor's degree |
| Career choices: Current situation | 80.9% know what career field to study<br>9.1% deciding between 2-3 careers in one area of study<br>7.2% deciding between 2-3 careers in different areas of study<br>2.9% do not know what to study |
| Possible certifications | SolidWorks: 3D CAD design software (computer-aided design) for modeling parts and assemblies<br>-Certified SolidWorks Associate (CSWA)<br>-Certified SolidWorks Professional (CSWP) |
| Add-on activities | Students participate in national and international contests, among other activities<br>Teachers participate in courses, trainings, congresses, among other activities |
| Generation 2015-2018 (most recently graduated) | 413 graduates, out of which:<br>332 continue their studies (80%), out of which:<br>244 choose engineering or science career (73%) |
| Generation 2015-2018 (most recently graduated) first 3 universities chosen: | Instituto Tecnológico de Querétaro (ITQ)<br>Universidad Politécnica de Querétaro (UPQ)<br>Universidad Tecnológica de Querétaro (UTEQ) |

## 12.8 Conclusion

Local and national workforce development more than ever needs to take intentional and strategic steps. Mexico has not been successful all through time to establish lasting solutions to foundational problems in its education system. Education is a cornerstone for development and needs to improve faster than ever. Solutions can be derived from successful programs around the world, or they can be small local initiatives that prove themselves successful and then are escalated to a level with greater impact, like the Engineering Bases Program.

Also, education in general and STEM education in particular should promote in the long run a different way to understand the world and its phenomena and should work toward creating meaningful solutions to old and new problems, with a social and inclusive vision. A global trend is pursuit of UNESCO's Sustainable Development Goals. Pursuit of these goals is mandatory in the following sense: the problems that the goals address cannot be ignored. The vision of STEM in the technopolis is to continually execute a powerful cycle for responsible investment, visionary entrepreneurship, and more and better employment.

## References


Alianza para la Promoción de STEM (2019). Visión STEM para México. Retrieved August 2, 2019 from https://movimientostem.org/wp-content/uploads/2019/02/Visio%CC%81n-STEM-impresio%CC%81n.pdf.

Centro de Análisis para la Investigación e Innovación CAIINNO (2018). Índice nacional de ciencia, tecnología e innovación 2018. Retrieved August 2, 2019 from https://www.caiinno.org/wp-content/uploads/2018/12/INDICE-2018.pdf.

Congreso General de los Estados Unidos Mexicanos (2019). Ley General de Educación, Nueva Ley publicada en el Diario Oficial de la Federación el 30 deseptiembre de 2019. Retrieved February 15, 2020 from http://www.diputados.gob.mx/LeyesBiblio/pdf/LGE_300919.pdf

Colegio de Estudios Científicos y Tecnológicos del Estado de Querétaro (n.d.). Información estadística básica de inicio de semestre febrero-junio/2019. Retrieved August 2, 2019 from http://cecyteq.edu.mx/SGC/EstadisticaBasica/Archivos/Inicio_FebJun2019.pdf.

Consejo de Ciencia y Tecnología del Estado de Querétaro CONCYTEQ (2017). Programa estatal de ciencia, tecnología e innovación, Querétaro 2017-2021. Retrieved August 2, 2019 from http://www.concyteq.edu.mx/concyteq/apropiacion/centro/9.

Consejo Nacional de Ciencia y Tecnología CONACYT (2016). Agenda de innovación de Querétaro. Retrieved August 2, 2019 from http://www.agendasinnovacion.org/?p=1123.

Consejo Nacional de Ciencia y Tecnología CONACYT (2017). Informe general del estado de la ciencia, la tecnología y la innovación. Retrieved August 2, 2019 from http://www.siicyt.gob.mx/index.php/transparencia/informes-conacyt/informe-general-del-estado-de-la-ciencia-tecnologia-e-innovacion/informe-general-2017.

Consejo Nacional de Ciencia y Tecnología CONACYT (2019). Padrón del Programa Nacional de Posgrados de Calidad. Retrieved August 2, 2019 from http://svrtmp.main.conacyt.mx/ConsultasPNPC/padron-pnpc.php.

Consejo Nacional de Evaluación de la Política de Desarrollo Social CONEVAL (2018). Evolución de la pobreza y pobreza extrema nacional y en entidades federativas, 2008, 2010, 2012, 2014, 2016 y 2018. Retrieved August 2, 2019 from https://www.coneval.org.mx/Medicion/Paginas/PobrezaInicio.aspx.

Consejo Nacional de Población CONAPO (2019). Proyecciones de la población de México y de las entidades federativas 2016-2050, Querétaro. Retrieved August 2, 2019 from https://www.gob.mx/cms/uploads/attachment/file/475447/22_QUE.pdf.

Estrella, V.(2019). Report Querétaro is highlighted for economic growth. El Economista. January 18, 2019. Retrieved August 2, 2019 from https://www.eleconomista.com.mx/estados/Queretaro-destaca-en-crecimiento-economico-20190118-0014.html.

Friedman, L., & James, S. (2007). Afterschool advantage: Powerful new learning opportunities. Retrieved August 2, 2019 from https://www.informalscience.org/sites/default/files/2014-06-24_2007_Afterschool_Advantage-Powerful_New_Learning_Opportunities.pdf.

Fundación México Estados Unidos para la Ciencia FUMEC (2018). Programa Bases de Ingeniería. Retrieved August 2, 2019 from https://www.fumec.org/Educacion-Innovacion.

Foro Consultivo Científico y Tecnológico, A.C. FCCyT (2014). Ranking Nacional de Ciencia, Tecnología e Innovación. Retrieved August 2, 2019 from http://foroconsultivo.org.mx/libros_editados/ranking_2013.pdf

Global STEM Alliance (2016). STEM Education Framework. The New York Academy of Sciences. Retrieved August 2, 2019 from https://www.nyas.org/media/13051/gsa_stem_education_framework_dec2016.pdf.

Gobierno del Estado de Querétaro (2016). Plan estatal de desarrollo Querétaro 2016-2021. Retrieved August 2, 2019 from http://www.queretaro.gob.mx/BS_ped16-21/.

Hernández, A. (2016). La migración interna de la población con educación superior en México 2010–2015. Retrieved August 2, 2019 from https://www.gob.mx/cms/uploads/attachment/file/232093/06_Almejo_Hernandez.pdf.

Innovación en la Enseñanza de la Ciencia INNOVEC (2019). Sistemas de Enseñanza Vivencial e Indagatoria de la Ciencia (SEVIC). Retrieved August 2, 2019 from http://innovec.org.mx/home/index.php/sevic.

Instituto Mexicano para la Competitividad, A.C. IMCO (2018). Índice de competitividad estatal: Un puente entre dos Méxicos. Retrieved August 2, 2019 from http://imco.org.mx/indices/un-puente-entre-dos-mexicos/resultados/entidad/22-queretaro.

Instituto Nacional de Estadística y Geografía INEGI (2015). Panorama sociodemográfico de Querétaro. Retrieved August 2, 2019 from https://www.inegi.org.mx/app/biblioteca/ficha.html?upc=702825082321.

Instituto Nacional de Estadística y Geografía INEGI (2017). Anuario estadístico y geográfico de Querétaro 2017. Retrieved August 2, 2019 from https://www.datatur.sectur.gob.mx/ITxEF_Docs/QRO_ANUARIO_PDF.pdf.

Instituto Nacional de Estadística y Geografía INEGI (2019). Encuesta nacional de ocupación y empleo, indicadores estratégicos, primer trimestre de 2019. Retrieved August 2, 2019 from https://www.inegi.org.mx/programas/enoe/15ymas/default.html#Tabulados.

ManpowerGroup (2018). Talent sufficiency survey. Retrieved August 2, 2019 from https://go.manpowergroup.com/hubfs/TalentShortage%202018%20(Global)%20Assets/PDFs/MG_TalentShortage2018_lo%206_25_18_FINAL.pdf.

NICHE (2019). 2019 Best schools ranking. Retrieved August 2, 2019 from https://www.niche.com/k12/the-science-academy-of-south-texas-mercedes-tx/#rankings.

Observatorio Mexicano de Innovación OMI (2017). Informe técnico del OMI: Estado de la innovación en México. Retrieved August 2, 2019 from https://omi.economia.gob.mx/Pages/Detalle-documento.aspx?IDP=19.

Organisation for Economic Co-operation and Development OECD (2016). Programa para la Evaluación Internacional de Alumnos PISA 2015; Resultados México. Retrieved August 2, 2019 from https://www.oecd.org/pisa/PISA-2015-Mexico-ESP.pdf.

Organisation for Economic Co-operation and Development OECD (2018). Strong foundations for quality and equity in Mexican schools. Retrieved August 2, 2019 from https://spec.sep.gob.mx/web/wp-content/uploads/2018/11/OECD-Mexico-Schools-Report-FINAL.pdf.

Profesor Chiflado TM (2019). Profesor Chiflado. Retrieved August 2, 2019 from https://www.profesorchiflado.com/.

Project Lead the Way, Inc (2019). PLTW. Retrieved August 2, 2019 from https://www.pltw.org/.

Rapeepisarn, K., Wong, K., Fung, C., Depickere, A. (2006). Similarities and differences between "learn through play" and "edutainment". *School of Information Technology*. Murdoch University Australia.

Secretaría de Desarrollo Agrario, Territorial y Urbano SEDATU, Consejo Nacional de Población CONAPO, Instituto Nacional de Estadística y Geografía INEGI (2015). Delimitación de las zonas metropolitanas de México. Retrieved August 2, 2019 from https://www.gob.mx/conapo/documentos/delimitacion-de-las-zonas-metropolitanas-de-mexico-2015.

Secretaría de Desarrollo Sustentable SEDESU del Estado de Querétaro (2018). Anuario económico Querétaro competitivo 2018. Retrieved August 2, 2019 from http://www.queretaro.gob.mx/sedesu/contenido.aspx?q=0P7NpIeTMww3HR5p8WDNH+uEt5B3ifl94/GBSxbgl7w=.

Secretaría de Educación Pública (2017a). Modelo educativo para la educación obligatoria. Retrieved August 2, 2019 from https://www.gob.mx/cms/uploads/attachment/file/207252/Modelo_Educativo_OK.pdf.

Secretaría de Educación Pública (2017b). Principales cifras del sistema educativo nacional 2016–2017. Retrieved August 2, 2019 from https://www.planeacion.sep.gob.mx/Doc/estadistica_e_indicadores/principales_cifras/principales_cifras_2017_2018_bolsillo.pdf.

Secretaría del Trabajo y Previsión Social (2019). Estadísticas del sector: salario diario asociado a trabajadores asegurados en el IMSS por entidad federativa 2019. Retrieved August 2, 2019 from http://www.stps.gob.mx/gobmx/estadisticas/302_0057.htm?verinfo=2.

Smilor, R. W., Gibson, D. V., & Kozmetsky, G. (1989). Creating the technopolis: High-technology development in Austin, Texas. *Journal of Business Venturing, 4*(1), 49–67.

United Nations UN (2015). 17 Sustainable Development Goals (SDGs) of the 2030 Agenda. Retrieved August 2, 2019 from https://www.un.org/sustainabledevelopment/.

Valero-Matas, J. A., Valero-Oteo, I., & Coca, J. R. (2017). El desencuentro entre ciencia y educación; un problema científico-social. *International Journal of Sociology of Education*. Retrieved August 2, 2019 from http://www.redalyc.org/articulo.oa?id=317153363002.

Vance, F., Nielsen, K., Garza, V., Keicher, A., & Handy, D. (2016). Design for success: developing a STEM Ecosystem. University of San Diego. Retrieved August 2, 2019 from https://stemeco-systems.org/resource/design-success-developing-stem-ecosystem/.

# Chapter 13
# Verbal and Mathematical Literacy Education and STEAM in the Technopolis of São Carlos, Brazil

**Cristiane Chaves Gattaz, Silvia Rocha Falvo, and Paulo Estevão Cruvinel**

**Abstract** How can training verbal and mathematical literacy educators of school-age children make a difference in one's hometown? From a regional development perspective, this chapter describes the potential of a "viral" transfer of primary and secondary STEAM education in Brazil, specifically in São Carlos, a leading technopolis of that country, also known as "SanCa Hub". Supported by the dense network of eight science centers[1] of the Federal University of São Carlos; other educational institutions; science, technology and innovation policies; influencers; and stakeholders, a verbal and mathematical literacy training ecosystem is being developed for literacy educators of São Carlos and region. The chapter also outlines an action profile to motivate these professionals to develop in their students a refined logical-mathematical ability in a transdisciplinary way, using STEAM methodology, and to "push" this system to the market. This case reveals the importance of the artistic and technological axis in the learning and teaching of Portuguese and Mathematics, reducing the gap between school and reality and promoting innovation practices for regional and national economic improvements based on technology development in strategic fields of interest considering the Society 5.0 model (Cabinet Office, Government of Japan, n.d.). To provide full context of the regional ecosystem, the authors include a description of the role of the local agriculture industry cluster, the adopted transdisciplinary education philosophy, how the technopolis and Triple Helix models relate to one another in the virtuous cycle

[1] Agriculture Science, Biological and Health Sciences, Science in Management and Technology, Science and Technology for Sustainability, Hard Science and Technology, Humanities and Biological Sciences, Education and Human Sciences and Science of Nature.

C. C. Gattaz (✉) · P. E. Cruvinel
Embrapa Instrumentation – Laboratory for Precision Agricultural Inputs Application, São Carlos, SP, Brazil

S. R. Falvo
São Carlos School, São Carlos, SP, Brazil



C. Zintgraff et al. (eds.), *STEM in the Technopolis: The Power of STEM Education in Regional Technology Policy*,
https://doi.org/10.1007/978-3-030-39851-4_13

of this city and region, and the lessons learned during the past 20 years regarding what today we call STEAM methodology.

## 13.1 Introduction

São Carlos, a Brazilian city, located in the geographic center of the State of São Paulo, distant about 150 miles from its capital, has an expressive agriculture industry cluster, which drives a multi-sector network building in agriculture related fields, inspiring relevant STEAM education experiences and programs especially in the context of verbal and mathematical literacy, with the support of government and industry leaders, leveraging real community needs. Since its foundation, this inland city has been built in alignment with the agriculture and education historic process. The first school for boy's verbal and mathematical literacy emerged in 1858 and for girls, in 1862, as the result of concerns about the formal education in that period. Due to growth of coffee production, a relevant product for local and national economy in the XIX century, São Carlos developed itself rapidly, gaining prestige among the other Brazilian cities[2].

In 1904, new local primary and secondary schools were created, reaching, nowadays, around nine local elementary schools applying STEAM education experiences in the context of verbal and mathematical literacy[3], as stated by Orlando Mengatti Filho, the Secretary for Education of São Carlos. The economic relevance and strong urban infrastructure of this city promoted the installation of banks, labs, manufacturing industry and commerce. In 1952, the first high school of engineering was inaugurated, linked to the University of São Paulo (USP[4]). In 1956, Sérgio Mascarenhas and Yvonne Primerano Mascarenhas left Rio de Janeiro to the city of São Carlos, where they became important figures in the history of the emergence of the Institute of Physics and Chemistry of São Carlos (IFSC/USP), still in the School of Engineering of São Carlos (EESC/USP), which at the time was headquartered in the present building of the Center of Scientific and Cultural Diffusion (CDCC/USP[5]).

[2] Some information about São Carlos history was retrieved from: https://www.youtube.com/watch?v=yHW5kKZX6fQ

[3] Name of the local public elementary schools: E.M.E.B. Angelina Dagnone De Melo; E.M.E.B. Arthur Natalino Deriggi; E.M.E.B. Carmine Botta; E.M.E.B. Prof. Afonso Fioca Vitali; E.M.E.B. Prof. Antonio Stella Moruzzi; E.M.E.B. Profª Dalila Galli; E.M.E.B. Profª Janete Maria Martinelli Lia; E.M.E.B. Profª Maria Ermantina C. Tarpani; E.M.E.J.A. Austero Manjerona; Name of the local private elementary school: São Carlos School

[4] USP: <http://www.saocarlos.usp.br/welcome-to-usp-sao-carlos/>. In 1986, the Institute of Advanced Studies *(Instituto de Estudos Avançados)* (IEA) was inaugurated (http://www.iea.usp.br/en/)

[5] CDCC / USP is located in the building of the former Società Dante Alighieri in the city of São Carlos-SP. It was created in 1980 by professors from USP, especially Dietrich Sciel (1940–2012), as well as primary and secondary educators of the city. The CDCC's main mission is to establish a system of didactic-pedagogical support for schools, their teachers and students, as well as to promote scientific and cultural diffusion activities.

The entrepreneurial vision of this couple of scientists transformed the city, leading it to be recognized as the Capital of Technology.

In 1960s, emerged the Federal University of São Carlos (UFSCar[6]). This city also received two research unit centers of the largest research corporation in agriculture in Brazil named Brazilian Agricultural Research Corporation (EMBRAPA)[7], a federal research center for science, technology and innovation[8], professional teaching schools[9], a private university[10], a virtual state university[11], a science and technology park[12] and large corporations[13], becoming an important agribusiness center in the fields of optics, new materials and instrumentation for milk, sugar-cane, orange, chicken, meet and corn, leveraging real community needs through educational programs (Gattaz et al. 2019).

Consistent with the high concentration of universities and research centers, São Carlos' citizens include a great number of scientists and researchers who also represents significantly the government and industry stakeholders for regional development: one Ph.D. researcher for every 100 inhabitants, average ten times greater than the proportional index of Brazil. Thanks to the research centers, São Carlos also boasts another important brand: the annual average of patent registrations is 14.5 patents per 100 thousand inhabitants. In the country, this ratio is 3.2 patents per 100 thousand inhabitants. The city also has 39 undergraduate courses and more than 250 companies are high technology.

As states the City Mayor Airton Garcia, such characteristics given to the city as the Capital of Technology, is also known as "SanCa Hub", and awake in different stakeholders and policy support the desire to invest in this region, benefiting not only the

---

[6] UFSCAR and its research center named Institute for Advanced and Strategic Studies (*Instituto de Estudos Avançados e Estratégicos*) (IEAE/UFSCar) – (<https://www2.ufscar.br/foreign-visitor/about-ufscar>; < http://www.ieae.ufscar.br/>)

[7] Embrapa Instrumentation (*Embrapa Instrumentação*) (<https://www.embrapa.br/en/instrumentacao>); Embrapa Southeast Livestock (*Embrapa Pecuária Sudeste*) (<https://www.embrapa.br/en/pecuaria-sudeste>)

[8] Federal Institute of Education, Science and Technology of Sao Paulo (Instituto Federal de Educação, Ciência e Tecnologia de São Paulo) (IFSP)

[9] National Service for Industrial Training (*Serviço Nacional de Aprendizagem Industrial*) (SENAI); Trade Social Service (*Serviço Social do Comércio*) (SESC); National Service for Trade Training (*Serviço Nacional de Aprendizagem Comercial*) (SENAC); Technology School (*Faculdade de Tecnologia*) (FATEC); Technical School (*Escola Técnica*) (ETEC)

[10] Centro Universitário Central Paulista – UNICEP <https://www.unicep.edu.br/instituicao>;

[11] Virtual University of the State of Sao Paulo (*Universidade Virtual do Estado de São Paulo*) (UNIVESP): https://univesp.br/institucional

[12] Technology Park (*Parque Tecnológico de São Carlos*) (Parqtec): <http://parqtec.com.br/quem-somos/>

[13] e.g. Electrolux (<https://www.electroluxgroup.com/en/>); Faber Castell (< http://www.fabercastell.com/>); International Institute of Ecology (*Instituto Internacional de Ecologia*) (IIE) <http://www.iie.com.br/index.php/component/content/category/77-mega-content-demo>

economy but also the citizen's quality of life. As a result, the city became the third inland city in Brazil that best combines the promotion of innovation with quality of life, dense interaction among stakeholders, incentive policies and economic development. In other words, simultaneously to the development and growth of the agriculture industry cluster in São Carlos, the Human Development Index of the city is significant – 0.85 (Census 2010 IBGE, 2019) and the solid and consistent relationship of notable influencers, hundreds of historical, cultural, scientific, technological, educational and innovation institutions bring a family-oriented connection among all players.

However, the Brazilian reality related to non-literate children completing their school life and the difficulties of a mathematical teaching system are themes very debated by primary and secondary educators, pedagogical coordinators and university community. There is a consensus that both the traditional teaching method, in which the educator is the sole holder of knowledge, is related to the difficulty of the students understanding the mathematical concepts contribute to students not pursuing education and careers in the discipline (Fonseca and Cardoso 2005). This idea is not new. Between the end of the 1800s and mid-1900s, driven by the development of psychology studies on learning and human development, by criticism of traditional pedagogy, and in opposition to the way curriculum content was imposed on students, many educators started to re-claim the active participation of students in the learning process. In this way, these proposals rescued Athenian principles of education by valuing the previous experience of the student and his knowledge prior to school learning. It revealed to re-capture these pedagogical perspectives (Gadotti 2000). This process was happening as new barriers arose; for example, such as the need of new skills of educators, new school structure, political change, new awareness, and the desire for program signature sponsors.

These social and educational problems also receive high attention of São Carlos universities and some of its representative "inspirators"[14], such as the present work that sought to chart a way to solve the presented challenge locally, reducing barriers using STEAM education with support of Federal, State and Local Policies in Education and Science, Technology and Innovation (S,T&I). These policies were designed to survive across political administrations and change, avoid broad-brush solutions (e.g. computers in all schools), help define problems, attract industry leaders, intentionally include educators and parents to students in strategy development, be cross sector, bring early adopters/pilots/wins; and cooperate with other industry clusters.

This process was thought to happen in the relationship of all stakeholders in primary and secondary education and agriculture industry cluster development leading to wealth and improved quality of life in the most leading technopolis of Brazil - São Carlos.

And so, it did. A minicourse was designed to integrate the learning and/or teaching issues of verbal and mathematical literacy educators, with that work performed by a higher educational institution and designed in context of the innovation ecosystem of São Carlos.

[14] Some of the most recent representative influencers of the city of São Carlos are introduced.

The verbal and mathematical literacy educators of the present case adopted STEAM education as defined by a movement led by the Rhode Island School of Design (RISD), whose president and designer, John Maeda, is its greatest enthusiast[15]. Widely adopted by institutions, corporations and individuals, STEAM methodology integrates science, technology, engineering, arts and mathematics. The RISD movement adds both Art and Design to STEM. Amazingly, the acronym STEAM gives rise to a word already existing in English whose meaning is "vapor" (Michaelis 2009), which, in a broader sense, can be considered as vaporizing, bringing innovative ideas - a break in paradigm (Fonseca and Bopprê 2013). This is consistent with this methodology's purpose, to stimulate students' participation in the learning process, preparing them for life and for the workforce pipeline.

The transfer process of the methodology of this case to many Brazilian cities until 2018 contributed to the gap reduction of the Brazilian Basic Education Agenda (2011–2021), considering the OECD indicators and their Education 4.0 paradigm: building better educators; strengthening early childhood education; educating a twenty-first century workforce; raising quality in secondary education; and maximizing federal impact and capitalizing on Brazil's education action lab (Bruns et al. 2012; OECD 2018).

## 13.2 Verbal and Mathematical Literacy Education and Agribusiness Challenges

São Carlos city is a benchmark in both advanced science, industry, and agribusiness education, which, in many cases, acts as a gateway for work in these sectors. Its solid training of students in primary (ages 6–14) and secondary schools (ages 15–17)[16], in professional courses, and of educators as multiplying agents for the agricultural segment, is a reference to all other agriculture cities in Brazil. Students in primary and secondary grades, mainly students over 10 years old, learn about nanotechnology, precision agriculture, rural basic sanitation, soil and water management and conservation, post-harvesting in fruits and vegetables, chemistry and biomass technology, meat and milk cattle breeding, sheep farming, and forage crops, among others.

In 2013, the city had an opportunity to improve the quality level of verbal and mathematical literacy education in those schools by means of a formal commitment assumed among the Brazilian Federal Government, Federal District, States, Cities and society, named PNAIC[17]. The main objective of this national program is to

[15] STEM to STEAM initiative of the Rhode Island School of Design (RISD) https://www.risd.edu/academics/public-engagement/#support-for-steam

[16] The Brazilian Educational System is structured in three groups: from 3 to 6 years old (childhood education – until first grade); from 6 to 14 years old (primary education – from first to ninth grade) and 15 to 17 years old (secondary education).

[17] The Brazilian National Pact for Literacy in the Right Age (*Pacto Nacional pela Alfabetização na Idade Certa*) (PNAIC)

ensure that all children are verbally and mathematically literate by the age of 8, at the end of the third year of Primary Education (ages 6–10) (BRASIL 2014).

Constituted by an integrated set of actions; materials; and, curricular and pedagogical references available by the Ministry of Education (MEC), this political and educational program has as its main axis the continued formation of verbal and mathematical literacy educators. The selected educators also received federal grants for training including teaching materials. São Carlos volunteered its participation in this program to contribute to the training and professional development of its respective literacy educators based on the following four pillars of action:

- Continuing education for educators and their study counselors;
- Production of materials for teaching and dissemination of educational technologies;
- Application and creation of systematic evaluations;
- Social mobilization towards education.

Considering the great potential of São Carlos as a technopolis and with the purpose of improving efficiency in the verbal and mathematical literacy of the educational system, the school educator and administrator Silvia Falvo, one of the co-authors of this chapter, delivered to local and regional education managers and educators a minicourse entitled "What adds more: working STEM or STEAM?". This event occurred during an initiative from PNAIC named "II Colloquium on Literacy Practices in the Initial Courses and Seminar PNAIC/UFSCar 2014", at UFSCar.

The minicourse trained educators on the importance of the STEAM methodology as a resource of verbal and mathematical literacy using project-based learning method (PBL), consistent with the Rhode Island School of Design (RISD) method, and psychopedagogical interventions[18]. In addition, this training also interacted with the agribusiness education curriculum applied in the primary and secondary schools, and professional courses, integrating students who were all over six years old, with most over ten years old, in specific agriculture projects[19], in partnership with local, state, federal and international actors – such as companies (e.g. Faber Castell, Electrolux, Microsoft, Apple, Toradex, Cargill, Alura, Oracle, Delivery Much, Serasa Experian), accelerators (e.g. Center for Development of Emerging

[18] Psychopedagogy is an educational practice for effective learning considering emotional, cognitive and social aspects used in the fields of Education, Business and Healthcare.

[19] National Robotics Competition: mathematical solutions for industrial problems (http://www.cemeai.icmc.usp.br/EduSCar/cemeai-2/). Program of Rational Use of Water (*Programa de Uso Racional da Água*) (PURA): local government program in partnership with the main water treatment and distribution company (Sabesp) for water consumption consciousness in local and state schools (https://www.educacao.sp.gov.br/pura). Home Garden Project is held by the partnership between schools, UFScar, the Secretariat of Agriculture and Environment, and the company Electrolux (http://www.uac.ufscar.br/saude/projeto-horta-na-escola). Embrapa & Schools: program creates an atmosphere for learning and awareness of the interfaces of science and technology with the agricultural sector and the environment. (https://www.embrapa.br/en/embrapa-escola); SancaThon Future Farms challenges participants to create solutions to modernize agriculture. (http://www.eesc.usp.br/portaleesc/index.php?option=com_content&view=article&id=5022&Itemid=164). Technovation Summer School for Girls promotes the insertion of girls (10–18 years old) in hard science and technology (https://www.icmc.usp.br/noticias/4091-computacao-tambem-e-coisa-de-menina-aprenda-a-fazer-aplicativos-na-usp)

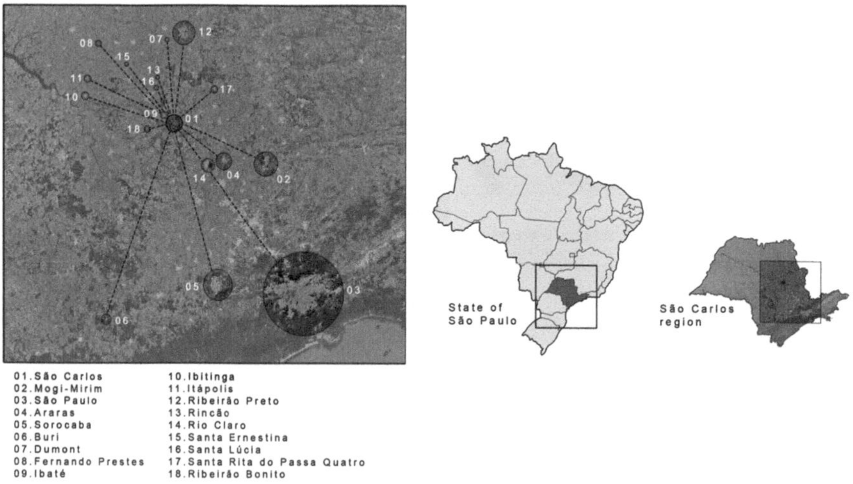


**Fig. 13.1** STEM or STEAM minicourse initial offering locations

Industries - CEDIN, ONOVOLAB, AgTech Garage), universities (e.g. UFSCar, USP), associations (e.g. Arts Association of São Carlos, Association of Parents and Friends of Disables of São Carlos, Brazilian Service to Support Micro and Small Enterprises (SEBRAE), ABStartups, Brazilian Society of Computing), research centers (e.g. Embrapa Instrumentation and Embrapa Southeast Livestock) and local government (e.g. Secretary of Education, Secretary of Science & Technology, Secretary of Agriculture, City Council). The interaction among the educators and students within this training process contributed with the offering of tools that could be used in other vulnerable and challenging situations related to agriculture (e.g. health, poverty, security) (Santos et al. 2019).

The goal was to present innovative ideas and activities that teachers could use in turn to awaken in students an enthusiasm for learning, using real world situations that mobilized them to a transdisciplinary knowledge, including agribusiness challenges.

Initially, the minicourse was offered to 20 educators from other 17 cities[20], in the year of 2013–2014, as shown in Fig. 13.1, São Carlos Region (Source: Elaborated by Caio de Souza Ferreira, Laura Mastrodi and Raquel Canola Zacour de Azevedo).

Each one of the participants applied the approach to their students from 6 to 10 years old, also engaging with their parents, the content and methodology presented in the minicourse, and brought feedback for continued learning through workshops, such as "Mother's Day Workshop". In such context, it was proposed to verbal and mathematical literacy educators the construction of a tissue jewelry holder, using the French Lining technique[21], as shown in Fig. 13.2 (Source: Elaborated by Silvia Rocha Falvo), integrating technology and art as means for the learning process.

---

[20] These cities range from 10 to 200 miles away from São Carlos city.

[21] It is a technique for making jewelry boxes (https://www.1stdibs.com/furniture/decorative-objects/boxes/jewelry-boxes/origin/french/?production-time-frame=0_6-weeks&price=[1%20 TO%20,100,000]&sort = popular).

**Fig. 13.2** Teaching while building using the french lining technique

During the construction of this object, several mathematical concepts (arithmetic and geometric) were addressed, such as flat and spatial geometry fields, measurement systems and financial education. It was also explored how these concepts could be approached with children of the fifth grade (age 11). Students were engaged in soil and water management, conservation, and post-harvesting activities with fruits and vegetables. The comments of educators, students and parents, and the results seen from these activities, indicated that students became more efficient in applying the mathematical concepts being taught. The communication level among the educators, the students and parents involved increased, as well as their motivation and skills practice. Feedback also revealed that a classroom filled with information technology is extremely important, but that manual creativity cannot be neglected (Oliveira 1996; Ramal 2002). In addition, cultural heritage - coming from parents, grandparents, and friends – had great value in teaching and mediating the construction of the present knowledge (Chalita 2001). During the PNAIC classes, the importance of the technology in learning as a counterpoint to manual creativity; the valorization of achievements also without technology, and the craft itself were emphasized.

According to the R&D Chief Officer at Embrapa Instrumentation, José Manoel Marconcini, and considering the British school – Brockwood Park School method as a reference[22], STEAM methodology turned the student to become more comprehensively prepared for the workforce pipeline and for the contemporary demands in the agriculture field and related fields, including in communities of high vulnerability. The goal is to prepare the citizens for life and not only for a positive recognition in the evaluation systems.

Neuroscience was also significant in applying STEAM education in the present case, because it allowed students to understand how their brains work and how they learn. In such context, art played an important role in the sense of having freedom to choose and to seek education, under the guidance of an educator, as quoted by André Gide (apud. Weinberg 1972, p. 20) "*art is born of constraints, lives by*

[22] Brockwood Park School method: http://porvir.org/escola-britanica-incentiva-jovens-se-descobrirem/

*combat, and dies through freedom*." In this sense, students were encouraged to find this artistic aspect of all the activities they developed, from a trip to the supermarket to a traditional history lesson.

In this sense, this minicourse indicated the importance of diversity; transdisciplinarity; cooperation and debate among those involved in the teaching and learning process. Paulo Renato Orlandi Lasso[23], an Educator at Paula Souza Center[24], emphasizes the importance of this approach as he states:

> Transdisciplinarity is very interesting in professional training, because it broadens the view, makes people be able to think "outside the box", see what happens around them in the society, in different contexts of social groups. It is fundamental for innovation, because innovative people need to understand the demands, the desires, and the expectations of several groups different from theirs. It allows that you do not keep focused only in your own field, but you transit through other areas. Nowadays, environmental sciences, social sciences, culture, art in general form a broader, more open, and more integrated mentality. It is essential for innovators to be able to think about the needs of these several groups, about the desires of other people.

In such context, interdisciplinary projects added value to Mathematics teaching as they showed students that knowledge and concepts do not exist individually, but they coexist, and interrelate. It is one of the many ways to take the scholars to a more meaningful learning, to a multiplicity of languages, as Antunes mentions:

> We are not born with stimulated abilities and definitive competencies. Intelligence is a biopsychological potential, an ability to solve problems and to create ideas. If, on the one hand, we inherit traces of the intelligences we have, on the other hand, it is the school's duty to stimulate them with vigor, opening up to the human being all the multiplicity of possible languages to use. (Antunes, 2007. p. 49)

Each one of the trained educators empowered others using the same methods and this "viral" transfer process in the following years until 2018 was a success, reaching 5.570 cities in Brazil, which more than 90 percent of them are small and rural cities.

## 13.3 The Agriculture Industry Cluster's Role in the STEAM Technopolis

São Carlos is one of the most representative participants for the development and growth of the agriculture industry cluster in the State of São Paulo, which stands out in the national agricultural sector for its great relevance to the agribusiness of the country. Amazingly, the São Paulo metropolitan area, with a population of 21 million, has contributed to 35% of the Brazilian Gross Domestic Product (GPD). In 2016, the GPD of São Paulo' agribusiness reached R$ 276 billion, 7.4% more than

[23] Bio info: http://lattes.cnpq.br/4488756073018870; interviewed on Apr 24th, 2019.

[24] Paula Souza Center (*Centro Paula Souza*) (CPS) is a governmental center of the State of São Paulo, linked to the Secretary of Economic Development. Present in 321 cities, the institution administers 223 Professional Schools (Etecs) and 73 state Colleges of Technology (Fatecs), with more than 297 thousand students in professional courses of technology in intermediate and high levels. More information, please visit <https://www.cps.sp.gov.br/>.

in 2015. Such result represented 13.8% of the total GDP of São Paulo and 18.7% of the GDP of Brazilian agribusiness last year. In addition, the area of influence of this state far surpasses its geographical boundaries, extending throughout all the other Brazilian states and several countries in South America, characterizing it as the most important poles of development in the Southern Hemisphere (CEPEC-IdN 2013).

As seen before, the high educational level, the potential to attract investments in the development, the absorption of innovative technologies and the quality of life are the main aspects of São Carlos that strengthen the high concentration of high-tech companies in agricultural innovation ecosystems. In addition, the generation and acquisition of knowledge of its universities, such as UFSCar and USP, in cooperation with research centers, such as EMBRAPA, Technology Park and private, public and rural high schools highlight the uniqueness of this technopolis. Amazingly, the territorial distance among these institutions is around three miles from each other, as shown in Figs. 13.3 and 13.4, São Carlos Region (Source: Elaborated by Caio de Souza Ferreira, Laura Mastrodi and Raquel Canola Zacour de Azevedo).

In collaboration with these most representative S,T&I stakeholders, the Academy of Brazilian Science (ABC)[25] and the São Paulo Research Foundation (FAPESP)[26] have contributed significantly to the agriculture industry cluster growth considering STEAM methodology as the main education resource in São Carlos. Research and Development (R&D) funding through many engineering research centers[27], programs of research, innovation and dissemination centers[28] has sponsored small business programs[29] for community development - as described in sect. 2 of this chapter - in alignment with the educational, scientific and technological policy strategies that support STEAM development - described later in this chapter. The R&D Chief Officer at Embrapa Instrumentation, José Manoel Marconcini[30], states:

> From the point of view of some facts of the research, we do not have a challenge that is monodisciplinary. A single facet does not solve even the society's challenges because they do have transdisciplinary forms. It is necessary to perceive social groups where technology will be deployed, know who will give value to this technology and how it would be used and incorporated in daily routine. The same applies in the city areas of health, education, industry, logistics and transportation, environment, climate changes … It has to have a harmony among various social actors and this harmony has to be orchestrated for the absorption of technology in society. It is necessary the connection, leveling of information at the appropriate level and orchestration / harmony, beyond what is beautiful. When we think about transdisciplinary, we need a set of people, each one with its specialty and competence, with a free spirit to build bridges with other areas and cannot be limited by its reality, i.e., it is important to create the environment for this to happen with orchestration. Art teaches how this should be done.

[25] Brazilian Science Academy (*Academia Brasileira de Ciências*) < http://www.abc.org.br/?gclid=EAIaIQobChMI75nTlLju4QIVEYWRCh2NEgWYEAAYASAAEgIFDvD_BwE>

[26] http://www.fapesp.br/en/about

[27] http://www.fapesp.br/cpe/home

[28] http://cepid.fapesp.br/en/home/

[29] Small Business Innovation Research and Business Research Support Program from the FAPESP (*Programa FAPESP Pesquisa Inovativa em Pequenas Empresas - PIPE and Programa de Apoio à Pesquisa em Empresas - PAPPE Fapesp*): http://fapesp.br/pipe/pappe_pipe/4/

[30] Bio info: http://lattes.cnpq.br/5373845785326215; Interviewed on Apr 24th, 2019

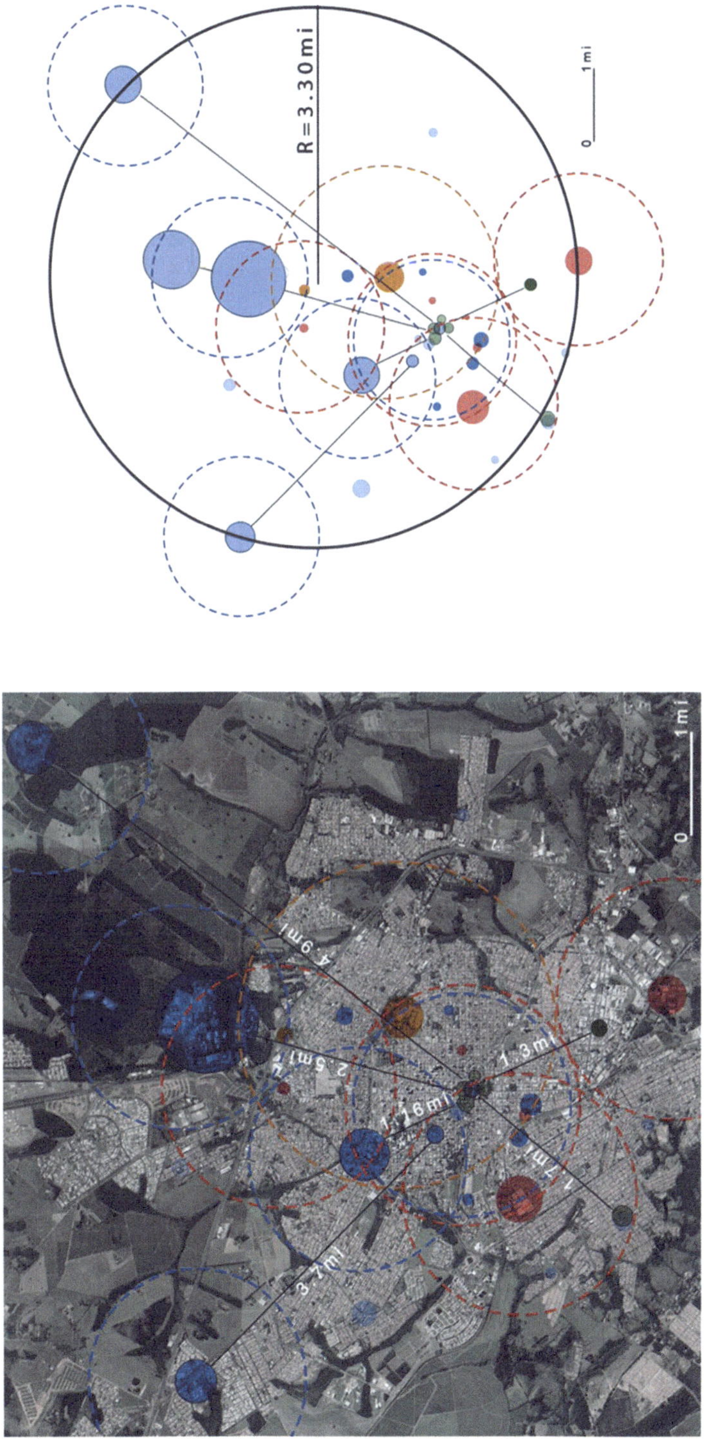


**Fig. 13.3** The concentrated technopolis in the São Carlos region

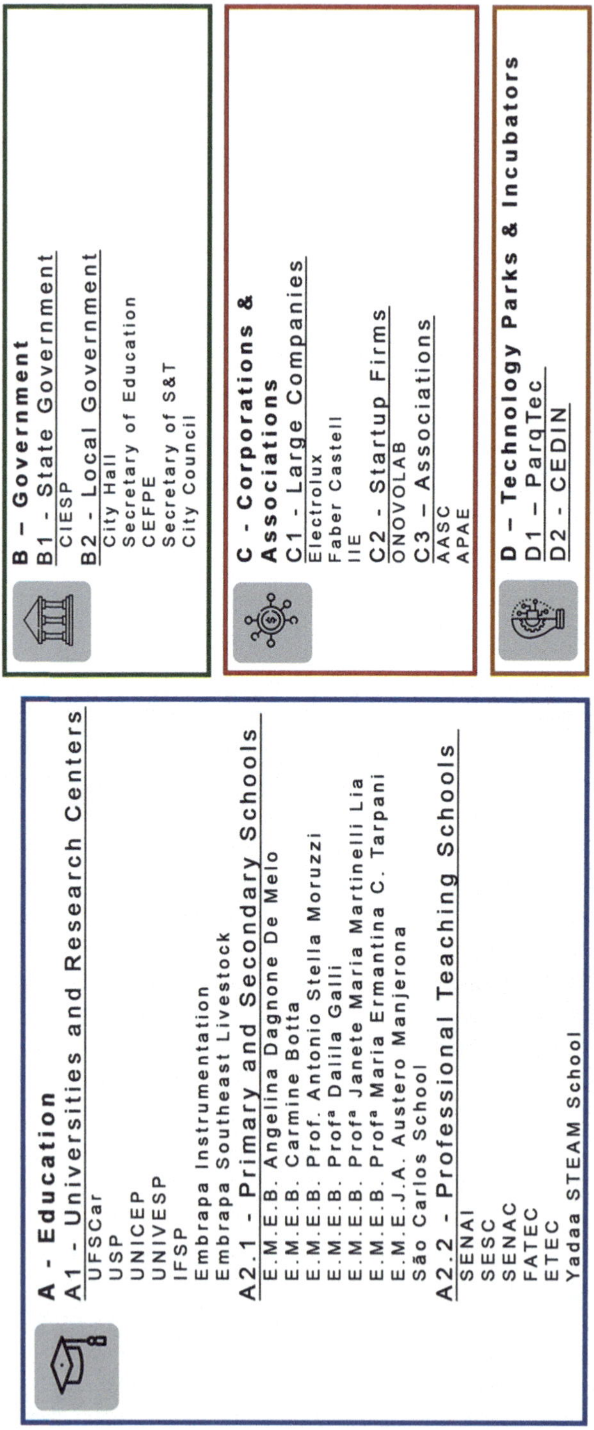


**Fig. 13.4** Actors in the São Carlos agriculture technopolis

However, the development of STEAM as a tool within the workforce pipeline is considered long term, as stated by Ubiraci Moreno Pires Corrêa[31], Council Member of the Center of Industries of the State of São Paulo (CIESP)[32]:

> The emergence of these new programs usually reflects on large industries, because there are relatively small wage masses. Large companies can implement many things with additional costs that do not affect the final price. On the other hand, small and medium-sized enterprises cannot afford to get involved with additional costs. They need to stick to the basic costs. After many of the new programs are used by large companies, they start to overflow to the small ones. However, this fact happens in medium and long term.

So research institutions have been supporting the use of STEAM for the development of this cluster in a shorter term through the influence of the networking of great "inspirators", represented as researchers from different fields, such as Natural History, Oceanography, Electrical Engineering and Electronics, Chemistry, Chemical Engineering, Biology, Bio-Geosciences, Plant Sciences, Environment Engineering, Physics, Biotechnology, Hydrobiology, Geography, Collective Intelligence and Complex Systems, Journalism, Art, Business, and Communication. These actors are also entrepreneurs; high-tech small business mentors; innovation network managers and communication professionals; federal, state and local public policy managers; socio-cultural and civic professionals of the institutions mentioned in this chapter, with active experience in the local community. One of these great "inspirators", Wilma Regina Barrionuevo[33], a Research Scholar at USP Campus São Carlos and the EduSCar Program of the UFSCar, states:

> We seek to awaken in the student the vision of the wonder that science is. In addition, when we begin this approach, we seek to show to students what they gain with it. What do they gain when they dominate content or, likewise, they dominate a language or express themselves better? It is important that they know that when we have information, this adds value… and the world opens, the opportunities appear and the students increase their skills in perceiving and choosing what is the right opportunity for them.

Related to the school system challenges, she also states:

> The student, since a child, learns to use cell phones, the computer, which have a more rapid and ephemeral language. He does not worry about if it is true or not. It is a great challenge for schools. The teacher has not yet mastered this new language, but the students already know. On the other hand, the teacher has historic, analytic and critical content that the students do not have. Therefore, it is important to integrate this knowledge.

In addition, José Dalton Cruz Pessoa[34], a Researcher at Embrapa Instrumentation, adds the following to this challenge:

---

[31] Interviewed on May second, 2019.

[32] Center of Industries of the State of São Paulo (*Centro das Indústrias do Estado de São Paulo*): a civil society of private law that aims to support the businessmen of São Paulo and represent them within the society and the Brazilian government. <http://www.ciesp.com.br/saocarlos/sobre/>

[33] Bio info: https://bv.fapesp.br/pt/pesquisador/5733/wilma-regina-barrionuevo/; Interviewed on Apr 16th, 2019

[34] Bio info: https://bv.fapesp.br/pt/pesquisador/3624/jose-dalton-cruz-pessoa/; interviewed on Apr 16th, 2019

> It is necessary to make an effort "to push" to the market the transdisciplinary educational approach, and not to do otherwise. If it "is pulled" by the market, we will have professionals prepared at the last minute. If the education system pushes well-trained professional educators, from the beginning, we will have greater capacity to increase Brazilian's productivity and be more competitive in several areas.

Due to this growing network-based communication of these "inspirators", the balance of "ground up" and "top down" strategies has continuously being reached in the interaction of existing educational and S,T&I policies, programs and infrastructure in support of STEAM education for regional economic and technology development. In this sense, connections among all innovation stakeholders have been continuously evolving and contributing in a more powerful way to a more agile and effective city management and development. This dense network relationship of the research universities; large and start-up technology firms; federal, state, and local government; and support groups[35] is best practice in supporting STEAM methodology in education, specially verbal and mathematical literacy education, as well as their practices and the study and management of the regional development and quality of life with federal impacts.

Figure 13.4 lists many of the actors in the São Carlos agriculture technopolis.

## 13.4 Education Philosophies, Regional Policies, and Adopted Methods

### *13.4.1 Philosophical Perspective*

The project-based learning efforts for improving verbal and mathematical literacy by means of STEAM education and for meeting agribusiness challenges presented in sect. 2 of this chapter have required skills of the science of complexity, collective intelligence and the understanding of real world issues. As cited by the Nobel Laureate Herbert A. Simon in his keynote speech, in the year of 2000, at the Integrated Design and Process Technology (IDPT) Conference:

> Today, complexity is a word that is much in fashion. We have learned very well that many of the systems that we are trying to deal with in our contemporary science and engineering are very complex indeed. They are so complex that it is not obvious that the powerful tricks and procedures that served us for four centuries or more in the development of modern science and engineering will enable us to understand and deal with them. We are learning that we need a science of complex systems, and we are beginning to construct it [...]

In such context, the guarantee of access to the knowledge needed for citizenship, the recognition of the students as citizens, and the respect and valuation of their previous knowledge – culture, language, diverse experiences – were key points for the learning process (Laplane et al. 2019). Paulo César de Camargo[36], a Research Scholar at UFSCar, illustrates this perspective emphasizing the importance of complexity aligned to collective intelligence as follows:

---

[35] Chamber of commerce, venture and angel capital, IP lawyers and other business professionals

[36] Bio info: https://www.escavador.com/sobre/3932072/paulo-cesar-de-camargo; Interviewed on Apr 16th, 2019

> Today, if I take a cell phone and make a question, I will have an answer. If it is not enough, I activate my network and I can exchange information. If I have a well-defined goal, behavioral rules and a good communication system, I create a collective intelligence and from this emerge solutions that a single subject will not be able to create. If we could do this with people, similarly to bees for example, it would be very interesting.

Besides, the constant social, cultural and professional transformations helped lead to the development of project-based learning and psychopedagogical skills of fundamental importance; and Mathematics greatly contributed in this context, since it is one of the sciences that permeates all areas. In this sense, Sérgio Henrique Vannucchi Leme de Mattos[37], a Research Scholar at UFSCar, states:

> Learning itself no longer serves. The knowledge we need is how to learn to learn in a world of uncertainties.

It was up to the school to make students aware of the need to what they were learning and to the context in which they lived.

In the mathematical literacy for example, there is a paradox in the learning and in the teaching process, as states Medeiros (1987):

> If it is true that mathematics permeates human activities, what is wrong with their teaching? Mathematics is present in the economic news of the newspaper and TV, music, painting, cooking recipes and nature in general. We live in a world of numbers represented everywhere. The human body itself already confers the experience of a spatiality. (pg. 22)

In this sense, making difference in mathematical literacy involved changes of paradigms. First, it was necessary to replace the traditional teaching method, in which only the teacher assumed the roles of holder and communicator of knowledge, for a pedagogical methodology that promoted the construction of knowledge together.

The conversion of scientific knowledge into school knowledge constituted another fundamental point for the success of mathematic teaching practices, as "doing and thinking the theoretical mathematician were not subject to direct communication to students" (Fonseca and Cardoso 2005, p.67). It involved the construction of contextualized and transdisciplinary knowledge and competence. Transdisciplinary education played a key role in the present case due to its importance in preparing learners for life in society[38]. George Kozmetsky, in his 2000 address to the Society for Design and Process Science (SDPS) community, stated that "technology continues to shrink the world. There is no choice other than to participate in the global community. Science and technology is too precious as a resource to be restricted from drawing the world together. That is what the 21st Century is all about." (Ertas et al. 2007). This conversion is also supported in solving agriculture challenges in the industry. Paulo Sergio de Paula Herrmann[39], a Researcher at Embrapa Instrumentation, states that:

---

[37] Bio info: https://bv.fapesp.br/en/pesquisador/98619/sergio-henrique-vannucchi-leme-de-mattos/; Interviewed on Apr 16th, 2019

[38] In 2007, the Institute for Transdisciplinary Collaborative Research & Education (T–CORE) is proposed as a new virtual research and development infrastructure housed within The Academy of Transdisciplinary Learning and Advanced Studies (TheATLAS)

[39] Bio info: https://bv.fapesp.br/en/pesquisador/86901/paulo-sergio-de-paula-herrmann-junior/; interviewed on Apr 16th, 2019

> This holistic view of transdisciplinarity is an important exercise that needs to happen and it is connected to a complex moment that we are living. The young people in education need to be open and it is necessary to gather efforts in this view, integrating different skills. We have information, but we need to have educational training.

In addition, the main President-Director of the Technological Park Foundation of São Carlos (Science Park), Sylvio Goulart Rosa Júnior, states that:

> The innovation phenomenon in São Carlos is the result of a sequence of lucid decisions taken by countless actors in the last decades. The dedicated and persistent work of these pioneers has resulted in a sophisticated science, technology and education complex within their Universities, Research Centers and Startups. The multidisciplinarity of this complex is the characteristic and strength of the Capital of Technology that has become an example and laboratory for the Brazil of the future.

Another paradigm that need to be considered by mathematics educators is the need of interconnection between Mathematics and Portuguese[40] Language, as it contributes to the development of reading, comprehension and production skills of any text in different situations. Thus, Fonseca and Cardoso (2005), presented some tips, such as explain and write, in informal language, the mathematical results; help students master the tools of reading, i.e., to understand the meaning of mathematical symbology; and, always contextualize to the applied situation.

### 13.4.2 Education Methods

In São Carlos as well as in other Brazilian cities, initiatives of the STEM movement became more noted in 2015, one year after the present case, with the use of "maker" culture and has been officially introduced into the local policies since then. As a result of many of these initiatives, the STEM International Cooperation Program[41] was launched in January 2015 by researchers from Brazilian regions including São Carlos as a joint initiative of the British Council through the Newton Fund, and the CAPES Department of Basic Education[42], in collaboration with The New Talents Program[43], which sought to answer questions such as how to improve educator training and motivation? How to instigate students to develop interest in science? How to

[40] Native language in Brazil.

[41] STEM International Cooperation Program *(Programa de Cooperação Internacional STEM)*: The action aims at the exchange of knowledge between educators and researchers to encourage curricular innovation and the creation of new strategies in the training of educators of basic education in the areas of science and mathematics. More information, please visit < http://www.capes.gov.br/educacao-basica/programa-de-cooperacao-internacional-stem>

[42] CAPES - Coordination for the Improvement of Higher Education Personnel, foundation of the Ministry of Education (MEC).

[43] The New Talents Program *(Programa Novos Talentos)*: aims to support proposals for extracurricular activities for educators and students in basic education, such as courses and workshops, aimed at disseminating scientific knowledge, at improving and updating the target audience and at improving the teaching of sciences in the Brazilian public schools. More information, please visit: < http://www.capes.gov.br/educacao-basica/novos-talentos>

overcome the difficulties pointed out in the different social contexts in Brazil, boost educator and student motivation, and overcome the challenge of creating a curriculum in science that is relevant to students' lives? (Nogueira-Ferreira et al. 2016).

By means of the dense relationship of the innovation stakeholders of São Carlos and its history described earlier in this chapter, the city has responded continuously to these questions giving attention to more formal initiatives with measured impacts.

More recently, the City Councilor, Roberto Mori, inspired the creation of the Arts Association of São Carlos[44]. The President of this Association, Fernando Cesar Crnkovic[45] also supports the presence of Art for STEAM development in all innovation stakeholders of São Carlos. He states Art not only as a means for a more ludic learning process, but also life skills development in professional education, as follows:

> Art is not a social whim nor a mere fun. It is a need. It covers the aesthetic aspect, which helps a lot to conquer the spectator, initially, in a sensorial way – but, above all, it is a form of contact with spirituality, of happiness search – with all abstraction and subjectivity that it includes. Art cannot have limits and, by essence, it must reach everyone.

Vanderlei Salvador Bagnato[46], also one of the inspirators of São Carlos city, who is a member of the Academy of Sciences for the Developing World and the National Academy of Sciences (USA), and the only Brazilian member of the Pontifical Academy of Sciences, gathered five Nobel Laureates (William Phillips, Eric Cornell, David Wineland, Serge Haroche, Dudley Hershbach), in honor of the renowned physicist Daniel Kleppner, to give medals to the students of the best Brazilian public schools placed in the Brazilian physics competition – the most leading challenge of the city's agriculture industry cluster. It was a great emotion for the students and educators as well as for the Nobels themselves.

The most recent initiative is the EduSCar[47] network, which is the result of a partnership among the engineering research centers, programs of research, innovation and dissemination centers of FAPESP, National Institutes for Science & Technology[48], Education Council of São Carlos region and the Secretary of Education of São Carlos, besides researchers from federal and state universities such as UFSCar and USP who are interested in improving the education level of the country.

EduSCar is a program developed for managing the application of two innovative tools for literacy education: Leon and Ludo Educativo. Leon is a STEAM digital tool of online reading for students of the fourth grade with increasing levels of complexity during one year. Experiments with this tool showed a very high level of literacy in this 12-month period. Now this tool is being implemented in all public elementary schools of the local and state network of São Carlos.

Another tool, Ludo Educativo has already 2.5 million accesses around the world. This digital tool is made up of several interactive games made for people from

[44] More information about Arts Association of São Carlos, please access: http://www.aasaocarlos.com.br/v2/

[45] Interviewed on Apr 16th, 2019

[46] Bio info: https://bv.fapesp.br/en/pesquisador/766/vanderlei-salvador-bagnato/

[47] http://www.cemeai.icmc.usp.br/EduSCar/

[48] http://inct.cnpq.br/

2 years to 70 years or more. While games are highly engaging and motivating, they do not bring violence. On the contrary, during different stages of these games, students will always be doing some action of environmental or social benefit. For example, throwing the paper in the trash, helping an old man cross the street, sanitizing his hands and food. While having fun, he will be taking a lesson in citizenship. This same tool will allow classroom evaluations using devices such as mobile phones and tablets. The Regional Director of Education, Débora Blanco, explains that only in the region of São Carlos, hundreds of thousands of sheets of paper are used annually to carry out the evaluations, according to her saying: "if we multiply these numbers throughout the network, there will be millions of paper sheets annually, generating a huge environmental impact on the production and subsequent disposal of this paper". She concludes," it is also important to think about the economy that digital evaluations will generate"[49].

Another initiative of STEAM education in Brazil was Yadaa[50] - the first school that used this methodology. It was founded in São Carlos in April 2014, and already owns plus eight franchises scattered throughout the state and 3 other more states in Brazil and more than 25 partner schools.

Art in coworking spaces[51], studios and in the streets are gaining more importance in the community, more maker studios begin to be created, STEAM skills are transferred by technical schools, new format of libraries come to exist (Experimentoteca[52]) by centers of scientific and cultural dissemination in the universities, among numerous other initiatives that strengthens the competitive ranking of the city of São Carlos as leading technopolis in Brazil.

## 13.5 How Do Policymakers in the São Carlos Region View Local Policies and How They Support K-12 Educational Goals for the Local Agriculture Industry Cluster?

In the two last decades, the development of STEM movement has been one of the priorities of educational, scientific and technological policies focused on innovation in Federal, State and Local Government in Brazil, especially by the Ministries of Education and S, T, I & Communication. The "inspirators" of São Carlos - described

[49] Interview with local news: Partnership with UFSCar resulted in a learning monitoring application. <http://www.regiaoemdestake.com.br/site/2019/05/16/sao-carlos-escolas-estaduais-da-cidade-e-regiao-vivem-experiencia-inedita-de-avaliacao-na-proxima-semana/>

[50] More information: <https://www.yadaa.com.br/sobre-nos>

[51] Most recent innovation coworking space: ONOVOLAB (http://colsaocarlos.com.br/): a campus of innovation in a former tissue factory of 1908.

[52] The Experimentoteca is a Science Laboratory that aims to rationalize the use of experimental material, in the same way that a public library facilitates the access of a large number of publications to an extensive public, in a loan system without costs to the user. (http://www.cdcc.usp.br/experimentoteca/index.html)

in earlier section - were more aware of this movement by means of their relationship with the countries that were already leading this movement. Silvia Falvo, the co-author of this chapter, was one of the leading school educators who helped introduce this movement into public and rural schools in the city in the present case.

The National S, T & I policy has been supporting STEM development in areas related to basic sciences by given their importance as a generator of knowledge, which is essential to national sustainable development, with direct impacts on the scientific, technological and innovation fields. This support has happened from the adoption of STEM curriculum in "Action Plans" aiming at Brazilian expansion and consolidation in scientific fields, as well as in the solution of problems specifically to each following theme for the development, autonomy and national sovereignty: Aerospace and Defense, Water, Food, Biomass and Bioeconomics, Social Sciences and Technologies, Climate, Economics and Digital Society, Energy, Strategic Minerals, Nuclear, Health, Convergent and Enabling Technologies (MCTIC 2016). This policy was replicated to state and local government. So, the development of these themes and basic sciences has supported the growth of the agriculture industry cluster in São Carlos and has motivated the use of STEM movement in primary, secondary and professional schools as a new educational paradigm. As stated by the City Mayor Airton Garcia[53]:

> Through projects and partnerships, we are fostering the development of the city with the most efficient and modern technologies applied in the construction of public policies. Also, education is a government priority, with investments above that provided by law. In 2018, the City of São Carlos invested 27.75% of the education budget. To further expand the network, we are investing in new partnerships, in technology absorption and especially in network training. As I always say: only with education will we improve Brazil. Education is investment.

The Secretary for Education of São Carlos, Orlando Mengatti Filho[54], supports the program by stating that "professional education is not like a building that is destroyed, or when the Secretariat term is over, the Government term is over ... It is intelligence, it is what is in the hearts of people. It is a network that beats, demands, but when you also have good initiatives and are not only demagogues, there is an answer."

This policy also aims at presenting methodological guidelines that guarantee the presence of what is ludic in the child's learning and respect for their universe, taking into account their way of thinking and building knowledge. The City Council, for example, supports this goal by adding music into the school curriculum and activities in joint with the industry cluster. Roberto Mori Roda[55], the City Councilor of Sao Carlos, is the main influencer for achieving this goal. Banda Marcial Faber Castell is a classic project that lasted for 23 years since 1985. It was a musical band in joint with the largest and oldest corporation of the city – Faber Castell[56]. Besides many other significant initiatives including the Association of Parents and Friends

---

[53] Bio info: https://pt.wikipedia.org/wiki/Airton_Garcia_Ferreira. Interviewed on Apr 24th, 2019

[54] Bio info: http://www.saocarlos.sp.gov.br/index.php/secretarias-municipais/educacao.html. Interviewed on Apr 25th, 2019

[55] Bio info: https://camarasaocarlos.sp.gov.br/vereador/?ent=70792&p=detalhe&id=31; < http://robertinhomori.net/site/index.php/minhasbandasminhavida > interviewed on Apr 23rd, 2019

[56] More info: < http://www.fabercastell.com/company/about-us>

of Disables of São Carlos (Associação de Pais e Amigos dos Excepcionais de São Carlos - APAE São Carlos)[57], in 2002, he also developed a law for promoting music in primary and secondary schools[58]. He states that:

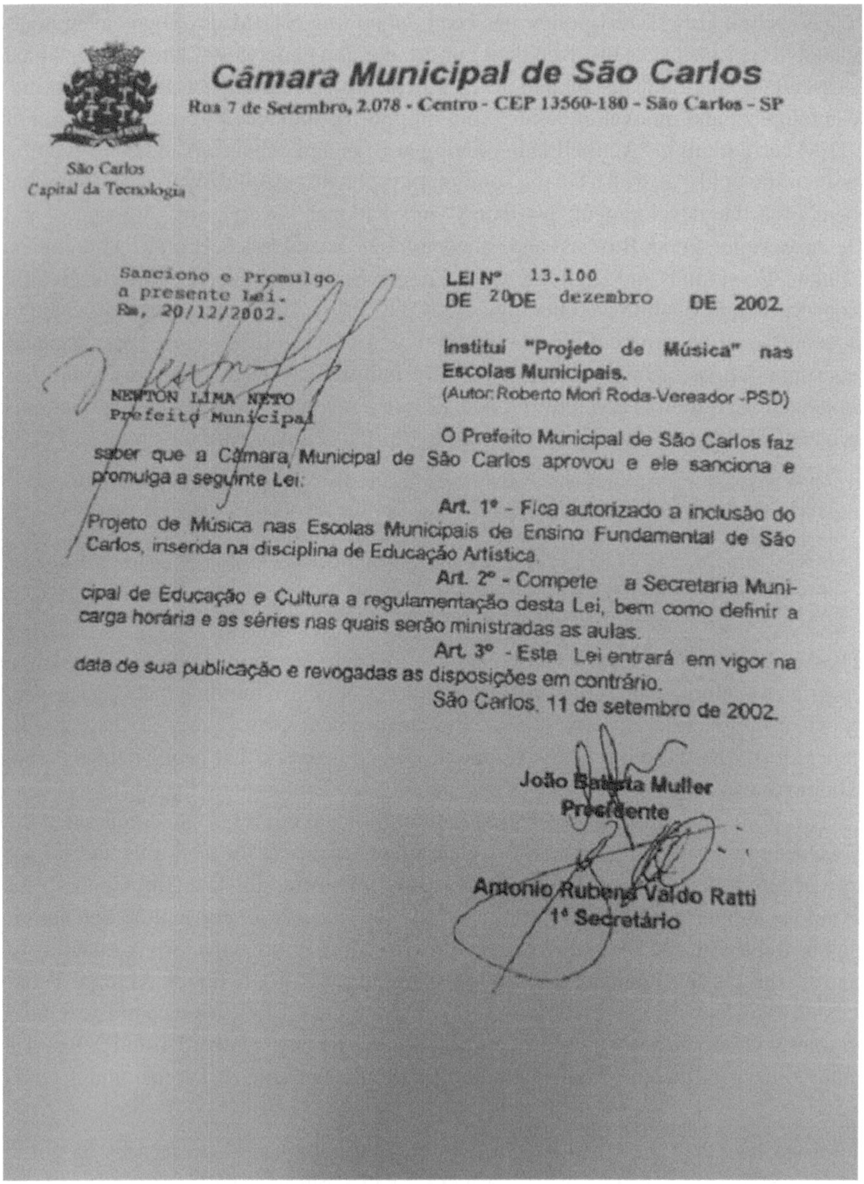

**Câmara Municipal de São Carlos**

Rua 7 de Setembro, 2.078 - Centro - CEP 13560-180 - São Carlos - SP

São Carlos
Capital da Tecnologia

Sanciono e Promulgo a presente Lei.
Em, 20/12/2002.

NEWTON LIMA NETO
Prefeito Municipal

LEI Nº 13.100
DE 20 DE dezembro DE 2002.

**Institui "Projeto de Música" nas Escolas Municipais.**
(Autor: Roberto Mori Roda-Vereador -PSD)

O Prefeito Municipal de São Carlos faz saber que a Câmara Municipal de São Carlos aprovou e ele sanciona e promulga a seguinte Lei:

Art. 1º - Fica autorizado a inclusão do Projeto de Música nas Escolas Municipais de Ensino Fundamental de São Carlos, inserida na disciplina de Educação Artística.

Art. 2º - Compete a Secretaria Municipal de Educação e Cultura a regulamentação desta Lei, bem como definir a carga horária e as séries nas quais serão ministradas as aulas.

Art. 3º - Este Lei entrará em vigor na data de sua publicação e revogadas as disposições em contrário.

São Carlos, 11 de setembro de 2002.

João Batista Muller
Presidente

Antonio Rubens Valdo Ratti
1º Secretário

Music Project in Local Schools at São Carlos. Prepared by Roberto Mori Roda.

[57] More info: <http://www.apaesaocarlossp.org.br/>

[58] More info: http://www.robertinhomori.net/site/index.php/2015-10-21-12-54-38/escola-livre

> The music brings to children the opportunity of good and healthy things. Music brought me a greater opportunity and it happened through a public school with the fanfare of 70s. I worked at Faber Castell, a multinational company that also projected my practice with music. This year, I will make 44 years of uninterrupted service with martial bands and fanfares. The lack of music school in the city, awoke in me the desire to create it. I elaborated a law so that there was discipline of music in schools. It was an authorization law, an indication, because I could not interfere in the city's budget. However, it allowed that a public policy was established.

So, STEAM movement became more noted than STEM in this training by introducing Liberal Arts as a discipline. The local government of São Carlos and region, also provided the verbal and mathematical literacy educators with the experience from the first to fifth grade in both public and private schools, and to whom would "virally" transfer to other cities around the country.

The Secretary for Environment, Science, Technology & Innovation of São Carlos, José Galizia Tundisi[59] brings evidence on how these policies have been core for the support of the interaction between STEAM development and the thinking and the economic and technological development of the city of São Carlos taking the "super smart society" model (Society 5.0[60]) as the main reference (JAPAN 2016), as follows:

> Who thinks the city? There are few people thinking of the city as a public system. The city is a system. You cannot separate education, environment, science, technology, industry, economy, health ... To treat this set as a system, you cannot treat as simply individual processes and compartmentalized problems. One must understand it as a complex system with its different interfaces, including the support of Artificial Intelligence. Thus, it takes an interdisciplinary and transdisciplinary action, considering the economic, environmental aspects, but mainly the social aspects that must permeate these processes. For such, we are creating an Applied Artificial Intelligence Center, adding scientists from different specialties (mathematics, engineers, agronomists, sociologists ... and thinking about São Carlos for 2119. - 1st city in Brazil with a scenario for 100 years. The Brazilian public system is not in a position to do this kind of study, but it can be supported by institutions that are not embedded in the public system, but which can contribute to the design of future public policies with a more strategic vision and with greater consistency. The cognitive thinking I needed to create super-brains, add music, paint, colors, all the experiences available to children and young adults. This requires a differentiated school with new teaching processes, new exposure processes, new approaches for a broader vision and brains with greater ability to make increasingly consistent connections. With super-brains, you change the economy and society.

As a result, these policies support the STEAM movement to help facilitate changes in the agriculture industry cluster and sustain the technopolis.

---

[59] Bio info: https://bv.fapesp.br/en/pesquisador/801/jose-galizia-tundisi/. Interviewed on Apr 24th, 2019

[60] Society 5.0 is a human-centered society that balances economic advancement with the resolution of social problems by a system that highly integrates cyberspace and physical space. Society 5.0 was proposed in the fifth Science and Technology Basic Plan as a future society that Japan should aspire to. It follows the hunting society (Society 1.0), agricultural society (Society 2.0), industrial society (Society 3.0), and information society (Society 4.0).

## 13.6 Recent Developments

The present case brought numerous developments that sustain the STEAM development and virtuous cycle of the agriculture industry cluster of São Carlos: new policies, programs, educational tools, educational excellence, schools and startups, and knowledge transfer centers.

The Secretary for Education of the city created the Center for the Training of Education Professionals (CeFPE), with activities in partnership with universities. It was inaugurated as a public policy of continuous training for improving the quality of education professionals in the city, teachers, directors, food producers, educational agents, administrative assistants and other network professionals. The City Mayor and the Secretary for Education of the city state that "the creation of the CeFPE raised in 2017 the average of the Basic Education Development Index for 2020 (IDEB-2017), the main quality parameter of national education".

This move by the city's Secretary of Education represented a major and recent increase in commitment to STEM approaches in public schools. On the other hand, STEM education is already a reality in private school and NGOs in São Carlos. In private schools, for example, robotics, games, 3D, programming and related activities have been the main themes applied in the agriculture industry cluster challenges described earlier in this chapter, considering that knowledge of these applications contributes significantly for the development of students, such as improvement of logical reasoning, and ability to solve problems. Besides, it is possible to use these abilities in labor market, which progressively adopt these concepts.

## 13.7 Conclusion

This case reveals the importance of reconciling the value of knowledge with the value of student engagement as a strategy to address the demands of a world in continuous development. Important contributions have been made into personal and professional development within all participant actors in the described case, as follows.

Today schools are yet incoherent institutions. In the moment of "transmitting knowledge" everyone is considered as equal to learn, however different, when evaluated. The STEAM methodology encourages the respect of one's individualities, culture and knowledge.

The students' participation in the learning process shows that knowledge arises from questioning subjects, who are not satisfied in living with incomprehension of things; nor to assume as unquestioned truths that has not yet been understood. Transdisciplinary projects that make use of STEAM technology create the opportunity of development of new ideas and creativity. It remains in the practices of educators and psycho-pedagogues enabled in this methodology the continuous effort to encourage students to perform in Mathematics; to create students capable of doing new things - creative, inventive and discoverers. In order to train people who can experience and not accept everything that is offered to them.

To trainers, the elaboration and application of the agriculture minicourse provided a unique experience - the visualization of mathematics teaching, in practice; the most intimate contact with the literacy educators; the transmission of ideas applicable to them; the return given by them of the importance of activities of this type; their interest in the STEAM methodology; their verbalization, that they do not have training and experience in this field.

The result was a self-empowerment; the pleasure of contributing to the formation of the Brazilian educator; the pleasure of being able to teach how to teach; the pleasure of learning to teach; the pleasure of preparing something in a format that motivates learning; the pleasure of presenting innovative ideas for the training of educators. The pleasure of empowering. The pleasure of training educators eager to be trained; trained to be able to train others.

In addition, the dense network relationship of the agriculture industry cluster led by the local inspirators was also powerful in STEAM development for regional development. The geographical setting provided a family-oriented connection among all players and increased the sense of community for absorbing new STEAM talents in a more agile and effective manner in the workforce pipeline.

These evidences keep promoting the use of STEM methodology as formal practice within regional technology policies for innovation and increasing the demand for new applications on STEAM methodology, giving also more opportunity to less high-technology industry regions.

**Acknowledgements** We thank Airton Garcia, José Galizia Tundisi, Orlando Mengatti Filho, Roberto Mori Roda, Ubiraci Moreno Pires Corrêa, José Manoel Marconcini, Paulo Sergio de Paula Herrmann, José Dalton Cruz Pessoa, Fernando Cesar Crnkovic, Paulo Renato Orlandi Lasso, Paulo Sérgio Camargo, Sérgio Henrique Vannucchi Leme de Mattos, Wilma Regina Barrionuevo, and Sylvio Goulart Rosa Júnior for their extraordinary support in being interviewed for bringing evidences that illustrate the validation of the content described in this chapter. We thank Caio de Souza Ferreira, Laura Mastrodi and Raquel Canola Zacour de Azevedo for developing the figures of this chapter as well as bringing knowledge on how geographical settings influence innovation and regional development. We also thank Débora Garcia and Nilce Chaves Gattaz for their great effort and valuable writing and translation skills.

## References

Antunes, C. (2007). *Novas maneiras de ensinar novas formas de aprender*. Porto Alegre: Artmed.

BRASIL. Secretaria de Educação Básica. Diretoria de Apoio à Gestão Educacional (2014) PACTO Nacional pela alfabetização na idade certa: Apresentação/Ministério da Educação, Secretaria de Educação Básica, Diretoria de Apoio à Gestão Educacional. Brasília: MEC, SEB.

Bruns, B., Evans, D., & Luque, J. (2012). Achieving world-class education in Brazil: The next agenda. direction in development; human development. World Bank. Retrieved August 1, 2019 from https://openknowledge.worldbank.org/handle/10986/2383.

Cabinet Office, Government of Japan (n.d.). Society 5.0. Retrieved August 1, 2019 from https://www8.cao.go.jp/cstp/english/society5_0/index.html.

CEPEC- Centro de Pensamiento en Estrategias Competitivas y Inteligencia de Negocios–IdN (2013). Ranking de ciudades latinoamericanas para la atracción de inversiones. Universidad del Rosario. Informe Oficial. Retrieved April 4, 2019 from https://web.archive.org/web/20170119073048/http://www.urosario.edu.co/urosario_files/9d/9d96f884-d433-45a8-947b-4e9877596f63.pdf.

Chalita, G. (2001). *Educação: A solução está no afeto*. São Paulo: Editora Gente.

Ertas, A., Gatchel, S., Rainey, V., & Tanik, M. M. A. (2007). *Networked approach to transdisciplinary research & education*. The ATLAS Publishing (TAM), 3(2),1–12.

Fonseca, M.C.F.R, &. Cardoso, C.A. (2005). Educação matemática e letramento: Textos para ensinar matemática e matemática para ler o texto. In: Nacarato, A. M., Lopes, C. E. (Org.). Escritas e leituras na educação matemática. (p. 63–76). Belo Horizonte: Autêntica.

Fonseca, M., & Bopprê, V. (2013, March 7). Ciências e matemática com uma pitada de arte. PORVIR. Inovações em Educação. 2013. Retrieved February 1, 2015 from: http://porvir.org/ciencias-matematica-uma-pitada-de-arte/.

Gadotti, M. (2000). Perspectivas atuais da educação. *São Paulo em Perspectiva, 14*(2). https://doi.org/10.1590/S0102-88392000000200002.

Gattaz, C. C, Bernardes, R. C., & Cruvinel, P. E. (2019). Leveraging digital knowledge ecosystem framework implementation: Case study: Aligning knowledge management and innovation goals for agricultural aerial pest control. Encyclopedia with Semantic Computing and Robotic Intelligence, online ready.

JAPAN. Government of Japan (2016) The Fifth science and technology basic plan. Retrieved June 1, 2019 from https://www8.cao.go.jp/cstp/english/basic/5thbasicplan.pdf.

Laplane, L., Mantovani, P., Adolphs, R., Chang, H., Mantovani, A., McFall-Ngai, M., Rovelli, C., Sober, E., & Pradeu, T. (2019). Why science needs philosophy. *Proceedings of the National Academy of Sciences of the USA (PNAS), 116*(10), 3948–3952.

MCTIC-Ministério da Ciência, Tecnologia e Inovação (2016). *Estratégia Nacional de Ciência, Tecnologia e Inovação 2016–2022*. Brasília: MCTIC.

Michaelis. Dicionário de Inglês Online. Melhoramentos (2009).. https://michaelis.uol.com.br/moderno-ingles/busca/ingles-portugues-moderno/steam.

Nogueira-Ferreira, F. H., Mello-Carpes, P. B. M., Forte, C. M. S., & Souza C. M. de M. (Orgs). (2016). Caminhos de um programa de educação científica: Relatos e produtos. UFU, Uberlândia, MG/ UNIPAMPA, Uruguaiana, RS/ UECE, Itapipoca, CE/ FURB, Blumenau, SC/UFV, Viçosa, MG/ UFG, Catalão, GO/ UFC, Fortaleza, CE/ UFMS, Cuiabá, MT/ UFPR - Matinhos, PR/ UFERSA, Mossoró, RN/ UFF, Santo Antônio de Pádua, RJ/ UFOPA, Santarém, PA/ UNEMAT, Cáceres, MT/ UFRN, Natal, RN/ IFBA, Salvador, BA: CAPES/ British Council/Fundo Newton.

OECD (2018). *Education at a glance 2018: OECD indicators*. https://doi.org/10.1787/eag-2018-en.

Oliveira, V. B. (Org.). (1996). *Informática em psicopedagogia*. São Paulo: SENAC.

Ramal, A. C. (2002). Educação na cibercultura: Hipertextualidade, leitura, escrita e aprendizagem. Porto Alegre: Ed. Artmed.

Santos, D. B., Leichsenring, A. R., Menezes Filho, N. A., & Mendes-Da-Silva, W. (2019). Income distribution and duration of poverty-level employment. In W. Mendes-Da-Silva (Ed.), *Individual Behaviors and Technologies for Financial Innovations*. Cham: Springer. https://doi.org/10.1007/978-3-319-91911-9_6.

Weinberg, K. (1972). *On Gide's PROMETHEE: Private myth and public mystification*. Princeton: Princeton University Press.

# Part III
# Cases of National Context

# Chapter 14
# Tracking STEM Education Development in China: National, Regional, and Local Influences

**Guolong Quan**

**Abstract** In China, STEM education is viewed as essential to the cultivation of talent in scientific and technological innovation. This chapter describes national, regional, and local policies that drive changes in teaching in schools, with an emphasis on novel STEM experiences. At the national level, the basic strategy is driven by scientific and technological innovation, and an increase in the number of budding connections to regional and local ecosystems can be seen in the developed regions of Shanghai and Zhejiang Province, where STEM education shows early signs of convergence with industrial development. The STEM Education Innovation Center, bolstered by creative competition in science and technology, is the primary vehicle for the regional promotion of STEM education and for the development of an ecology compatible with scientific and technological innovation. As the economy and industry mature, the Department of Education must lead the way in establishing and continuously improving regional STEM policies, stabilizing expectations and demands regarding the subjects taught and the quality of STEM programs. Finally, the development of school-based STEM education in cooperation with regional industrial clusters must become the norm. This chapter examines all of these influences, describes how they interact, and provides examples of how they unfold in Shanghai and Zhejiang Province.

## 14.1 Introduction

In recent years, China's GDP growth rate has remained at around 6.8%. In the new period of national economic development, scientific and technological innovation strategy is regarded as an important means to develop both traditional and emerging industries and to promote the real and the digital economy. For example, to promote "Internet Plus," the new technologies and models can be used to transform tradi-

G. Quan (✉)
Jiangnan University, Wuxi, China
e-mail: quangl@jiangnan.edu.cn



C. Zintgraff et al. (eds.), *STEM in the Technopolis: The Power of STEM Education in Regional Technology Policy*,
https://doi.org/10.1007/978-3-030-39851-4_14

tional industries; deepening the "Internet Plus Government Services," it can be improved of the ease of work for enterprises and the masses; vigorously optimizing the innovative ecosystem, the enthusiasm of the whole society innovation can be attracted. As another example, to deepen the reform of science and technology management system, to strengthen the main position of technological innovation of enterprises, to expand the policy of increasing fund for R&D costs and reducing deduction ratio to all enterprises, and to make policy measures to support the in-depth development of innovative entrepreneurship. The implementation of the corresponding policy has increased the turnover of technical contracts by more than 30% over the past year (China News Network 2019b).

In the new situation, the outlines of the National Medium- and Long-Term Science and Technology Development Plan (Ministry of Science and Technology of the PRC 2006) and the National Medium- and Long-Term Education Development Plan (Central People's Government of PRC 2010), as well as personnel training to meet the needs of scientific and technological innovation and industrial development, have been promoted as important strategic steps. A series of outlines, reports, and plans have emerged to point the way for national science and technology development and education (Ministry of Science and Technology of the PRC 2006; State Council of the PRC 2017; China News Network 2018; National Institute of Education Sciences 2017). In recent years, the Ministry of Education has issued a series of policy documents focused on the promotion of STEM work (e.g., Ministry of Education 2016b: 2017a, b). In particular, after the launch of the STEM Education 2029 Action Plan, the eastern provinces initiated STEM teacher training, STEM curriculum development, and other local, national, and international cooperative promotions. All these efforts are aimed at providing sufficient scientific and technological innovation support and human resource reserves for China's economic development and industrial take-off in the future.

## 14.2 Key Points

- China's economy is on the upswing, and maintaining economic growth, developing sunrise industries, and targeting productivity are important goals. The development of industry based on artificial intelligence (AI) technology has become an important point of China's future economic development.
- Scientific and technological innovation is expected to drive industry take-off. Government departments have introduced corresponding policies for the development of science and technology on one hand and education on the other, to guide these twin engines of future economic development.
- Scientific and technological innovation and the cultivation of innovative talents are two important goals for the introduction of science and technology and education policies, which provide impetus for the development of various industries through the employment of talents.

- STEM education, led by the education sector, has begun to take root in some schools, and pilot work is being conducted with a view to extending it across the entire country. Further development faces the challenges of insufficient teachers, poor teacher quality, poor environmental conditions, and lack of learning resources and facilities.
- STEM promotion, which takes place mainly through the use of international mainstream project-based learning and through local and international partnerships between schools and education companies, provides support for personnel training needs. School-based STEM teaching is faster, whereas STEM courses and study training offered by education companies is of higher quality.
- Regional STEM education is being actively developed, with each region imprinting its own particular signature. Examples include the establishment of the STEM Education Collaborative Innovation Research Center in Jiangsu Province, the STEM Education Theme Conference held in Zhejiang Province, the development and release of outlines, organizational innovation competitions, and so on. Shanghai and Zhejiang have begun active communication between the education sector and the industrial sector.
- STEM education is not yet sufficiently integrated with or guided by the needs of regional economic and industrial development. Over the short term, it is not always easy to see the economic benefits of STEM education in China. Nevertheless, it is of utmost importance to long-term economic development.

## 14.3 STEM Industry Clusters in China

With the development of China's economy and the establishment of science and technology innovation strategy, a number of new industries began to emerge, and many traditional industries are eager to upgrade as well. The purpose of STEM education is to train sufficient innovative talent to develop both new and older industries.

### *14.3.1 Driving China's Economy and Industrial Development With Scientific and Technological Innovation*

At the 2018 National Assembly, Chinese President Xi Jinping stated the important assertion of "in grasping scientific and technological innovation, we have grasped the 'bull nose' of the overall development" (CNR 2018). Developing the "bull nose" refers to building and producing based on the key elements preferentially. The Chinese government has put forward a three-step strategy to enable China to enter the ranks of innovative countries by 2020, to be at the forefront of newly innovative countries by 2030, and to become the world's leading technology and innovation powerhouse by

2050. To achieve these goals, the government has taken on the long-term, systematic, and strategic project of promoting the cultivation of innovative talent, implementing forward-focused personnel training, and developing innovation-oriented basic education programs and infrastructure.

The National policy of science- and technology-driven economic development not only encourages mass entrepreneurship and public innovation, but also significantly guides the development of pillar industries. According to the 2019 China Industrial and Informatization Development Trend Outlook report (Sadie Think-tank 2018), future industrial hotspots for the development of industry and informatization include the following areas.

- Information technology (IT) innovation
- Software and information services
- AI
- Virtual reality
- Cloud computing
- Big data
- Digital creativity
- Industrial technology innovation
- Digital economy
- Intelligent manufacturing
- The mobile Internet of Things
- Radio management and application

That said, the importance of human resource development to industry and informatization innovation should not be overlooked. Talent training in the new era requires nontraditional skills, including in the areas listed below, with an emphasis on integration and innovation skills based on professional and subject-related accomplishment.

- Information
- Innovation and design
- Professional knowledge integration
- Computational thinking
- Innovative thinking
- Design thinking
- Critical thinking

### *14.3.2 Technology Policy Focuses on Key Areas for Industrial Development*

The outline of the National Medium- and Long-Term Science and Technology Development Plan (2006–2020) describes the major scientific issues and pioneering technological directions that must be addressed as China undergoes national and social development, pointing out the future direction of cutting-edge technologies (Ministry of Science and Technology of the PRC 2006). In 2017, the new generation of the Artificial Intelligence Development Plan issued by the State Council recognized AI as a new engine of economic development. AI development will result in the reconstruction of all aspects of economic activities, create a need for new intelligence in both macroscopic and microscopic fields, and lead to the emergence of new technologies, products, industries, corporate forms, and business models. AI development will also trigger major changes in the economic structure; revolutionize human production, lifestyle, and thought patterns; and achieve an overall leap in social productivity. As China's economic development in the new normal deepens, the task of supply-side structural reform will become very arduous. In order to re-energize China's economic development, we must speed up and deepen the application of AI and cultivate and strengthen the AI industry.

The form and scale of AI-driven industrial development is very promising. The first step is to establish AI technology standards, service systems, and an industrial ecological chain. If a number of the world's backbone AI enterprises are cultivated in China, a core AI industry of more than 150 billion yuan would drive the related industry scale by more than 1 trillion yuan. Then, a new generation of AI could be widely used in such diverse areas as intelligent manufacturing, intelligent medicine, smart cities, intelligent agriculture, national defense, construction, and other fields. A core AI industry of more than 400 billion yuan could drive the related industry scale by more than 5 trillion yuan. AI could be further applied to the broadly expanding areas of production, social governance, national defense, and construction. Through the application of core technology, key systems, support platforms, and intelligent applications to the entire industrial chain and high-end industrial clusters, a core AI industry of more than 1 trillion yuan could drive the relevant industry scale by more than tenfold.

The New Generation Artificial Intelligence Development White Paper prepared by the China Electronics Society, which explores the industrial boundaries of the new generation of AI, divides the industrial categories and application scenarios of AI and studies relevant investment and financing characteristics and trends (China Electronics Society 2017). Table 14.1 shows the current core industrial chain of the new generation of AI at three different levels. However, China's industrial cluster network is still in development. Although the economic and technological development is basically completed, further improvement is needed in the industrial

**Table 14.1** Current core industrial chain of the new generation of artificial intelligence

| Level | Technology area | | | |
|---|---|---|---|---|
| Application level | Drones | Intelligent driving | Intelligent robots | Intelligent education |
| | Intelligent finance | Smart medicine | Intelligent security | Intelligent search |
| Technical level | Speech recognition | Text recognition | Picture video recognition | |
| Base level | Smart sensors | Smart chips | Algorithm models | |

agglomeration of coordinated development, industrial and entrepreneurial innovation links, innovative cooperation mechanisms, and so on. With AI technology as the industrial cluster leading in development, upgrading through the network approach and the "government-market-cluster" governance system are also needed (Zhao 2019).

The new generation of the AI development plan and the above-mentioned White Paper (China Electronics Society 2017) clearly state that high levels of AI innovation talent and teams should be nurtured in the following ways:

- Support and cultivate leading AI talents that show potential for development
- Enhance the training of professional and technical personnel in basic research, applied research, operations, and maintenance of AI
- Prioritize training of compound talents, most particularly longitudinal compound talents with AI theory, methods, technology, products and applications; as well as horizontal compound talents that master AI as well as economics, social sciences, management, standards, law, etc.
- Constantly explore the views expressed within the new compound talent education and training mechanism

### *14.3.3 Follow-Up of STEM Policy*

In order to implement the innovation-driven development strategy, the State Council and the Ministry of Education introduced policy outlines and documentation regulations focusing on personnel training and human resource reserves. With this in mind, the China Academy of Educational Sciences established the STEM Education Research Center (Xinhua News Agency Network 2017). In the same year, given the rapid development of AI technology, the State Council issued the New Generation of the Artificial Intelligence Development Plan, which suggested ways to implement AI-related courses in primary and secondary schools[1] and to gradually promote programming education (Observer 2017). Moreover, AI textbooks for schools have been developed and used.

The STEM Education Research Center's White Paper argues that STEM education in China is steadily developing in terms of theoretical research and educational policies and practices, including the initiation of regional STEM pilot schools and teacher training, the establishment of STEM professional classrooms and maker spaces in pilot schools, project-style learning in the classroom, and more (National Institute of Education Sciences 2017). As a regional example, the Education Bureau (EDB) in Qingyang District, Chengdu, incorporated STEAM education into its concept of the "School of the Future," considering it important to the educational ecosystem of the future. In May 2015, the EDB launched the Innovation Course Experiment Pilot Project (Education Management Information Center of Education Ministry

[1] Unless otherwise specified, henceforth "schools" refers to primary and secondary schools.

2017). By 2017, there were a total of 15 pilot schools in the region, all prioritizing STEAM for project implementation. In 2016, the first Create the Future project was organized to promote the exploration of new technology. This project not only offers a platform for exchanges, displays, and competitions among pilot schools, but also encourages pilot schools, teachers and students to create opportunities to earn honors and awards. Effective promotion of the STEAM curriculum is guaranteed by the policy and system behind the "Experiment Schools for Innovation Course."

In contrast, the Qingyang District's STEM Pilot School is not well connected to the local industrial cluster. Qingyang's STEM practice reflects the national trend, in which the initiatives of China's lower education administrative departments respond directly to the spirit of the national education policy, indirectly to the development of the country's economic industry, and are somewhat removed from local industrial development. Moreover, STEM is explored from the bottom up (Education Management Information Center of Education Ministry 2017), meaning that promoters pay more attention to the immediate interests of teachers, while integration with local industry is seen as a luxury. Also, within about three years, the national policy began to seek to move STEM education forward.

Another hindrance to the integration between local STEM initiatives and industry is the fact that in various parts of China, industrial clusters are themselves in the developmental stage. Although economic and technological development zones are forming and high-tech industrial zones are at the core of industrial agglomeration areas, inter-industry cohesion is lacking, industrial and corporate innovation links are weak, and both innovative cooperation mechanisms and guidance for industrial cluster organizations are lacking (Zhao 2019). The deep-value orientation in the less-developed regions is different, which makes it challenging to connect regional STEM education with the development of industrial clusters. Most importantly, there is no communication or collaboration between the leaders of the two sectors.

As a point for comparison, let us look at the parallel situation in the United States. At the state and municipal levels, the education system of the United States has a great amount of independence. At other levels, the American education sector fosters a closer relationship between parallel departments of education in the domestic regions and municipalities. This facilitates intersectoral communication and collaboration within a given region. This close relationship is best reflected at the national level. STEM measures of regional and urban areas are more directly influenced by the national education policy, and then respond to the development of the national economic industry.

## 14.4 Education Philosophies and Methods

Education policies that meet the needs of economic and industrial development and can stimulate the development of science and technology and are important for the implementation of science, technology, and innovation driving strategies. The cultivation of innovative talents is fundamental to ensuring that the strategic goal can be achieved. Furthermore, reasonable scientific educational ideas and methods are essential to furthering the driving strategy of science and technology innovation and rejuvenation in both science and education.

## 14.4.1 Philosophies of STEM Policy

The outline of the National Medium- and Long-Term Education Development Plan (2010–2020) was introduced with the goals of modernizing education, forming a learning-oriented society, strengthening human resources, and fully implementing quality education as a strategic theme in education reform. Since 2015, the Ministry of Education has issued a number of policy documents to actively promote the development of STEM education. The documents presented in Table 14.2 indicate that increasingly more attention has been paid to breaking down subject-related and occupational barriers, and focused on professional and comprehensive compound personnel training. This is directly manifested in the increasing emphasis on interdisciplinary learning, construction of comprehensive laboratories and educa-

**Table 14.2** STEM-related documents issued by the ministry of education since 2015

| Release date | Policy document | Main content |
|---|---|---|
| September 2015 | Guiding opinions on comprehensive and in-depth advancement of Education during the thirteen five-year plan period | Explore new education models such as STEM and maker education<br>STEM concept first proposed |
| June 2016 | The thirteen five-year plan for education informatization | Actively explore interdisciplinary learning, STEM education models, and their applications<br>Improve students' information literacy |
| July 2016 | Opinions on further improving the equipment work of ordinary schools under the new situation | Support the exploration and construction of comprehensive laboratories, educational maker spaces, and other educational environments<br>Encourage multifunctional technical transformation of existing classrooms<br>Adapt to the learning needs of students |
| February 2017 | Compulsory education primary school science curriculum standard | Advocate interdisciplinary learning methods<br>Suggestion that teachers can deepen STEM education in their teaching practices |
| July 2017 | New generation artificial intelligence development plan | Set up AI-related courses in schools<br>Gradually promote programming (STEM) education) |
| October 2017 | Guideline for comprehensive practical activities in schools | Comprehensive practical activities are compulsory courses stipulated in national education and general high school curriculum programs, and set up in parallel with subject courses. This is an important part of the basic education curriculum system |

tional maker spaces, attention to students' information literacy, comprehensive practical skills, programming skills, and other capacity development.

All of the STEM-related education policy documents presented in Table 14.2 require project-style teaching. Multifaceted, project-style teaching and learning can take place within an interdisciplinary model, and content can be learned and understood from a multidisciplinary perspective, increasing students' knowledge integration and migration ability. Project-style teaching is also conducive to students' understanding and solving problems in real situations and increases their receptiveness to new memories and experiences. Finally, project-style learning requires more cooperation and exchanges, and hence is conducive to teamwork, cooperation, and the development of complementary sets of skills.

A good education is one that helps students learn not only to ask and answer "why" and "what," but also to rise to the more practical challenge of "what to do." Project-style learning cultivates students' autonomous STEM accomplishments, innovation abilities, and the compound talents required by the policy documents cited above. More importantly, project-based teaching follows a personnel training model that combines production, learning, and research in the early stages of basic education. It not only supplants the traditional "cramming" style of basic education — instead infusing students with a sense that learning is vibrant and dynamic — but also implants the modes of production, study, and research into students' thinking and habits from an early age.

In project-style learning, subject-type teaching can be highlighted through the methods of project research and implementation. The relevant knowledge is embedded through a certain subject, and the students' creative thinking and problem-solving abilities are cultivated through the analysis, inquiry, program design, and implementation of a certain problem. The STEAM class analysis treated in the China STEAM Education Development Report (Education Management Information Center of Education Ministry 2017) fully reflects these characteristics of project-style learning. This framework consists of a real situation, key problems, a design scheme, an implementation plan, an improvement plan, and a product-exchange display. Students not only lay the foundation of specialization in solving the specific problem, but also apply integration ability to find the solution. Over time, their ability to solve practical problems through the comprehensive application of multidisciplinary knowledge improves naturally, and their practical ability and innovative spirit grow naturally. This prepares them to approach scientific research from a transformational perspective in the future.

In June 2017, the first successful symposium on the project teaching method (project-based learning) and its integration into the primary and secondary school curriculum was held in Beijing (Sina Education 2017). The ultimate goal of the symposium was to promote the reform and practice of deep learning in schools over a background of IT.

In 2016, the Ministry of Education and 11 other departments jointly issued their opinion "On the promotion of study travel for primary and secondary school students." Study travel, more commonly known as "research travel," focuses on out-of-school educational activities (field trips) that combine research-based learning and travel experiences. The opinion emphasizes that

1. in order to give students room for positive growth and all-around development, interesting, systematic, intellectual, and scientific activities suited to their physical and mental characteristics, their acceptance level, and their practical needs should be incorporated into the field trip; and
2. depending on local conditions and regional characteristics, students should be guided off the campus and into new environments in order to expand their vision, enrich their knowledge, increase their understanding of society, be close to nature, and participate in daily-life experiences in new ways.

While the educational and practical nature of research travel are of utmost importance, safety and public welfare are also very high priorities. The opinion points out that research travel will be included in primary and secondary education teaching plans, and that all localities should build research travel bases that respond to local conditions, standardize their organizational management, and explore sound financing mechanisms (Ministry of Education 2016a). The opinion suggests that in the future, more elements of scientific innovation will be incorporated into the research travel curriculum, and that disciplines such as physics, chemistry, biology, geography, astronomy, and engineering can be experienced and understood through active exercises, children's social interactions, and contact with nature and society. To innovate basic education, it is also necessary to

1. update the concept of parent education,
2. innovate the design of schools,
3. secure support for the government education policy, and
4. expand the supply of social education resources.

Only then can a large number of innovative scientific and technological talents be cultivated in support of the goal of "building a world science and technology innovation powerhouse by 2050" (Wu 2018). Support for these four aspects of STEM teaching and learning activities is exemplified in the specific cases of STEM education in Shanghai, Zhejiang, and Hubei.

### 14.4.2 Philosophies and Methods of STEM Teaching

In STEM research, design-based learning and project-oriented learning addressing real problems are recommended as teaching methods to cultivate students' STEM literacy (Wang 2015). In actual implementation, teachers have explored many methods, including:

- inquiry learning that focuses on the process to inquiry something around a set of questions of a problem,
- cooperative learning that solves problems by teams or groups based on a problem or project,
- lecture-based learning that conduct learning with lectures predominantly,
- autonomous learning that students learn by himself or herself based on the materials that designed for STEM learning,

- cooperative inquiry learning that conducts according to inquiry learning process by groups or teams,
- project-based learning that the project is the basis to learn for students, aiming at the project completing successfully by students' effort and teachers' guidance.
- problem-based study that requires to solve the problem with one's own thought,which is the important target, and
- design-based learning that focuses on students' design based on the requirements of real situation or the given task, which make manifests using relevant knowledge and information.

Among these, inquiry-based and cooperative learning are the teaching methods most commonly used, accounting for 68.7% and 60.4% of teachers, respectively. The proportion of teachers who use the international STEM education common in project-based, problem-based, and design-based learning was at least 20 percentage points lower. This is especially true of design-based learning, which was used by only 25% of teachers (Education Management Information Center of Education Ministry 2017). A search of STEAM classroom teaching modes revealed that the common elements of STEAM classroom teaching are real situations, practical problems, design schemes, improvement plans, implementation plans, and product exchange and sharing. For the corresponding teaching cases, see China STEAM Education development report at page 108 (Education Management Information Center of Education Ministry 2017).

The concept of "project plus practice" is embodied in STEM courses. According to the China STEAM Education Development Report released in January 2017 (Education Management Information Center of Education Ministry 2017), STEM courses in China include both subject development courses and comprehensive practice courses. Subject development (i.e., subject practice) courses focus on a particular subject while integrating knowledge from other disciplines. Comprehensive (i.e., innovative) practical courses completely break the boundaries between disciplines and integrate multidisciplinary knowledge. Both types of courses are oriented to solve problems in real situations and aim to improve student comprehension through problem solving. At present, STEM courses are mainly school-based, community-based, comprehensive practical activities, and the like. As an example of the above the STEM course developed by the Shanghai STEM Plus Research Center has a duration of about 10 hours per semester. Unlike traditional courses, it applies project-based STEM teaching methods, focuses on STEM literacy, and uses exploratory processes, making it a useful supplement to normal curriculum teaching in schools.

There are two reasons relating to curriculum resources for which the STEAM education curriculum was established. One is to translate foreign STEAM education curriculum resources into Chinese and then put them into practice in the country. Examples of the use of STEAM curricula from abroad include the STEM course translated by the team leaded by Professor Zhongjian Zhao from East China Normal University, the STEM course created for the "Huagong Star" brand by Wuhan "Chuang-Qu-Tian-Di" Education Consulting Co., Ltd. based on the MIT Fab Lab laboratory project "How to make (almost) anything", and The STEM course that shared cooperatively between the Education board of Zhejiang Provincial and the

Indiana State of the United States. The other is that domestic institutions and teachers do not yet have a clear understanding of or experience with expensive foreign courses and need special training to use them to advantage. The domestic curriculum with which teachers are most familiar is generally dominated by technical learning courses and uses technical tools for design and realization. Moreover, although there are many kinds of courses in existence, there is a lack of continuity and a lack of guidance from the upper echelons.

The implementation of STEM teaching methods with projects, problems, and design at their core means that schools should cooperate with non-educational institutions. This opens up space for K-12 schools to work with other departments in industry, government, and nonprofit organizations to support STEM education and teaching. Schools can also receive solidarity, volunteers, mentors, and funding from other sectors they cooperate with. This is particularly true when STEM content is aligned with industrial clusters that are important to cities/regions, which promotes cooperation. However, in seeking out multi-sector cooperation, STEM education is subject to many other factors, such as industry competition, business costs, balance of interests, company secrets, industry control, education systems, and so on. In the current trend within STEM development, there is no perfect combination of STEM education and regional industrial cluster development. The STEM services of the Shanghai STEM Cloud Platform and the STEM Plus Research Center, supported by multiple departments, have achieved good results while working with schools to train for family participation, yet they have not developed close links with any industrial clusters. As the market allocation of talent comes closer to the needs of industry, the time and cycle of talent growth as well as the evolution of industrial clusters can both benefit.

Of course, schools also receive grants, donations, or assistance from business, or educational companies to aid in STEM education. Government provides grants to STEM schools through education policies and projects. The example of Zhejiang Province reflects the situation of government input. (See part IV below.) And enterprises or educational companies, out of concern for education or the need for future development, contribute to or fund STEM education in schools in some form. For example, on April 25, 2018, Asahi Hui Group joined Hangzhou Mark Middle School education group to hold the Asahi Hui City Laboratory project donation ceremony. Through the generosity of the Asahi Hui Group, Mark Experimental Middle School, which has offered multiple school-based courses to develop the architecture for the school-based STEAM curriculum, now has a new platform for enhancing comprehensive student literacy, as well as a new chapter for STEAM education. The Asahi Hui City Laboratory project, in line with future directions in urban planning, anticipates the shape the city will take in the future. The project, supported by advanced technology, reflects the deep integration of technology and education achieved through project-based STEM learning. Still, basic education curriculum reform must be given the necessary support. An example of such support is Xu Hui Group's Future City Laboratory project in Mark Experimental Secondary School, which is set to produce the world's science and education products in Zhejiang and may even be promoted on the national market. Just over a year ago, the Asahi Hui group's Future City Laboratory-Global A-STEM Classroom donation ceremony was held at the Tongzhou campus of Beijing Polytechnic University's affiliated secondary school (Zhongxu Future Education 2018). However, this ceremony is also limited to corporate donations to schools, and does not reach the level of business collaboration with schools.

## 14.5 Policies That Support STEM Programs

As can be inferred from the concept and method of STEM education, STEM projects are mainly implemented by primary and secondary schools. Although the schools do not directly connect with or serve regional industrial clusters, they respond directly and positively to the directives of higher education departments and follow the spirit of Ministry of Education policy. Moreover, policies and documents in the education sector have affected the informal education industry. In developed regions, educational companies typically cooperate with schools to carry out STEM activities or training. Therefore, at this time school-based STEM projects and company promotions are the two main avenues for the development of STEM education, and they complement each other.

The 2017 White Paper points out some urgent problems with the development of STEM education, including the mismatch between personnel training and market demand, the lack of high-quality STEM teachers, and limits to the implementation of STEM programs. If these problems cannot be solved in a timely manner, it will be difficult for STEM programs to benefit the Chinese market with a full range of innovative talents. As STEM education progresses, the problems still existing for its implementation must constantly be revisited. Three examples of such problems are listed below.

1. All kinds of STEM education and training institutions spring up, but few can really secure the cooperation of scientific research institutions and training enterprises.
2. The courses categorized as STEM are really just traditional courses in existing disciplines, courses that have always existed. What is needed are new courses based on instructional designs that apply STEM philosophies, theories and models and integrate content across multiple disciplines.
3. In the absence of STEM curriculum standards, the answers to key questions regarding what to teach, how to teach, and how much to teach are vague, uncertain, and unable to identify the desired results (China News Network 2019a).

If the most important goal of STEM education is to build talent for the future development of the national economy and industry, then STEM education must inevitably be examined from the perspective of economic development and industrial demand. From the perspective of both scientific research institutions and STEM enterprises, a variety of problems in STEM education remain to be overcome. However, judging from the Ministry of Education's promotion policy and from local education sector initiatives of 2018, China's STEM education development model is well under way. How the education sector acts in the future will determine whether the impact of the college entrance examination will be reined in, and whether cooperation with industry will become more unrestricted, to the benefit of long-term development.

STEM education in China consists of two important elements. First, STEM courses are hosted in schools. In accordance with the document of the Higher Education Department, schools teach STEM through curriculum planning, autonomous or cooperative curriculum construction and implementation. Second, STEM courses and teacher training services are provided by educational companies. Educational companies actively seek to offer STEM courses and training services for the sake of their own survival and development. Often, they provide schools with courses, teaching materials, teachers, research, and related services, while offering STEM learning and training services to students' families. These family-oriented services are generally independent of the services provided to the schools.

At present, STEM education has three main focal points: (1) professional training of STEM teaching teams; (2) introduction and development of STEM curriculum resources; and (3) creation of positive environments and conditions for STEM education – that is, schools and enterprises prepare the conditions for the implementation of the STEM project/business and determine their own programs and implementation plans according to their own circumstances (China Education News Network 2017).

The following examples more specifically reflect the current situation of STEM implementation in China.

### 14.5.1 Progress in STEM Led by Education Sectors and STEM Action in Zhejiang Province

The National Medium- and Long-Term Education Development Plan (2010–2020) clearly states that it is necessary to focus on improving the quality of education and enhancing the capacity for sustainable development around the strategic goals of education reform. The program calls for strengthening key areas and weak links through the implementation of a number of major projects. In 2016, STEM education was integrated into the national strategic development policy. In the same year, the Ministry of Education's Education Informatization Thirteen-Five plan "clearly stated that the conditional areas should actively explore the use of IT in the creative space" (Ministry of Education 2016b), interdisciplinary learning (STEAM education), creative education, and other new educational models. In 2017, the Ministry of Education issued the standard of science curriculum for compulsory primary schools, advocating interdisciplinary learning methods and advising teachers to try STEM education in teaching practice. In 2017, The China Academy of Educational Sciences STEM Education Research Center officially released the China STEM Education White Paper. It pointed out that some progress has been made in STEM education, especially in educational practice, theoretical research, and education policy.

Many cities actively explore ways to promote STEM education. Zhejiang Province, Shenzhen City, and Chengdu City have issued special documents, vigorously promoted STEM courses, and carried out pilot school declarations and work in STEM teacher training. Many schools have established STEM professional classrooms or maker spaces to promote project-based learning in classroom teaching. The documents released by these cities are listed in Table 14.4 in the later section.

The China STEM Education 2029 Action Plan, launched in May 2018, is an important measure for actively promoting the development of STEM education. The significance of this plan lies in its efforts to bring STEM education to the widest possible range of student groups by training them in scientific thinking, innovation, inquiry, innovation consciousness, and complex problem solving (International Daily 2018). Through the plan, 1000 schools have been selected, guided, and cultivated to explore STEM practice. A research force has been organized to help schools to train a group of backbone teachers, build a number of special projects and schools, design the mode of promoting economic development in different regions. The content of the action plan emphasizes:

- Promote the top-level design of STEM policies to help implement innovation-driven strategy
- Introduce and implement a smooth plan for STEM talent training. The STEM course teaching system must be perfected in order to effectively tie STEM education to elementary, middle and high schools. Optimize the attractiveness, quality, and emphasis on science of STEM activities in order to allow students to participate safely in STEM activities. Emphasize multivariate evaluation, formative evaluation, and learning process evaluation in personnel training.
- Build up and strengthen STEM teaching staffs. As the planners, designers, organizers, and implementers of STEM activities, teachers are the key to the success or failure of STEM education. Teacher development should include interdisciplinary requirements, teamwork, training, participation in learning, participation in practice, etc. (Sohu Network 2019).
- Establish and improve the long-term cooperation mechanism for active, consensus-based participation, exchange, cooperation, and investment. Fully mobilize institutions across the whole society in creating an integrated STEM innovation ecosystem. Alongside efforts by government, schools, manufacturers, and high-tech enterprises, encourage social organizations (such as Children's Palace and Youth Palace, science and technology museums, and digital media enterprises) to open spaces to provide a broader learning platform. Further integrate innovative personnel training, engineering technology education, and entrepreneurship education into the framework of the National Science and Technology Management Platform. In future local STEM platform action plans, such an ecosystem will undergo a complex and iterative process, and slowly build up.
- Promote the success of STEM education on a larger scale: Share excellent practice cases; provide quality resources; introduce first-class STEM experts; organize special trainings, project discussions, field guidance, and other activities. Combine STEM education with school reform, maker education, community practice, etc. by further expanding the scope of the pilot, widening the path of social participation, and strengthening advocacy and funding.

#### 14.5.1.1 Initiatives of Zheijang Province

In September 2017, the Department of Education of Zhejiang Province launched the recommendation of seed schools for STEM education projects in its schools. Relying on the Seed School project, participants learned from the successful experience of STEM education in Indiana and promoted in-depth implementation of the Zhejiang–Indiana Primary and Secondary School Curriculum Translation Project, with the goal of innovating the talent training model and improving the quality and level of STEM education. Ultimately, 15 schools in the province were identified as seed schools, and 15 as breeding schools for STEM projects (Department of Education of Zhejiang Province 2017).

In March 2018, the Department of Education of Zhejiang Province organized 15 overseas training programs in the dynamics of educational reform for kindergarten, primary, and secondary school teachers. The educators were sent to countries with well-developed basic education programs to learn advanced educational teaching concepts and practices, with a view toward enhancing teaching and management as practiced in the province's schools (Department of Education of Zhejiang Province 2018a).

- The program in *Germany* provided special training in industrial manufacturing informatization technology for key teachers in secondary vocational education. Trainees received up-close experience with advanced teaching and management in Industrial Manufacturing 4.0, which concerns information development in German vocational education. They got a first-hand feel for the underlying vocational education philosophy, the teaching environment, and the appeal of the teaching methods, and they came away with an understanding of the cutting-edge knowledge, technology, and skills required for industrial manufacturing IT.
- The content of the advanced training program that participants attended in the *United States* includes the same case and practice paradigm as the course in Germany, as well as strategies for STEM teacher training, the mechanism and means of STEM resource construction, the technology and method of STEM evaluation, etc.
- All training funds are supported by provincial financial departments.

April 2018 both the first STEAM education conference and the inauguration ceremony of the STEAM Education Synergy Innovation Center in Zhejiang Province were held (Zhejiang Province Teaching and Research Network 2018). In his speech, the Director of the Teaching and Research Department of Zhejiang Provincial Department of Education Xuebao Ren said that the purposes of actively promoting the STEAM project are twofold: to increase awareness of the significance and value of the STEAM project, and to practice project strategy.

In a keynote conference report, Mengzhou Xu, Deputy Director of the Business Administration and Research Department of Zhejiang Provincial Party School, detailed the advantages of, challenges to, and measures of high-quality development for Zhejiang's economy over the background of increasingly fierce global competition.

A second keynote report, by Feng Zhang, Deputy Director of the Teaching and Research Department of Zhejiang Provincial Department of Education, reviewed the early development of STEAM education in the province. He believes that the overall goals of the STEAM education promotion strategy are to consolidate the science and mathematics curricula of small and medium-sized primary schools, strengthen the mastery of technical application, effectively promote research-related learning, and cultivate students' innovative spirit and practical ability. STEAM education is to complement the child's basic education with hands-on, practical learning while promoting the reform of teaching methods and improvements in basic curriculum learning (China Education Online Zhejiang Station 2018).

Sub-forums for AI, scientific inquiry, and technology and engineering design were arranged at the meeting. Influential university experts in the field and teaching and research practitioners were invited to make reports. The meeting also arranged for an online parade of STEAM education practices to showcase the achievements of STEAM education exploration in schools in Zhejiang Province, and to refine the practical characteristics and working ideas of STEAM education in schools. The online component garnered nearly 10,000 live visits (China Education Online Zhejiang Station 2018).

#### 14.5.1.2 STEM Education Cooperation Between Zhejiang Province and University in U.S.A.

In June 2018, the Department of Education of Zhejiang Province signed a memorandum of understanding with Kean University of Union, New Jersey. According to the agreement, the two sides will engage in STEM curriculum development, teacher training, sister school pairings, academic seminars, and other cooperative projects (Department of Education of Zhejiang Province 2018c). With the approval of the Ministry of Education, Kean University and Wenzhou University partnered to formally establish Kean University-Wenzhou in 2014. To further deepen educational cooperation, Department of Education of Zhejiang Province prioritizes friendly relations between Zhejiang and New Jersey and gives full play to the bridging and linking roles of Kean University-Wenzhou.

#### 14.5.1.3 STEM Education Cooperation Between Zhejiang Province and State of U.S.A.

In October 2018, the Department of Education of Zhejiang Province and the Indiana Department of State renewed a memorandum of understanding on educational cooperation. Later that year, they jointly held the Zhejiang–Indiana Educational Cooperation Exchange and the opening ceremony of the Zhejiang–Indiana STEM Instructional Research Center in Hangzhou. Five schools in Hangzhou have signed sister school cooperation agreements with three Indiana schools.

Since entering into friendly relations in 1987, Zhejiang and Indiana have jointly established more than 50 sister-school partnerships. More than 500 teachers from Zhejiang Province have participated in training in Indiana. In the field of STEM education cooperation, the Department of Education of Zhejiang Province, and Indiana State University jointly set up the Zhejiang–India STEM Course Translation Project. Through this project, the government of Zhejiang province invites about 30 teachers and experts from Indiana State for STEM course teaching each summer. More than 1500 teachers have been trained.

In the same time period, at the second STEM Education conference in Jiangsu Province, the Jiangsu STEM Education Collaborative Innovation Research Center officially released the first provincial basic education STEM Course Guidance Outline, which standardizes and guides the healthy, high-quality development of STEM education (Suzhou Industrial Park Teacher Development Center 2018).

In December 2018, the Education Board of Zhejiang Province issued a three-year action plan for provincial educational informatization (2018–2020) to speed up the digital transformation and development of education in the province and to modernize education. The plan calls upon STEAM courses to cultivate students' core accomplishments, enhance learning inquiry, and build a diversified collection of curriculum resources to promote the new ecological development of education in line with the needs of "digital Zhejiang." These goals serve the needs of the current major industrial clusters in Zhejiang Province, as well as emerging and future economic agents (Department of Education of Zhejiang Province 2018b).

The initiative in Zhejiang Province links STEAM education promotion and talent training to the development of enterprises, industries, and economies in the region. Many education sectors provide the impetus for this trend, particularly at the regional level. The Department of Education of Zhejiang Province is a primary promoter of STEAM education. Most important is the Education Department's communication and contact with field experts and industry leaders, which has moved STEAM education in Zhejiang Province forward. The communication between the education sector and sectors in related industries provides the conditions for further synergistic development of regional industries and STEM education.

### *14.5.2 STEM Total Solution Led by Enterprises in Shanghai*

Scientific research and educational enterprise cooperation is a bright spot in STEM development. Under the influence of STEM-related education policy, educational enterprises actively prepare STEM products based on market demand and cooperate with scientific research institutions to develop STEM companies. Some STEM enterprises have had a good impact and reached a certain scale. Their products and services have played a positive role in supporting and promoting STEM education in schools, and have also met the needs of families for STEM education to a certain extent.

The Shanghai STEM Cloud Center, supported by the Shanghai Association of Science and Engineering and East China Normal University, was established in 2014. With the support of geoscience research institutions, the Cloud Center fills many unmet needs by providing schools, institutions, teachers and students with STEM services, most especially top-level project design, course development, the hardware required for the courses, teaching materials, and high-quality teacher training (Shanghai STEM Cloud Center 2017). The ultimate goal of the teacher training component is to enable teachers to design STEM courses on their own, so that STEM programs actually land in schools.

The Cloud Center's services encompass the fields of life sciences, material science, technology and design, earth and environmental sciences, and social and behavioral sciences. Its STEM project is grounded in the development of students' STEM literacy, the construction of a STEM laboratory where youth can form a creative exchange community, and STEM research literacy training courses. Students' comprehensive STEM literacy is monitored and evaluated in order to improve the growth of STEM teachers; on the basis of the findings, a STEM education model suitable for use in schools is then designed.

School-oriented STEM service is the core business of the Cloud Center, which has reached more than 400 schools to date, with a 98% school user renewal rate. This success can be attributed to its effectiveness in truly landing STEM education in the school and its attention to the needs of relevant parties in the school. The Cloud Center accomplishes this first by collaborating with the school on a deep level. The school is first investigated, and then top-level STEM education design is tailored to the current situation and the needs of managers. Secondly, the Cloud Center continuously conducts course iterations to meet the school's need for curriculum updates and upgrades, and to satisfy its teaching and research requirements.

From its inception, the Cloud Center's work has focused on teaching and research. It offers more than 300 courses, and more than 30 full-time teachers are responsible for curriculum development. At the same time, the Cloud Center absorbs excellent courses designed independently by school teachers. It also meets the needs of principals, students, and other multi-users. In addition, like most institutions getting entry of school, the Cloud Center is in cooperation with channel merchants to gain entry into the school. Because of the high gross margin and continuity relative to hardware product services, the Cloud Center is able to maintain good, cooperative relations with channel operators and to take some initiative.

The services implemented at the Shanghai STEM Cloud Center include support and assistance from a wide range of sources, including the Shanghai Association for Science and Technology (such as Tongji University Science and Technology Association, Shanghai Shipyard Science and Technology Association, Baosteel Group Co., Ltd. Science and Technology Association), the Shanghai Sci-Tech Museum, and the Shanghai Youth Science Society. The STEM Education Service at Cloud Center has been recognized by families and schools, which has changed the attitude of customers and attracted the attention and support of the government and all sectors of society. Table 14.3 presents a few examples of four types of changes initiated by the Shanghai STEM Cloud Center in schools that take advantage of its services.

**Table 14.3** Changes regarding STEM education in Shanghai

| Change in parent's view of STEM education | Innovation of school education model | Support of government education policy | Supply of social education resources |
|---|---|---|---|
| STEM literacy is necessary for children's future development | Practical courses are carried out in a planned manner | Implementation of STEM courses is actively advocated and encouraged | Respond and support |
| STEM learning is different from traditional knowledge learning | Project-based teaching and learning is applied to help children integrate knowledge | STEM courses and activity modes/ arrangements are specified | React and incorporate |
| | STEM courses are integrated into traditional classes | | Coordinate and carry out |

The Shanghai STEM Cloud Center's overall approach to STEM education has become an important force in the development of basic STEM education in Shanghai. The choice, orientation, and implementation of Shanghai's education product companies cannot be separated from the city's strategic role in using scientific and technological innovation to promote the regional economic growth and urban development. In April 2019, over the background of smart city construction, SAIL, a Shanghai organization promoting AI, was instrumental in inaugurating the Shanghai Artificial Intelligence Development Alliance, the World Artificial Intelligence Innovation tournament, and the first AI pilot application scene in Shanghai. Dignitaries from Shanghai's municipal government, academia, a Consulting Expert Committee, and other experts and leaders attended the events. The Alliance and tournament were launched to create a new plateau of intelligent development with industry and a new ecological environment in Shanghai. The functions of the Alliance are to gather global AI talent; absorb AI technology, industry, applications, research, investment, and financing; and to invite other enterprises and institutions to join in promoting the application of AI using cutting-edge core technology.

At the same time, the brand recognition enjoyed by the tournament allows the Alliance to explore the formation of a long-term AI innovation ecosystem. The resources of enterprises and experts gathered together secure the professionalism and international influence of the events. Gradually, a pattern of "government guidance, enterprise operation, academic interaction" formed, which can be explored to create a new ecology for Shanghai (Shanghai Center for Economic and Information Development Research 2019, Launched…, para. 3). Over such a background, the market and prospects of Shanghai's STEM companies will improve. With the support of policy and the organization of departments, the development of the regional economy and of science and technology policy is bound to gradually connect with the regional STEM forces to form a common force.

### 14.5.3 *Educational Policy Leading Complementary Teaching and Enterprise Training*

In its actual implementation, China's STEM policy and action plan are characterized by (1) a comprehensive curriculum and interdisciplinary integration, (2) practical problem solving through project-based activities, (3) stimulation of curiosity and development of hands-on skills, and (4) interactive learning that fosters teamwork. STEM education as implemented by educational enterprises is of higher quality than that provided in schools (Create a Headline 2018). In the future, formal school-based STEM education will increase in scale and talent training effect.

#### 14.5.3.1 Fast Organization of STEM Instruction in Schools

STEM projects, guided by education policy and promoted by the education sector, are easy to organize, scale up, and implement quickly in regional and urban schools. Since 2014, the State Council and the Ministry of Education have issued a number of policy documents, and various localities have introduced policies to cooperate with STEM education, which actively promote the development of STEM education (Create a Headline 2018), as shown in Table 14.4.

In 2017, the China Academy of Educational Sciences established the STEM Education Research Center in Beijing to implement innovation-driven development strategy and play the leading role in promoting scientific and technological innovation and improving national competitiveness.

The China STEM Education 2029 Action Plan, described earlier, plays a role in this organization of STEM schools. The plan provides a platform for policy, theory, research, practice, strategy, content and evaluation. The Action Plan started with a total of 79 pilot schools, 228 seed schools, and 76 seed teachers.

Governments provide three types of economic support at all levels to promote STEM education in schools: subsidies according to number of students; reasonable remuneration for teachers (the preferred method of Shanghai, Shandong, and Guangxi provinces); and a separate subsidy for the school (the preferred method in Xi'an and Zhejiang provinces).

However, while the market capacity is huge and the market drivers are clear, in-school STEM teaching has not emerged as a benchmark. Due to the lack of uniform standards, STEM education is carried out in a variety of ways in different regions and schools, including compulsory courses, elective courses, interest classes, laboratories, and 3D printing. In the procurement of STEM-related services, independent school choice is extremely strong. Educational enterprises have difficulty bringing their services into schools, resulting in delays in the development of STEM training in schools. Table 14.5 is a collation of the ways STEM education is currently carried out through several of the more common methods.

**Table 14.4** STEM policy documents and number of schools affected in major cities since 2014

| Issue date | Development area | Policy file name* | No. Schools affected |
|---|---|---|---|
| January 2014 | Beijing | Guiding opinions on doing a good job in after-school service for primary and secondary students | 984 |
| September 2014 | Shanghai | The standard of science curriculum in primary schools of compulsory education | 1141 |
| September 2015 | Jiangsu | Guiding opinions on school construction of STEM education projects in Jiangsu Province (for trial implementation) | 243 |
| June 2016 | Shandong | Guiding opinions on the construction of creative spaces in schools in Shandong Province | 200 |
| October 2016 | Shenzhen | Guide to the construction of creative education practice rooms in schools in Shenzhen (for trial implementation) | 300 |
| November 2016 | Xi'an | Implementation of project in Xi'an city to promote the maker education practice room in 500 schools before 2020 | 500 before 2020 |
| April 2017 | Henan | Outline of medium- and long-term Education development plan in Henan Province | 100 |
| August 2017 | Guizhou | Guiding opinions on the construction of maker spaces in primary and middle schools in Guizhou Province (for trial implementation) | 96 |
| November 2017 | Xiamen | Notice on further strengthening the work of science and technology education in schools | 250 |
| December 2017 | Guangzhou | The Thirteen Five-Year plan for the development of education in Guangzhou (2016–2020) | 155 |

*File names translated by the author

**Table 14.5** How STEM education develops in schools

| STEM education development method | Teacher source | Representative areas |
|---|---|---|
| Compulsory course | Education bureau unified training for school teachers | Zhejiang Province |
| Elective course | Out-of-school teachers, supplemented by teachers in schools | Beijing |
| Interest class (optional or not) | STEM plus research center's unified training of teachers in schools (approved by Shanghai Municipal Education Commission) | Shanghai |
| Maker lab | Teachers in schools, as a supplement to off-campus teachers | Xi'an<br>Zibo |
| 3D printing course | Unified training of teachers in schools throughout the province | Yunnan Province (procurement of 3D printing equipment for 6000 schools) |

#### 14.5.3.2 Effective Performance/Implementation of STEM Enterprise Training

Enterprise-based STEM projects are characterized by easy implementation, high-quality courses, and relatively thoughtful service. These characteristics complement those of the school-based STEM organization in strength and scale. Like the services of the Shanghai STEM Cloud Center, enterprise-based STEM curriculum systems have relatively large numbers of users and have produced good results. Parents show a high degree of satisfaction with enterprise-led STEM training. Parents generally believe that, like art training, STEM training can inculcate scientific and technological knowledge in children, which can spark their interest in related fields (see Table 14.6 for a comparison of parents' perception of the traits of art training versus STEM training). At the same time, STEM training can also hone children's hands-on ability, focus, and independent problem-solving skills. The visible results of STEM training—for example, the building of educational robots by students—translate to a high degree of satisfaction for both children and parents.

However, the market share of enterprise-based STEM training is still not as broad as examination-oriented education and art training. Table 14.6 expands on this observation. Art training is highly accepted by parents, with high demand, much help to students in progression, continues to be practiced by students after the program, etc., while STEM training has low acceptance by parents, is low profile, is not popular, etc.

This is the general state, but the Shanghai STEM Cloud Center aims to change this state. In the future, as the demand for STEM-educated talents increases in the labor market, the relationship between STEM learning and further education will also be strengthened. This will drive growth in the STEM training market, at a pace dependent on how many directions and paths of employment the government creates. In the past two years, this development has taken a more definite direction. For example, in 2017 Zhejiang Province added IT to the list of subjects covered by the

**Table 14.6** Comparison of parent perceptions of STEM training and art training

| Comparison category | | Art training | STEM training |
|---|---|---|---|
| Demand analysis | Acceptance | High | Low |
| | Result | Dominance (graduation, competition) | Invisibility (work, competition) |
| | Progression help | Large (many art specialty students, many art institutions) | Small (few tech specialty students) |
| | After-class sticky | Lots of practice | Little contact |
| Supply analysis | Popularity | High (nationwide) | Low (mainly in big cities) |
| | Site requirements | Low | High |
| | Number of teachers | Many | Few |
| | Course content | Partially standardized | Non-standardized |

entrance examination, and in April 2018 the Ministry of Education authorized 60 institutions of higher learning to offer robot engineering as a major.

The content of STEM training in enterprises is largely divided between robotics and children's programming. The educational robotics training market is well established, while children's programming belongs to the emerging market. Robotics training is difficult to scale up due to high operational costs. The competition is extremely fierce. As a result, cross-regional expansion is difficult and the profit level is not high. According to market research, currently more than 95% of small institutions in the market have fewer than ten direct stores; fewer than ten institutions have more than 50 stores. At the same time, the key success factors for large-scale expansion in robotics training institutions number five: site selection capability, shopping mall relationships, operation ability, capital strength, and management ability. In the industry's current situation, it is not easy to fully possess all five keys to success.

The introduction of education policy has boosted the children's programming market, and general technology and IT have been added to the college entrance examination as an optional subject in some province like Zhejiang. This notwithstanding, after all, the IT test is not a must. The role of programming policy in boosting the market is limited. No other province has followed Zhejiang's lead regarding adding IT to the entrance examination, and the number of students affected is very limited. More importantly, if we compare the content of children's programming with the technical subjects tested on the college entrance examination, we will find that the two are not closely related. Therefore, STEM education is still developing and requires further observation.

#### 14.5.3.3 Education Sectors and Policies Are Vital to Linking STEM Education With Regional Industry Development

STEM teaching in schools and STEM training offered by enterprises each have their own conditions, as well as their own advantages and characteristics. The advantages of the two approaches are complementary in efficiency and benefit. Both STEM initiatives are based on education policy. The school-based STEM teaching in Shanghai and enterprise-based STEM training in Zhejiang both show that:

- Education policy calls for the implementation of STEM courses in schools, and calls on enterprises to provide STEM services
- The education sector is to promote the formal implementation of STEM teaching practice in schools
- The education sector is likely to be a leader in connecting regional STEM education with the development of regional industries or industrial clusters.

STEM companies are better able to meet the needs of the STEM market and promote the development of formal STEM education when they are supported by STEM education policies. The STEM Education Conference organized by the Zhejiang Provincial Department of Education is a good example of a venue for such support.

The purpose of the education policies issued by the national and regional education departments is to meet the needs of talents and other resources in the development of economy and industry on both the national and regional levels. This means that the development of school-based STEM teaching and corporate-based STEM services depends on education policies and situations. The Education Department's education policy is what implicitly connects national or regional personnel training with the demand for industrial development. Logically, it follows that the implementation of STEM education in Zhejiang, Shanghai, Jiangsu, and other places is inevitably connected with the development of local industries in those areas. It is simply that a direct connection regarding desired demands and potential values has not yet developed.

Owing to the state of current relationships between STEM education and industrial development in many parts of the country, we can expect it to be a long time before the Education Department takes systematic action that has an effect on personnel training and industrial development. In summary, the important reasons for this are:

- The development of regional industrial clusters is not mature and is undergoing change
- There is a lack of effective communication between education sectors and economic development sectors
- the development of formal STEM education requires adjustments in the handling of high school and college entrance exams
- many schools are not structured to interact with industry and others from outside the school
- the role of STEM education in meeting the demand for industrial talents is not clear, and its effects is not easy to identify
- the timeframe, manners, and conditions under which STEM talents enter industrial clusters are facing challenges

## 14.6 Progress

The launch of the China STEM Education 2029 Action Plan signifies that STEM education is being implemented, expanded, and gradually given a role in the cultivation of innovative talents through concrete measures and projects. China's STEM education is gradually beginning to develop, most notably in urban areas, but it has not yet had a direct effect on the economy. Nevertheless, preparations are being made for the long-term cultivation of the scientific and technological innovation talents needed for economic and industrial development. This goal is embodied in the policy and direction of STEM education, the practice and promotion of STEM by schools and enterprises, the organization and leadership of education departments, the research and support of institutions, and the germination of connections between schools and industry.

### 14.6.1 Leaders of Educational Sectors Start to Organize STEM Development

STEM initiatives carried out in developed provinces in 2018 amply illustrate that STEM activities organized by education sectors are a powerful tool for practicing STEM education within the framework of education policies. The practical examples from Zhejiang Province and Shanghai city revealed that, in addition to education policy, organization and guidance from the Education Department can coordinate the communication and cooperation of relevant departments, attract the attention of stakeholders, and have an effect on publicity and promotion. It is this interdepartmental communication led by the education sector that increases the opportunities for interface between formal and informal STEM education and industrial clusters. As communication deepens, STEM education will have more prospects for development. The efforts of departmental leaders have accelerated this process.

The regional and/or local cases of Jiangsu, Zhejiang, and Shanghai also illustrate the current organizational approach to promoting STEM education in the education sector, making use of policy documents, cluster meetings, competition activities, and the like for enhanced impact. In these ways, the education sector plays a leading role in school STEM teaching and enterprise services, and its action plan provides guidance for further STEM practice activities.

The implementation of STEM education by the Ministry of Education, the STEM Education Research Center, and regional centers has an underlying set of ideas which can be summed up as "finding the right direction, focusing on bottlenecks, proposing strategies, and formulating plans." In the iterative actions of focus, strategy, and planning, regions will gradually find the path to development and the means that best suit their own situations. The respective advantages of school teaching and enterprise service will also play a significant role.

### 14.6.2 STEM Education Is Entering the Embryonic Phase of Connecting with Industry Demand

The cases of Shanghai and Zhejiang Province serve as an important sign of the development of STEM education in those regions. That is, representatives of departments tasked with regional economic and industrial development, along with industrial enterprise leaders, participated in the local STEM development conferences, which resulted in more industry information and industry dynamics influencing the development of STEM education. The introduction of talent demand, industry technology bottleneck, industrial cooperation modes, and industry development dynamics provides more information for STEM schools and enterprise services and gives more direction to the Education Department's guidance of STEM education.

There is also a need for mechanisms for effective communication between the education sector and the economic sector and industrial organizations. In terms of the essence and original intent of developing STEM education, the overall advancement of STEM education at the national and regional levels cannot be separated from the joint collaboration between the education sector and industrial organizations. Such partnerships are budding in Shanghai, Zhejiang, Jiangsu, and other places, and this is a good start. However, issues related to cost, organizational constraints, institutional frameworks, and collaboration models remain to be resolved. The direct interface between STEM talent training and industry development needs is profoundly impacted by the development of and changes in regional industry, the timeliness of personnel training, the possibility of K-12 students contributing to industrial development, changes in employment market allocation, the restriction of education management and industry rules governing the connection between the two sides, and so on. Before the emergence of STEM, the direct connection between enterprises and school education was more embodied in vocational education than in the K-12 stage. If the direct connection between enterprises and K-12 school education is to be considered inevitable in the future, it is necessary firstly to develop STEM education to the point of in-depth application.

## 14.7 Lessons Learned

### *14.7.1 Impact of Educational Policies on STEM Instruction in Schools And STEM Service Enterprises*

The policy documents of the Ministry of Education, the improvement of teaching conditions, the strengthening of financial support, and the adjustment of relevant policies like those regarding employment will broaden the scope of STEM education. For example, the current entrance examination system provokes resistance to STEM development. This might change with the reform of examination standards, which is also beneficial to the cultivation of scientific and technological innovation talents. For example, a policy adjustment, such as employment and registered residence, inspires more interest in the cultivation of STEM innovative talents and makes it a principal attraction in the talent training market.

### *14.7.2 Organization of Education Sectors Is Vital to Linking With Industrial Clusters*

STEM education was originally intended to be closely integrated with industry and with economic and social development. STEM education heavily emphasizes innovative thinking and practical operation in projects and activities. It requires students to integrate the knowledge they gain from STEM courses into real problem solving, so that teachers and students can have the characteristics and qualities that distinguish scientific and technological innovation talents. STEM education in China has inherited these essential requirements from its early education policies, yet it has not been able to develop them. Nowadays, the development of STEM education is an opportunity to raise the level of students' scientific and technological innovation ability. There is also a general trend to work jointly with industrial clusters. Such signs can already be seen in the trajectory of regional development.

The experience in regional STEM education shows that education sector leadership is an important liaison between economic management and the industrial sector on one hand, and the Education Department on the other. In terms of domestic education, the bottom-up response model determines the promotion of regional education and requires the Education Department to take the lead in connecting the education sector and regional industry. Only with such leadership can institutions, companies, schools, teachers, and students achieve a greater vision and find the space to practice and implement STEM teaching/learning and services. For example, during project-style learning, the problems of test pressure, course time, learning space, and other external conditions must be solved in order that the shortages in curriculum, materials, teachers, and teaching and research may be addressed successfully. For instance, exam-related pressure on students has been reduced through the proposal of "reducing the burden" and examination system reform. Overall, STEM education is on track as work progresses.

### *14.7.3 Meeting the Needs of Teachers and Students and Ensuring Effective Implementation*

A profound insight from STEM training in enterprises is that if corporate STEM initiatives and activities are closely adapted to the actual teaching and learning situations of schools, teachers, and students, then STEM education can develop healthily within the scope of the school unit (Zhihu 2017). If these needs are kept in mind, enterprises and schools can carry out STEM activities together in realistic environments and conditions.

It is beneficial for participants in school-based STEM teaching and enterprise-driven STEM training to observe the actual effects of their activities. STEM project schools in Zhejiang Province have seen their STEM initiatives begin to bear fruit. Stakeholders related to policy, organization, participation, implementation, and other topics can benefit from awareness to further enhance their own focus on STEM propulsion and participation. The 98% school user renewal rate enjoyed by the Shanghai STEM Cloud Center, and the positive response of the participants, can be attributed to the Cloud Center's having met the needs of schools, teachers, and students. Consequently, the number of its users continues to increase.

### *14.7.4 Linking to Regional Industrial Clusters and the Importance of Sample, Functioning Systems*

STEM initiatives in Shanghai, Zhejiang, and Jiangsu Provinces should be examined to identify both what they have in common and how they differ from each other. Beginning with Shanghai, the STEM Education Cloud Center and the STEM Plus Research Center provide holistic solutions to institutions, schools, teachers, and students. Regional education sectors interact to create an innovative ecosystem through sharing at conferences, development of alliances, innovation competitions, etc. In intelligent city development based on AI technology, the cultivation of scientific and technological innovation talents is receiving more and more attention, and STEM education is likewise playing a more important role in the province. Zhejiang and Jiangsu Provinces, in turn, have established the STEM Education Collaborative Innovation Research Centers. Zhejiang Province is more invested in overseas teacher training, whereas Jiangsu Province has made its mark by developing the Basic Education STEM Course Guidance Outline. In short, each region's approach to STEM education development has its own traits. This indicates both that the influential ideas in the STEM sector vary from region to region, and that STEM education must be developed regionally, in response to the salient characteristics of each individual region.

The overall promotion of STEM education with an eye toward synergy and openness points to the long-term strategy of driving economic development with science and technology innovation. For STEM education, which is dominated and influenced by the education sector, personnel training is the direct goal; the local economy and its industrial development are influenced by talent rationing policy and the talent market system. Innovation-oriented personnel training, supported mainly by school-based STEM education and enterprise-based STEM training, benefits local industries by improving the talent market. Education policy on personnel training also directly affects school-based STEM teaching and enterprise-based STEM training through its reaction to a direct or indirect talent market and an explicit or implicit economic market. Figure 14.1 graphically represents the relationship among sectors involved in STEM education.

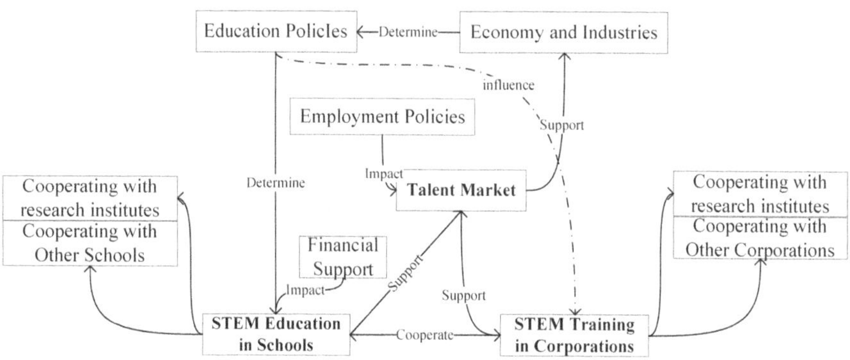


**Fig. 14.1** The relationship among industrial demand, educational policy, and STEM education

### 14.7.5 Building Policy, Systems, and Environments for STEM Education and Industrial Clusters

The trend toward technological and economic development and the demand for industrial development of regions in China is an important benchmark for the development of STEM education and the cultivation of innovative talents. However, neither of them is an absolute or direct point of reference, due to the influence from the change of regional industry, the timeliness of personnel training, the change of employment market allocation, the limitation of educational management and industry rules on the connection between the two sides, etc. That is, the main areas of regional STEM development and STEM talent training are not necessarily the only response to the development of regional industry. Talent market management and employment market institutions are also important and direct respondents.

Driven by the strategy of "Science, Technology, and Innovation Drive," the STEM policy and its Action Plan launched by the Education Department have finally put national economic and industrial development in their sights, albeit indirectly. The talent market and job market are the direct beneficiaries of current STEM promotion, which indirectly serve economic and industrial development. If there is a macroscopic and a microscopic direction in STEM education, then the national STEM policy, which indirectly points to the talent market and the job market, is the macroscopic one; STEM courses and learning that point to regional industries are microscopic. This requires the local economic and educational authorities to take the lead, with the specific needs of local economic and industrial development in mind, as they implement STEM curriculum design, teacher training, international cooperation, and other projects.

If the macroscopic direction of STEM education is characterized by nationwide policy and projects that influence economic and industrial development, then the microscopic direction is made up of specific projects or plans carried out on a national, regional, or citywide level according to the development needs of a specific or local industry. Simply put, the implementation of macro policies and projects is hard-pressed to promote STEM in ways that are closely integrated with industrial development needs, and it is not the most effective vehicle for combining the scientific and technological innovation needed by industries with STEM systems, mechanisms, content, strategy, and so on. However, with the promotion of STEM action and the development of the national economy and education, the development of local industry is bound to be compatible with education policy, with the direction of the country's economic and educational strategy. The microscopic orientation of education toward local economic development will also have time for concrete implementation. In other words, STEM promotion of the current macroscopic direction benefits the micro-direction with which STEM education is so closely integrated, and is also poised to receive the long-term direct economic benefits driven by scientific and technological innovation.

## References


Central People's Government of PRC (2010). Outline of national medium- and long-term education reform and development plan. Retrieved July 19, 2019 from http://www.gov.cn/jrzg/2010-07/29/content_1667143.htm. (In Chinese; translation by author).

China Education News Network (2017). China STEAM education development report. Retrieved July 19, 2019 from http://www.baidu.com/link?url=SzirUlGiYKiaIuaaFT8IoIxmkkp-r8PO-NJ0RagMAKCMLW8bgKDqIKLRXOb-ozTHFp3RCp1DOIxGC26-b555zK&wd=&eqid=81da5dd2000bd7ba000000035ce913aa. (In Chinese; translation by author).

China Electronics Society (2017). New generation artificial intelligence development white paper. Retrieved July 19, 2019 from http://www.cbdio.com/BigData/2018-02/26/content_5681858.htm (In Chinese; translation by author).

China Education Online Zhejiang Station (2018). First STEAM education conference of primary and secondary schools in Zhejiang Province successfully held, provincial STEAM Education Cooperative Innovation Center established. Retrieved July 19, 2019 from http://www.eol.cn/zhejiang/zhejiang_news/201804/t20180427_1597441.shtml. (In Chinese; translation by author).

China News Network (2018). China STEM education 2029 action plan launched in Beijing will train a group of innovative talents. Retrieved July 19, 2019 from http://www.chinanews.com/gn/2018/05-15/8514456.shtml. (In Chinese; translation by author).

China News Network (2019a). China STEM2029 Action Plan: High-quality STEM education institutions explore cultivation of innovative talents. Retrieved July 19, 2019 from https://baijiahao.baidu.com/s?id=1611937293456304711&wfr=spider&for=pc. (In Chinese; translation by author).

China News Network (2019b). Reading the government work report: China's economic transcript in 2018. Retrieved July 19, 2019 from http://www.chinanews.com/gn/2019/03-05/8771513.shtml. (In Chinese; translation by author).

CNR (2018). [Listen, Xi Jinping addressed on scientific and technological innovation] Grasping scientific and technological innovation, it has seized the "bull nose" of overall development. Retrieved July 19, 2019 from http://china.cnr.cn/news/20180528/t20180528_524248599.shtml. (In Chinese; translation by author).

Create a Headline (2018). The wind of policy has spurred China's STEM education. Who is better at school training and off-campus training? Retrieved July 19, 2019 from http://www.ctoutiao.com/857183.html. (In Chinese; translation by author).

Department of Education of Zhejiang Province (2017). Notice of the Office of the Department of Education of Zhejiang Province on publishing the list of seed schools and breeding schools for STEM education projects in primary and secondary schools. Retrieved July 19, 2019 from http://jyt.zj.gov.cn/art/2017/12/18/art_1532973_27485255.html. (In Chinese; translation by author).

Department of Education of Zhejiang Province (2018a). Notice of the Office of the Department of Education of Zhejiang Province on the selection of students for overseas training projects of primary and secondary school teachers in 2018. Retrieved July 19, 2019 from http://jyt.zj.gov.cn/art/2018/3/23/art_1532982_27487428.html. (In Chinese; translation by author).

Department of Education of Zhejiang Province (2018b). Notice on issuing the three-year action plan for education informatization in Zhejiang Province (2018–2020). Retrieved July 19, 2019 from http://www.zjedu.org/art/2018/12/19/art_275_35559.html. (In Chinese; translation by author).

Department of Education of Zhejiang Province (2018c). Zhejiang Province will work with Kean University in USA to promote STEM education. Retrieved July 19, 2019 from http://jyt.zj.gov.cn/art/2018/6/6/art_1543973_26234527.html. (In Chinese; translation by author).

Education Management Information Center of Education Ministry (2017). China STEAM Education development report. Retrieved July 19, 2019 from http://www.shaoerbc.org/book/2017steam.pdf. (In Chinese; translation by author).

International Daily (2018). China STEM Education 2029 Action Plan. Retrieved July 19, 2019 from http://www.guojiribao.com/shtml/gjrb/20180516/509014.shtml. (In Chinese; translation by author).
Ministry of Education (2016a). Ministry of Education and 11 other departments on promoting the study of primary and secondary school students. Retrieved July 19, 2019 from http://www.moe.edu.cn/srcsite/A06/s3325/201612/t20161219_292354.html. (In Chinese; translation by author).
Ministry of Education (2016b). Ministry of Education, Education Informatization, Thirteen Five-Year Plan. Retrieved July 19, 2019 from http://www.moe.gov.cn/srcsite/A16/s3342/201606/t20160622_269367.html. (In Chinese; translation by author).
Ministry of Education (2017a). Compulsory education primary school science curriculum standards. Retrieved July 19, 2019 from http://www.moe.gov.cn/srcsite/A26/s8001/201702/t20170215_296305.html. (In Chinese; translation by author).
Ministry of Education (2017b). Curriculum guidance for comprehensive practical activities in primary and secondary schools. Retrieved July 19, 2019 from http://www.moe.gov.cn/srcsite/A26/s8001/201710/t20171017_316616.html. (In Chinese; translation by author).
Ministry of Science and Technology of the PRC (2006). Outline of the national medium- and long-term science and technology development plan (2006–2020). Retrieved July 19, 2019 from http://www.most.gov.cn/kjgh/kjghzcq/. (In Chinese; translation by author).
National Institute of Education Sciences (2017). China STEM education white paper. Retrieved July 19, 2019 from https://wenku.baidu.com/view/f43a9834f68a6529647d27284b73f242326c3150.html. (In Chinese; translation by author).
Observer (2017). State Council: Primary and secondary schools should set up artificial intelligence courses to promote programming education. Retrieved July 19, 2019 from https://www.guancha.cn/TMT/2017_08_28_424678.shtml. (In Chinese; translation by author).
Sadie Think-Tank (2018). 2019 China industrial and informatization development situation outlook series report. Retrieved July 19, 2019 from http://www.cbdio.com/BigData/2018-12/24/content_5964043.htm. (In Chinese; translation by author).
Shanghai Center for Economic and Information Development Research (2019). Shanghai focuses on creating artificial intelligence innovation ecology to create new AI highland. Retrieved July 19, 2019 from http://www.sheitc.org/Articles/view/1803/19. (In Chinese; translation by author).
Shanghai STEM Cloud Center (2017). STEM Education's overall solution. Retrieved July 19, 2019 from https://wenku.baidu.com/view/cbf97d0566ec102de2bd960590c69ec3d5bbdb84.html?sxts=1551000430961. (In Chinese; translation by author).
Sina Education (2017). Seminar on the integration of project teaching methods and primary and secondary school courses held in Beijing. Retrieved July 19, 2019 from http://edu.sina.com.cn/zxx/2017-06-26/doc-ifyhmtek7779468.shtml. (In Chinese; translation by author).
Sohu Network (2019). The key to the success of STEM education: Teachers' professional development. Retrieved July 19, 2019 from http://www.sohu.com/a/250984881_534881. (In Chinese; translation by author).
State Council of the PRC (2017). A new generation of artificial intelligence development planning. Retrieved July 19, 2019 from http://www.gov.cn/zhengce/content/2017-07/20/content_5211996.htm. (In Chinese; translation by author).
Suzhou Industrial Park Teacher Development Center (2018). Notice on forwarding "Guidance outline (trial) of STEM course of basic education in Jiangsu Province." Retrieved July 19, 2019 from http://tdc.sipedu.org/Item/21841.aspx. (In Chinese; translation by author).
Wang, X. (2015). Study of the creator education model for STEM education. *China Educational Technology, 8*, 36–41.
Wu, S. (2018). Innovative talents come from innovative basic education. Retrieved July 19, 2019 from http://epaper.gmw.cn/gmrb/html/2018-09/23/nw.D110000gmrb_20180923_2-06.htm. (In Chinese; translation by author).
Xinhua News Agency Network (2017). Education research center of the Chinese academy of educational sciences established in Beijing. Retrieved July 19, 2019 from http://education.news.cn/2017-06/06/c_129626460.htm. (In Chinese; translation by author).

Zhao, L. (2019). The study on Chinese industrial cluster development path based on network forms of organization: Based on the experience of industrial cluster development in developed countries. *Science & Technology Progress & Policy, 36*(7), 56–70.
Zhejiang Province Teaching and Research Network (2018). Zhejiang's first primary and secondary school STEAM education conference & synergy innovation center inaugural ceremony held in Hangzhou. Retrieved July 19, 2019 from http://www.zjjys.org/jdxw/7055.htm. (In Chinese; translation by author).
Zhihu (2017). How to start a business based on STEM education and its business model in China. Retrieved July 19, 2019 from https://www.zhihu.com/question/50260939. (In Chinese; translation by author).
Zhongxu Future Education (2018). "Xuhui City Laboratory" donated to Hangzhou Mark middle school in Asahi World to promote the development of STEAM education in schools. Retrieved July 19, 2019 from http://www.chnfedu.com/index/article/id/82. (In Chinese; translation by author).

# Chapter 15
# Case Study: STEM Contribution in Indian IT Clusters

**Sang C. Suh, Hemanth Bandi, Jinoh Kim, and U. John Tanik**

**Abstract** This chapter provides a general introduction to STEM (Science, Technology, Engineering and Mathematics) education in India and how it has helped to establish IT clusters. Technopolis cities in India were especially affected by STEM education, which led to the emerging trend of integration of STEM education and technology policy in select regions in India. Policies were implemented to support STEM education in primary schools through Ph.D. level, in addition to Indian government development of STEM models to support twenty-first century technology development and education reform. This chapter concludes with common as well as unique aspects of STEM education in the technopolis of India.

## 15.1 About This Chapter

This present study emphasizes how STEM education contributed to the development of the technopolis cities in India, demonstrating the emerging trend of integration of STEM education and technology policy in select regions across the country. This chapter also explores the theories and models established by the Indian government since the 1980's as forecasted for twenty-first-century skills development, learning theories, and education reform. This chapter also summarizes the evolution of STEM education in India and how STEM initiatives have helped to establish IT clusters across the nation. Unlike most other case study chapters in this volume, which mainly show the growth of STEM secondary education as driven by regional policies, this chapter provides an overview of multiple policies with a majority of them at the national level, covering the full spectrum of education from primary school through the PhD level. A conclusion section highlights common themes,

S. C. Suh (✉) · J. Kim · U. J. Tanik
Texas A&M University-Commerce, Commerce, TX, USA
e-mail: Sang.Suh@tamuc.edu

H. Bandi
McKesson Corporation, Richmond, Virginia, USA



C. Zintgraff et al. (eds.), *STEM in the Technopolis: The Power of STEM Education in Regional Technology Policy*,
https://doi.org/10.1007/978-3-030-39851-4_15

summarizes views of STEM education particular to India, and suggests takeaways for STEM K-12 educators, STEM college educators, industry professionals, and non-profit leaders and policymakers.

## 15.2 Route of STEM Evolution

The Republic of India is the second most populous country in the world and the largest democracy. Over the past few decades, India's technology sector has experienced growth through the formation of information technology (IT) clusters due to large number of STEM graduates, low labor costs, and well-developed infrastructure. This spurred the government of India to build technology parks across different metropolitan cities (Somesh K Mathur, 2008). For instance, Sonner tech parks became IT clusters hosting offices of many companies in one place. STPI (Software Technology Parks of India) have experienced high growth over the last two decades which made Multi National Companies (MNCs) prefer India for outsourcing their IT services. This could be attributed to Indian students choosing a STEM education path as early primary school. Under the Right to Free Education Act, which went into effect on April 1, 2010, all children between the ages of 6 and 14 were bestowed by the government the opportunity to free and compulsory education as a fundamental right (The Right of Children to Free and Compulsory Education Act, 2009). As a result, 229 million students were enrolled in accredited urban and rural schools across the country in just three years.

IT jobs require a broad range of skills in problem-solving, computational thinking, statistical modeling, and abstractive design, which can be gained mostly from topics in the STEM discipline. To attract MNCs and create IT service clusters, India needed a huge number of technically skilled graduates. The government of India introduced various policies to promote STEM education in the 1980's. By then, STEM education in India was focused primarily on science and mathematics. The Indian STEM education system did not originally concentrate extensively on technology and engineering, which did not help graduating students to advance their technical skills. However, this emphasis changed rapidly in the last decade, as Indian education has become more of a practical research and development effort rather than theoretical. As practical considerations increased, one of the challenging tasks the government of India faced was to increase student enrollment in STEM courses at the secondary education level.

### *15.2.1 STEM Philosophies Served*

Many future trends forecasted that having a career in STEM would lead to the best jobs globally (Tulsi & Poonia, 2015). The government of India decided to initiate schemes to identify and promote scientific knowledge from the primary school level. Although the government of India implemented different strategies to improve STEM education while motivating students to enroll in scientific research courses, the lack of teaching staff has been a significant concern. To ameliorate that, many funding opportunities were initiated to train primary and secondary level teachers. One approach was to make sure secondary school teachers take a Computational Thinking test, which covers basic problem-solving skills, conditional logic, debugging, and systematic error detection.

Many traditional jobs have manpower being replaced by automation tools. Technology is transforming the world, and all countries are trying to become technological world leaders. To train students in skills that approach those that very large IT companies are looking for, there should be a curriculum covering the basics of computer sciences from primary school to university. Students who choose a STEM major when they enter college are more likely to succeed if they are already proficient in STEM skills. Students who were lagging, especially in math, may be hard hit during major STEM coursework (Benedict, 2018).

The All India Council for Technical Education (AICTE) recently decided to begin a training program for faculties of technical institutes. AICTE chairman Anil Sahasrabudhe said that Minister of Human Resources Development officially launched the new curriculum for the technical institutions faculty. (AICTE, 2018).

### *15.2.2 Education System in India*

The Indian system of higher education is the third largest in the world. The technical education system is comprised of 8462 institutions. Those institutions offer degrees in engineering and technology. 3524 institutions offer a diploma in engineering and technology, and they can teach 3.4 million students per year (Tulsi & Poonia, 2015). Students between the ages of 6 and 10 attend primary school (grades 1–5). Students 11–13 years of age attend upper primary school if they have passed all the exams in the lower grades. Both primary education and upper primary education are compulsory for all citizens in India. Students aged 14–17 receive secondary education, which includes most of the technical training institutes as well. Figure 15.1 explains the segmentation of the education level.

As stated above, enrollment in primary and upper primary education is rising, but the same cannot be said for secondary education. There are several reasons for these declines, including high tuition fees in private schools and lack of resources in public schools. The government has noted these statistical trends and has also started to improve the higher education system. With this goal in mind, the government is implementing many reforms and bringing in more IITs, NITs, and IIITs (Indian institutes of technology, National Institute of Technology, and Indian Institute of Information Technology, respectively). The Government of India also started e-learning in 2010 which helped to improve Information and Communication Technology in higher education (Astha Dewan, 2010).

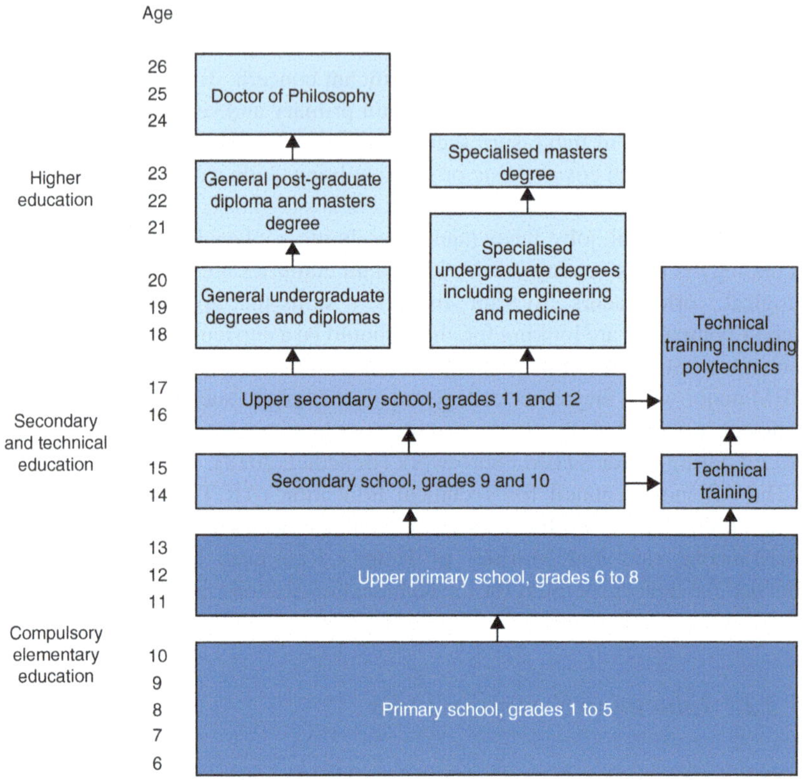


Note: Grade and age profiles vary across states and duration of higher education courses varies by discipline.
*Source*: Ministry of Human Resource Development.

**Fig. 15.1** Education system in India (Nanda, 2018)

## *15.2.3 Software Technology Parks of India*

Software Technology Parks of India (STPI), which provide internet and incubation services, was established by the Indian Ministry of Electronics and Information Technology in 1991 with the objective of encouraging, promoting, and boosting the software exports from India. STPI played a vital role in improving country's GDP.

### 15.2.3.1 Policies Support STPI

The policies discussed below are noteworthy in the ways they support STPI.

Maharashtra's Information Technology/Information Technology Enabled Services (IT/ITES) Policy-2015: The Government of Maharashtra announced its first IT Policy in 1998. It was followed by the Information Technology and Information Technology Enabled Services (IT/ITES) Policy-2003 and IT/ITES Policy-2009 to generate employment, increase efficiency and improve the quality of life. The IT/ITES Policy-2009 came into force from 29 August 2009. (Government of Maharashtra, Draft of Maharashtra's Information Technology/Information Technology Enabled Services Policy (IT/ITES), 2015). In Fiscal Year (FY) 2004, employment in India's IT sector stood at 830 K; by FY 2014, it had nearly quadrupled to 3 M.

The main objective of IT/ITES Policy-2015 is to preserve Mumbai's status as the country's top IT cluster, which would also enable the state government to overcome the problem of unemployment. Initiating this policy would also help enable Maharashtra as one of the preferred IT corridors in India.

Maharashtra State Innovative Startup Policy: This policy aims to drive economic growth and job creation in the coming years by preparing more entrepreneurs and designing more facilities for advanced technologies such as artificial intelligence, deep learning, machine learning, cloud computing, and the Internet of things.

As part of introducing this policy, the Government of Maharashtra shared its vision: to adopt a holistic approach "including establishing a network of incubators, cultivating entrepreneurial mind-sets among students [by developing infrastructure and framework for startups], connecting relevant stakeholders, simplifying the regulatory environment and making strategic investments to foster entrepreneurship across the state" (Maharashtra State Innovation Society, 2018). This could extend the creation of IT parks/clusters across the state, which in turn would help to improve the state's economy. This policy apparently helps Maharashtra to be one of the premier destinations for startups.

Information Technology Investment Policy 2015–20: The main objectives of this policy of the Government of Goa are to create world-class IT infrastructure in the state and to enlist a minimum of ten MNCs in establishing STPI. This policy also aims to promote a "modern, clean, green, eco-friendly working environment." Goa offers several incentives, including an industry-friendly state government and a peaceful social environment, to attract IT companies to establish IT parks in the State (Government of Goa, 2019).

## 15.3 Indian STEM IT Clusters

### *15.3.1 Bangalore IT Cluster*

Bangalore was the first city in India to develop an IT cluster, which has developed substantially since the liberalization reforms introduced in 1991. Most IT firms of Indian origin, including Infosys and Wipro, have their headquarters in Bangalore. The city has more state technology parks than any other state across the country, helping small firms and MNCs to open offices.

The top IT technology parks located in Bangalore are: Electronic City Phase I, Electronic City Phase II, Electronic City Phase III, Bagmane Tech Park, Manyata Tech Park, International Tech Park (ITPL), Prestige Group, RMZ Infinity, Ecospace Business Park, Global Village Tech Park, Embassy Tech Village, and Embassy Golf Links Business Park.

The growth of Bangalore's IT cluster has been a significant factor in the expansion of the city's population. This growth creates public problems such as high traffic and environmental pollution. In fact, Bangalore is no longer called the "Garden City" because so many of its trees were chopped down to build IT cluster infrastructure. Consideration is now being given to moving some companies to Tier II cities in Karnataka, such as Mysore, Mangalore, Belgaum, and Hubli-Dhanbad. Mangalore and Mysore have all the facilities needed to develop IT clusters in their cities. This will not only address problems in the Tier I city of Bangalore but will also drive much-needed employment and development in the Tier II cities.

### 15.3.2 Chennai IT Cluster

Chennai, a metropolitan city, the capital of Tamil Nadu state, is the largest industrial and commercial center of southern India. Chennai became an IT hub because of educational institutions, great infrastructure and software parks. The Tamil Nadu government implemented STEM as a pilot project across 320 schools in 2015-2016 and 2016-2017 fiscal years under the Rashtriya Madhyamik Shiksha Abhiyan (RMSA) scheme (more below) which produced skilled STEM graduates. The government invested in building IT Parks and Special Economic Zones that encouraged MNC's to choose Chennai.

The top IT technology parks in Chennai are DLF-IT Park (SEZ), Olympia Tech Park, SRM Tech Park, Shapoorji Pallonji Infocity (SP Infocity), TIDEL Park, ETL Infrastructure, IITM Research Park, Prince Info Park, and TCG Group.

### 15.3.3 Mumbai IT Cluster

Mumbai, formerly known as Bombay, capital of Maharashtra, is the most populous city of India, with 12.4 million as per the 2011 Census. Five Fortune 500 companies are in Mumbai. Policies such as Maharashtra State Innovative Startup Policy helped to improve IT architecture and created new jobs. Amravati, Nagpur, Aurangabad, Nashik, Bhiwandi, Pune, Solapur, and Kolhapur are Tier II cities in Maharashtra. Due to the population growth of Mumbai, the government of Maharashtra is focusing on attracting IT companies to locate in Pune, whose infrastructure and transportation facilities position it as an excellent location for IT service clusters (Jadhav, 2013).

## 15.4 Indian STEM IT Policies

The twenty-first century is marked by competition as countries around the world challenge each other through technological growth. To take the lead over other countries in the world, the Indian government has implemented many programs across primary and secondary school education to improve students' computational and interpersonal skills. Some of the programs are discussed below.

### *15.4.1 Rashtriya Avishkar Abhiyan*

In July 2015, Dr. A.P.J. Abdul Kalam, former President of India, launched Rashtriya Avishkar Abhiyan (RAA) under the auspices of the Ministry of Human Resource Development (MHRD) (Press Information Bureau, 2015). The objective of this program, which is in place in all primary and secondary schools, is to "encourage children to enter science and mathematics courses through exposure to/participation in science laboratories, technical competitions, exhibitions, and science museums" (Shastri Bhawan, 2015).

#### 15.4.1.1 Background

In 1964, the Kothari Commission reported that "technical learning should take place inside and outside the classroom, beyond textbooks" (National Education Commission 1964-66, 2015). In 1993, the Yashpal Committee expressed an interest in promoting "learning without burden" (Learning without burden, 1993). By that time, most of education in public schools was exam based, so students were merely preparing to pass examinations. That did not help students improve their aptitude and reasoning knowledge.

#### 15.4.1.2 Objectives

The RAA was conceived with the following goals (Shastri Bhawan, 2015)

- To motivate students to be engaged in STEM courses through innovation activities like experimentation, aptitude reasoning, model building etc.
- To make primary school students enthusiastic in STEM courses.
- To make use of advanced technology in STEM to teach school students.
- To create STEM clubs to facilitate student engagement with other technology leaders to help students understand scenarios in the real-world.

### 15.4.2 Atal Innovation Mission (AIM)

"Atal Innovation Mission (AIM) including Self-Employment and Talent Utilization (SETU) is the government of India's endeavor to promote a culture of innovation and entrepreneurship" (Ramanan, 2018). Under this initiative more than 10,000 schools are expected to establish technology labs by end of 2020. AIM has been structured to influence the STEM graduates by providing financial support in a sustainable manner.

### 15.4.3 Entrepreneurship Promotion

The Indian startup ecosystem has grown exponentially in recent years. Entrepreneurs Narayana Murthy, Azim Premji, Kiran Mazumdar-Shaw, and many others have paved the way for others to succeed on this path. The government of India is also supporting entrepreneurs through various ministries, such as the Department of Industrial Policy and Promotion, the Department of Science and Technology, the Department of Biotechnology, and the Ministry of Skill Development and Entrepreneurship. AIM's main purpose is to increase the number of entrepreneurs, which in turn will boost the country's economy. (Dubey et al., 2012) discusses the need of developing entrepreneurship programs for engineering students in India.

### 15.4.4 Innovation Promotion

AIM, a scheme promoted under NITI, is establishing Atal Tinkering Laboratories (ATL) in schools across India. The objective of this scheme is to increase interest, increase creativity, and engender imagination in young students and also to train a design mindset, computational thinking skills, adaptive learning skills, and physical computing skills (Ramanan, 2018). The All India Council for Technical Education (AICTE) is a national-level organization for technical education, headed by the Department of Higher Education, MHRD. In 2017, it was reported that up to 60% of engineering students in India had failed to secure post-graduation employment. Fewer than 1% of students in India participated in summer internships, and those internships can lead to stable jobs (Mckinsey report, 2018). To change this trend, the MHRD is planning a major revamp of India's STEM education system. All colleges and universities will need to adapt their curricula to the technology being used in the world and report these changes to AICTE. Once this is accomplished, most programs of higher education will be accredited by the National Board of Accreditation.

The MHRD recognized the need for entrepreneurship and started promoting it. In his budget speech for FY 2015-16, Arun Jaitley, then Minister of Finance and Corporate Affairs, announced the government's intention to establish AIM with an initial sum of rupees 5000 million and 10,000 million, respectively. This budget is allocated for the following three platforms.

- Atal Tinkering Labs
- Atal Incubation Centre
- Scale-up support to established incubators

## 15.5 Indian STEM Education Policies Emerging from IT and Economic Development Policies

### *15.5.1 Improvements in IITs, NITs, and IIESTs*

MHRD (Ministry of Human Resources and Development) has continuously worked to improve technical education in India by opening innovative technology institutions.

#### 15.5.1.1 Indian Institutes of Technology (IITs)

IITs are autonomous public institutes for higher education in India. They are created to be centers of excellence for training, research, and development in science, engineering, and technology. As of 2018, there were 23 IITs in the country, with a total of 11,279 seats (Manash Pratim Gohain, 2018). To produce good research facilities, the MHRD has scheduled a special training for Professors at IITs asking to develop and submit a separate curriculum by the end of every year. The main purpose of strengthening the IITs' curriculum is to bring graduates up to date on current industry technology. Many students were interested to participate in IITs due to their high education standards and available research resources (Ministry of Human Resource Development, 2017). Students also can meet and work with eminent professors and scientists. Funding opportunities and financial assistance is extended to deserving students.

IITs are the top-ranked technical colleges in India. IT companies have many expectations on STEM graduates, and questions have arisen about the talent of IIT graduates (Chowdhury & Ghosh, 2014). In a recent Interview, IT industrialist and Infosys co-founder N.R. Narayana Murthy said students leave the country because our nation "does not provide enough attractive opportunities for the brightest minds. The government in collaboration with IITs should take strict measures to stop this 'brain-drain' and encourage our talented engineers to work in our country by providing them with due remuneration, recognition and opportunity" (IIT-JEE, 2019).

#### 15.5.1.2 National Institutes of Technology (NITs)

The MHRD noticed that the success of technology-based institutions (IITs) has led to high demand for technical and scientific education. In 2002, the Ministry decided to start NITs instead of expanding IITs. By 2018, 31 NITs were offering engineering and technology-based courses at undergraduate and postgraduate levels

(National Institutes of Technology, 2017). Admissions to NITs are determined through the All India Engineering Entrance Examination, conducted by the Central Board of Secondary Education. The Central Government controls NITs and provides all their funding. The MHRD decided to upgrade 20 NITs to full-fledged technical universities. The teaching, training, and research activities of the institutes are periodically reviewed by the senate to ensure that educational standards are maintained.

#### 15.5.1.3 Indian Institutes of Engineering Science and Technology (IIESTs)

Allotted seats in IITs and NIT's may not be enough for students and MHRD recognized the need to create more institutes under the jurisdiction of IITs. IIESTs (Indian Institutes of Engineering Science and Technology) were thus established as one of the public institutions that provide technical education. The main purpose of these institutes is to improve student research skills.

#### 15.5.1.4 IIT Delhi's Thesis into Startups

In 2018, IIT Delhi made a historic move in giving doctoral students the option of starting a startup instead of writing a thesis. According to Director V. Ramgopal Rao, "It will be a platform to harness deep technology from blockchain to artificial intelligence via young companies" (UNC Enable, 2019). Entrepreneurship Development Cell has been launched to support the budding entrepreneur effort (Nanda, 2018). Students who choose this option will gain entrepreneurial knowledge and experiential insight under the guidance of prominent researchers. The startups the students initiate could cooperate with MNCs to establish additional firms in all regions, including Tier II and Tier III cities. In fact, IIT-Delhi had setup an incubation center in Haryana to accommodate several companies each year.

## 15.6 Takeaways

The era of IT service clusters in India started in the early 1980s. Since then, India has become a destination for all top MNCs to outsource their IT services. There is a great need to align STEM education policy closely with technology policy, in such a way that all graduate students have basic knowledge of computer science areas. There are many reasons for MNCs to participate in and help develop IT clusters in India, which have been discussed in this case study. STEM education in India has played a key role in establishing STPI (Software Technology Parks of India) across India, and the MHRD implemented many policies to promote STEM education (e.g., RAA, AIM, Make in India). The MHRD also opened many IITs to promote STEM education and increase activities in technology clusters all over the country.

While most of the policies are intended to promote the economic growth of the country, AIM endeavors to help entrepreneurs establish startups and expand the presence of IT clusters in Tier II as well as in Tier I cities.

A major item on the agendas of all the policies discussed in this case study is to promote STEM as early as the primary education phase, which helps students in problem-solving, computational thinking, statistical modeling, and abstract design, among other areas (Davies, 2011). The government of India and the MHRD have implemented many schemes to promote STEM in rural regions to expand on their success.

STPI, which started early in the twenty-first century, has developed IT clusters across the country. India is now a hub for IT services, perhaps the world's premiere hub, thanks to its large supply of STEM graduates, low compensation rates, low taxes, and good infrastructure. STPI offers all required facilities in a single place. Some of the key takeaways from this chapter include:

1. India has established itself as an IT powerhouse.
2. Indian leaders recognized the need to build a talent pipeline from primary to Ph.D.
3. Indian leaders put policies in place to encourage more STEM coursework by students. It has been successful as indicaed by 36% of growth in STEM enrolment from 2001 to 2010.
4. India still views STEM courses, for the most part, as isolated versus as integrated effort, which does not necessarily promote higher level initiatives of research, entrepreneurship, and critical thinking.
5. Faculty re-training at the K-12 and college levels is essential to promoting new levels of STEM success that lead to jobs after graduation.

## References

AICTE, "A Comprehensive Training policy for Technical Teachers", https://www.aicte-india.org/sites/default/files/TRAINING%20POLICY%20FOR%20TECHNICAL%20TEACHERS_BOOK%20%2812%29.pdf, 14th November, 2018.

Astha Dewan, International Conference on Technology for Education), "Scope of technology in higher education in India: A study", 2010, https://doi.org/10.1109/T4E.2010.5550105.

Davies, A. et al., "Future work skills 2020", Institute for the Future, University of Phoenix Research Institute, University of Phoenix, 2011.

Government of Goa, Information Technology Investment Policy 2015–20, https://www.goa.gov.in/wp-content/uploads/2016/05/Investment_Policy2015_FINAL_COPY_POLICY.pdf, p. 5, Accessed August 25, 2019.

Government of Maharashtra, Draft of Maharashtra's Information Technology/Information Technology Enabled Services Policy (IT /ITES) – 2015, https://www.maharashtra.gov.in/PDF/Web_Marathi_IT-ITES_Policy_2015.pdf, Page 1, Accessed August 25, 2019.

IIT-JEE, https://usaraj.com/iit-jee/, "What is IITs?", Para. 6, Accessed August 26, 2019.

Learning without burden, Report of National Advisory Committee appointed by the Ministry of Human Resource Development, 1993.

Maharashtra State Innovation Society (MSInS), Department of Skill Development & Entrepreneurship, Government of Maharashtra, Maharashtra Start up Innovation Policy – 2018, https://www.startupindia.gov.in/content/dam/invest-india/Templates/public/state_startup_policies/Maharashtra_State_Innovative_Startup_Policy_2018.pdf, Page 7, Accessed August 25, 2019.

Manash Pratim Gohain, https://economictimes.indiatimes.com/industry/services/education/11279-seats-being-offered-in-the-iits-in-2018-an-increase-of-291-over-last-year/articleshow/64483912.cms, published on June 6th, 2018.

Mckinsey Report, https://economictimes.indiatimes.com/jobs/only-6-of-those-passing-out-of-indias-engineering-colleges-are-fit-for-a-job/articleshow/64446292.cms, updated on 4th June, 2018.

Ministry of Human Resource Development, https://mhrd.gov.in/sites/upload_files/mhrd/files/document-reports/AR_2015-16%20ENGLISH.pdf, published on June 21st, 2017.

Narendra Jadhav, Role of Mumbai in Indian Economy, http://www.drnarendrajadhav.info/new-version/drjadhav-data_files/Published%20papers/Role%20of%20Mumbai%20City%20in%20Indian%20Economy.pdf, retrieved 25th August, 2013.

National Education Commission 1964-66, PB Works, Retrieved on 20th June 2015.

National Institutes of Technology | Technical Education | Government of India, Ministry of Human Resource Development, mhrd.gov.in. Retrieved 29th June, 2017.

Pallavi Dubey, T.S.S. Subramanian and Anupam Singh, IEEE International Conference on Engineering Education: Innovative Practices and Future Trends (AICERA), Need of promoting entrepreneurship at institution level for engineering students in India, 2012. https://doi.org/10.1109/AICERA.2012.6306741.

Prasanth K Nanda, https://www.livemint.com/Companies/ezoXVnfXJRnoUiMyiqRfmL/IITDelhi-to-help-PhD-students-convert-their-thesis-into-sta.html, 18th April, 2018.

Press Information Bureau, Government of India, Ministry of Human Resource Development, http://pib.nic.in/newsite/PrintRelease.aspx?relid=123120, Accessed on 09th July, 2015.

PK Tulsi and MP Poonia, IEEE Global Engineering Education Conference (EDUCON), "Building excellence in engineering education in India", 2015. https://doi.org/10.1109/EDUCON.2015.7096035.

R Ramanan Atal, "New India Challenges – A first step towards becoming an Innovation Nation", IEEE India Info. Vol. 13 No. 4, Objective, para. 1, Oct. - Dec. 2018.

Shajulin Benedict, "Incubation Centres – A Need for Successful Innovations via Entrepreneurs!", IEEE India Info. Vol. 13, No. 4, Pages 60 - 63, Oct - Dec 2018.

Shastri Bhawan, New Delhi, Government of India Ministry of Human Resource Development (Department of School Education and Literacy), https://mhrd.gov.in/sites/upload_files/mhrd/files/raa/Guidelines.pdf, 28th May, 2015,

Shauvik Roy Chowdhury and Koushik Ghosh, 2nd International Conference on Business and Information Management (ICBIM), Profile and practices of Indian premier institutes compared to global standards on the basis of QS World Universities Ranking 2013–14, 2014. https://doi.org/10.1109/ICBIM.2014.6970959.

Somesh K Mathur, "Indian Information Technology Industry: Past, Present, Future& A tool for National Development", 2008. https://pdfs.semanticscholar.org/c47e/1ad9ea94843365e1930980029e30501a2354.pdf.

The Right of Children to Free and Compulsory Education Act, published by Government of India Authority, http://righttoeducation.in/sites/default/files/Right%20of%20Children%20to%20Free%20and%20Compulsory%20Education%20Act%202009%20%28English%29.pdf, 26 August, 2009,

UNC Enable, https://www.facebook.com/permalink.php?id=1729158617379722&story_fbid=1884587608503488, Para. 1, Accessed August 26, 2019.

# Part IV
# Conclusion: Making Your Community a STEM Technopolis

# Chapter 16
# Intentional Integration of K-12 STEM Education With the Challenges of Cities: Do This, Avoid That, Here Are Tools

**Cliff Zintgraff**

**Abstract** In the volume *STEM in the Technopolis: The Power of STEM Education in Regional Technology Policy*, the editors brought together three kinds of chapters: (1) *foundational principles* of STEM education in the context of cities and regions: integrating with industry clusters; applying STEM education philosophy, theory, and models (pedagogy); digital equity; and quantitative measures; (2) *cases at city/regional level* reflecting these principles; (3) *cases with rich national context* that also reflect these principles. Focus was brought to these ideas especially at the primary and secondary (K-12) level. In this final chapter, a summary and guide to application is provided. The chapter is organized around the successes, barriers, strategies, tools and techniques that can be used to implement the ideas of this volume in one's local community. Details come from the volume's chapters and from the results of a public conference organized around volume development, led by the editors, and attended by most lead authors. Tools and techniques include the Eight Indicators of a Virtuous Cycle, the STEM Technopolis Model/Wheel, the STEM Technopolis Virtuous Cycle, technopolis digital equity strategies, categories for measurement, a cross-reference between common challenges and cases, and a process any committed stakeholder can follow to being to build a STEM Technopolis in their community.

## 16.1 Introduction

This final chapter of *STEM in the Technopolis: The Power of STEM Education in Regional Technology Policy* provides a short summary of the volume, with a main focus on how the ideas of the volume can be implemented. This chapter repeats only limited details from the prior chapters, and the reader is referred to those chapters for in-depth information. Rather, this volume focuses on the following hypothetical question from a reader: *How can I build a STEM Technopolis in my community?*

C. Zintgraff (✉)
The University of Texas at Austin, Austin, TX, USA
e-mail: cliff.zintgraff@utexas.edu



C. Zintgraff et al. (eds.), *STEM in the Technopolis: The Power of STEM Education in Regional Technology Policy*,
https://doi.org/10.1007/978-3-030-39851-4_16

To answer the question, this chapter presents results from a public conference on the volume's topic. At the conference, participants were asked to identify successes, barriers and strategies. Chapter authors then identified the most important items in each category consistent with their chapters and experience. Results are presented with a focus on strategies that amplify successes and address barriers. A summary is provided of the foundational ideas and models from the volume. A cross-reference is provided between this chapter's main themes and the volume's cases. Finally, a process is shared that readers can use to get started creating a STEM Technopolis in their community.

## 16.2 Strategies for Building a STEM Technopolis

On April 5–6, 2019, the volume editors held a conference at The University of Texas at Austin (UT Austin). The conference was hosted by the university's $IC^2$ Institute (Innovation, Creativity, Capital), a research unit within the Office of the Vice President for Research, and by the university's College of Education, and also partnered with the Society for Design and Process Science. The conference chair was Dr. Paul Resta of the UT Austin College of Education, and the conference design and sessions were led by the editors. Most lead authors attended, including those developing cases from Medellín, Colombia; San Antonio, U.S.; Querétaro, Mexico; São Carlos, Brazil; Fundão, Portugal; and India. Additional experts were invited to participate in panels.

Day 1 of the conference was open to the public. Foundational principles were shared, and cases were presented. Two working sessions addressed these questions:

- What successes already exist (as evidenced by the cases)?
- What are the barriers to policy creation or implementation?
- What successes should become policies?
- How do we overcome barriers?

Day 2 was reserved for author-only discussions, chapter-by-chapter reviews, and further attempts to address the first day's questions through the lens of the authors' cases. During a group exercise, successes, barriers and strategies were identified and ranked. Across those three categories, main themes emerged. Table 16.1 provides a summary of the individual successes, barriers and strategies that were

**Table 16.1** Successes, barriers and strategies from the STEM in the Technopolis author's conference

| Prior successes | Barriers | Strategies |
|---|---|---|
| Multi-sector network building (celebrations, open houses, meetings)<br>Leveraging real community needs<br>Buy-in from industry leaders | Teachers able to deliver program<br>*Herding cats*<br>Securing signature sponsors<br>School structure | Include parents and students in strategy development<br>Be cross sector<br>Avoid broad-brush (top-down) solutions<br>Policy that survives political change<br>Intentionally include educators |

identified and most agreed on by the authors. The themes are described below. Themes are expressed with a focus on ways to amplify wins and overcome the problems associated with delivering great STEM Technopolis experiences to students.

## 16.2.1 Strategy: Multi-Stakeholder Network Building

A robust network involving education, industry, government and non-profits is a powerful resource for STEM programs. Based on conference discussion among the authors, and on the content of the cases, no strategy is more urgent and necessary. A network is a source of content, advocacy, volunteers, mentors and funding. A robust group of stakeholders brings richness to teachers' and students' STEM experiences. Those experiences are consistent with and reinforce the STEM education pedagogies that have come to be understood as best practice (see Petrosino et al., this volume).

### 16.2.1.1 Multi-Sector Network Building

There was wide agreement that networks should span sectors, especially when programs leverage the strength of a city or region. Such programs are more likely to be effective and sustainable. Three kinds of networking building were identified: celebrations, open houses, and meetings. Celebrations are events where an organization's mission, sponsors, volunteers, educators, student participants, etc. are recognized for their contributions and/or success. Open houses give the general public an opportunity to see spaces and activities and engage with active participants. Meetings are the places where the hard work of communication and planning are done to propel programs forward—a place where all concerns can be addressed and win-win solutions sought.

Multi-sector network building is widely seen in this volume. Multiple instances are present in the Medellín, Colombia case, and in particular, the STEAM-LABS, STEAMakers, and Horizons programs (Roldán, this volume). San Antonio's CyberPatriot Mayor's Cup brings as many as one thousand people together for a program celebration (Sanchéz and Zintgraff, this volume), not to mention wide support for training and mentoring sessions. Taiwan's work on Fourth Industrial Revolution includes robotics competitions (Yang, Yang, Chou, Wei, Chen, and Kuo, this volume) that create numerous opportunities for open houses and celebrations.

### 16.2.1.2 Buy-In From Industry Leaders and Securing Signature Sponsors

Buy-in from industry leaders was successful in a number of communities as a way to build strong STEM education programs. In the highlighted cases of the volume in particular (those highlighted in Chap. 1), the participation of industry leaders has been essential to long-term and ongoing success. Medellín's innovation agency, Ruta N, convenes an industry leader forum monthly, and those leaders receive briefings on the development of STEM (in Medellín's case, STEAM) programs. San Antonio's CyberPatriot program is deeply supported by local industry leaders as documented in the case. Likewise, the Taipei/Taiwan pursuit of excellence in the Fourth Industrial Revolution has led to industry support for STEM education programs.

It must be noted that industry support was listed as both a prior success, and also as a barrier under the heading *securing signature sponsors*. The author observes that industry buy-in is sometimes difficult to achieve, but powerful when obtained. Vocal support, volunteers, mentors and funding often follow the support of industry leaders, consistent with technopolis observations about the role of influencers (Gibson and Butler 2015).

While individual industry leaders may have particular and noble intentions to help certain kinds of programs, support over time and at scale are most likely when activities advance an industry's or a company's well-established goals. Companies have business needs, like developing a workforce. They have branding needs, like helping a community understand the importance of their company and industry cluster to the community (financial impact, taxes paid, philanthropic impact, etc.). They usually have established philanthropic goals, and those goals generally involve working in the communities where they have significant workforce.

Appeals to the worthiness of one's STEM program are good, but it is usually most effective to consider potential signature sponsors as a market, with a STEM program offering value to sponsors, specifically the kinds of value noted above. One can then ask which sponsors will find an offering most appealing. Keep in mind this fundamental tenet of marketing analysis: No organization can serve every potential customer in a market. Pursue the sponsors for whom a STEM initative most naturally offers value, and be willing to make changes to sponsor packages that provide greater value to sponsors while remaining true to values and mission.

#### 16.2.1.3 Stakeholders: Intentionally Include Educators

Two items, one barrier and one strategy, were identified related to educators. The two are highly related. The barrier is related to teacher preparation and is covered in a subsequent section. The strategy is inclusion of educators as stakeholders, including classroom teachers. While perhaps less relevant for out-of-school programs, most programs operating at scale include some meaningful amount of teacher and/or formal-school-day involvement.

It might seem obvious that teachers are stakeholders, but educational reformers do not always behave in this way. Cuban (1986) studied decades of educational reform efforts in U.S. schools, dating back to the 1920s, and found that teachers are often not included in reform decision-making. A high-profile, somewhat over-the-top, but highly illustrative case came from no one less than Thomas Edison. He infamously said that film (new at the time) would soon replace books in the classroom. Film did not replace books. The expense, unreliability, limited availability, and general hassle of film was a poor substitute for the low-cost, flexibility, and wide availability of books. Books did a better job of reliably delivering information to students.

Teachers are professionals who best understand the moment-to-moment tasks required to help students learn. Educators know the structure and limits of schools. When efforts fail, and teachers were not consulted as stakeholders, they are not to blame. Alignment must occur across the full collection of stakeholders (Zintgraff and Hirumi in press).

#### 16.2.1.4 Include Parents in Strategy Development

Significant research highlights the major role of parents in students' education (See and Gorard 2015). Despite all the effort put into delivering great student experiences, student success often comes down to the support of parents. For parents who received a quality education and have received the ensuing benefits, passing on that value is a more natural task. Parents from underrepresented populations often do not have that advantage. Many such parents still understand and pass on the value of education. Overall, the active support of parents from underrepresented populations is a common barrier.

The involvement of parents can be seen in multiple cases in this volume. In Medellín's Engineering N program, parents attend a final science fair (Róldan, this volume). In San Antonio's CyberPatriot, parents attend the college fair and Mayor's Cup awards ceremony (Sánchez and Zintgraff, this volume). In Taipei, Taiwan, parents want to know how programs will affect college entrance (Yang et al., this volume; see more below). In Mexico's Movimiento STEM program, outreach to parents is a priority, and participation in meetings is encouraged (Rojas and Segura, this volume). In China, parent perception of STEM education experiences is a direct concern, and Quan (this volume) dedicates a table to the question of why arts training has received higher ratings than emerging STEM experiences.

A common theme is parent's concerns about how programs might affect student's preparation for high-stakes standardized tests that affect future college admissions. The topic itself is controversial, but regardless, the views of parents must be addressed. Concerns about these issues can be seen in cases from Taiwan and China (Yang et al., and Quan, all this volume). Preparation for testing is high in India. While current U.S. trends are moving toward less standardized testing, the amount of testing remains high and central to the education experience. The concern remains that robust STEM experiences must never be only activity, but also rigorous learning (e.g., Technological Pedagogical Content Knowledge (TPACK), Mishra and Koehler 2006).

#### 16.2.1.5 Strategies for Herding Cats

*Herding cats* is a metaphor that illustrates the difficulty of getting busy people with different goals to join together and focus on a particular mission. Herding dogs is easy, dogs want to herd. Cats and busy people usually do not. Getting people representing the full range of technopolis stakeholders to gather, advocate, plan, execute and celebrate regularly is a difficult task.

People skilled at consistently motivating diverse stakeholders apply a range of techniques. Connecting to deep community needs (as described in this volume) creates a strong foundation. Creating a vision helps people see that success is possible, and a detailed vision helps people see their own role and contribution. Consistent message delivery and opportunities to engage are essential. Effective facilitators get people to commit to deliverables on their own terms, and those terms become the basis for accountability. Effective facilitators engage people in the work at hand, and people engaged are more likely to show up than people simply spectating. Sharing the rewards, both monetary and psychological, provides impetus over time for multi-stakeholder network building.

Support groups, as identified in this volume's STEM Technopolis model, can be especially effective at gathering diverse stakeholders. Their relatively independent posture provides them advantages as a convener around issues. Gibson and Oden (2019) identified the important role these "small i" (Storper 2013, p. 100) institutions play in building networks for regional development.

## *16.2.2 Leveraging Real Community Needs*

There was strong agreement among authors that leveraging real community needs was a successful approach to building STEM programs. It is perhaps self-fulfilling to note the depth of support in this volume—working alongside the challenges of cities and regions was a criterion for case inclusion. Still, one can note the richness documented in the cases in this volume, which include: Medellín's STEAM education grounding in the city's priority industry clusters and their focus on culture change for youth; San Antonio's focus on its strong cyber security cluster; Taiwan on Fourth Industrial Revolution; Querétaro, Mexico (Rojas and Segura, this volume) addressing STEM workforce needs; São Carlos, Brazil (Gattaz, Falvo, and Cruvinel, this volume) seeing mutual benefits between STEM education and its agriculture sector; Fundão, Portugal (Aguiar and Pereira, this volume) transforming its future through ICT STEM education and company development; and Shanghai, China's adoption of national policy toward the local goal of education in ICT cloud development and the resulting economic impact (Quan, this volume).

### 16.2.2.1 Avoid Broad-Brush Solutions

This volume has focused mostly on policies, plans and programs at the city or city-sized regional level. Is there a role for top-down planning? Politics and culture differ from country to country, state to state, and among cities/regions. What common lessons can be drawn, and what common strategies can be shared?

The term *broad brush* catches the lesson of what not to do. Taking an approach that works in one place and copying it to another ignores the local setting, and unless one is very lucky, one is set up for failure. This lesson applies regardless of the local politics or culture. Top-down planning in a grassroots culture will not succeed, nor will exclusively participatory approaches when a culture of central planning dominates.

Many of the cases highlight grassroots approaches, but with a city playing a central or coordinating role. The cases of Medellín, San Antonio, São Carlos, and Fundão exhibit this characteristic. All four of these cases demonstrate the active involvement of mayors and senior city staff playing a coordinating or greater role. One sees mayors recognizing the powerful role STEM programs can play in the economic development of a community, the long-term education and development benefits, and consistency with the priorities of their electorates. Supporting education is smart for short- and long-term economic development and is almost always important to voters. Securing the support of mayors and city staff should be a significant strategy of any STEM program with the status or goal of operating at city scale. Securing such support requires alignment with city priorities.

#### 16.2.2.2 Not a Broad Brush: Local Adaptation of National Programs

At the same time, programs with national support are also a highlight of the volume. Strikingly, this characteristic is not seen only in locations one associates with dominant top-down planning. San Antonio's adoption of the national CyberPatriot program (run by a national association); Querétaro, Mexico's amplification of national government policies; São Carlos's use of national agriculture and STEM education policies; Fundão's building on national ICT priorities; application of Fourth Industrial Revolution policy in Taiwan; along with local adaptation of national policies and programs in China and India; are a major theme of the volume, and a strategy any STEM education advocate should consider.

The case from China, and in particular details of activities in Shanghai, are telling. The city has taken national policies and made them specific to Shanghai's focus on cloud development. The politics and details may be vastly different, but the principles are not so different from San Antonio's choice to robustly adopt CyberPatriot, consistent with the cybersecurity focus of the city, state and nation. In both cases, principles, policies, programs, resources and narrative are available as tools for building the local program.

#### 16.2.2.3 Policy That Survives Political Change

Every stakeholder in the technopolis can relate to the changing of political winds, and in the process, the killing of program momentum, and sometimes the end of programs themselves. Teachers have their own version of this phenomenon. In the author's home state (and likely throughout the country), teachers have a saying, that if an educational reform being introduced in classrooms is not a good one, wait three years, and another will take its place. Poorly designed programs and reforms should not survive, of course, but that is never the initial goal of STEM program designers.

Leveraging real, deep community needs is a powerful way to insulate against political change. Support runs deep in the volume in topics familiar to the reader by now: energy, ICT, health, cyber security, economic development, agriculture, advanced manufacturing, computer science, and cloud software development, to name a few. These are economic drivers, but in several cases, STEM policy is nothing short of a long-term response to crisis situations in communities (more below). Overall, programs in Medellín, San Antonio and São Carlos have demonstrably survived (in Medellín's case, it's the larger policy that has survived) across 15+ years of changes in local political administrations. It is only through deep roots to major issues of local import that such programs sustain.

## *16.2.3 Teachers Able to Deliver Programs*

### 16.2.3.1 Teacher Professional Development Done Right

The most common barrier expressed was the ability of K-12 teachers to effectively deliver the STEM programs developed. The author recognizes this as a very common concern (e.g., Spector 2014; Zintgraff 2013). Fidelity of implementation, often affected by teacher beliefs and training (Pajares 1992; Harris et al. 2009), are real concerns when innovative approaches to teaching and learning are brought into classrooms, and again when attempts are made to bring them to scale.

It is the responsibility of all stakeholders to find a way to build the right curriculum and get teachers trained on its delivery. Good instructional design, a process that requires understanding of the full situation, including existing teacher beliefs and of pedagogical practice, is essential. Time must be made for professional development, funding made available, professional development content designed, and technopolis stakeholders who can assist engaged at their points of strength. When these factors are in order, then teachers gain the responsibility of professional development to prepare for delivery.

### 16.2.3.2 School Structure

For developers of innovative STEM programs, a common barrier is the structure of the institution (K-12 school, college, etc.) that is required to deliver a program. As noted above, a quality instructional design considers all factors related to delivery. The setting for delivery is one important factor. A traditional K-12 school day can have seven class periods, each less than one hour, and minimal time for planning the co-teaching of interdisciplinary experiences.

Principals (rectors) have significant influence over the structure of a school. STEM professional development training often includes specific training for principals and other administrators. Longer classes that meet less often each week provide more opportunity for students to engage deeply in projects. Reducing class load gives teachers time for co-planning. Enough of these items, taken together, can begin to change the culture of a school.

There are options that do not require restructuring of the school day, as useful as that exercise is. Zintgraff and Hirumi (in press) incorporated in their Learner-Centered Design (LCD) Framework the idea of *strategies* versus *tactics* that teachers can deploy in classrooms. Strategies are more fully developed approaches (like project-based learning) that require sufficient space and time and a good structure to operate. Tactics, like posing problems in classrooms, driving discussion or reflection, or ad hoc work in groups, can be used by a skilled teacher in a pre-planned or ad-hoc manner within existing school structures. In doing so, teachers bring elements of the STEM Technopolis into the teacher's classroom. School structure creates limits but should not be seen as an absolute barrier. Many teachers in the same school introducing related tactics can become the impetus needed for systemic change.

### *16.2.4 Great STEM Experiences as the Opportunity in Crisis*

This theme was not identified during the author's conference, but your chapter author notes how some cases represent the opportunity in crisis. The Medellín 2021 plan, a direct reaction to the city's difficult times of the 1990s and early 2000s, is the foundation of the city's STEAM education efforts, with those efforts run by the same agency pursuing a broad front of innovations. Roldán (this volume) provides a rich description of the history that led to the energy for developing a knowledge economy in which everyone rides. The case of Fundão, Portugal also stands out, a rural city facing the often-intractable problem of negative net migration. Fundão successfully used a combination of traditional industry development, high-technology industry development, and STEM/workforce education to change the direction and the narrative. With less immediate intensity, many volume cases demonstrate how global challenges experienced in local settings can be used to build great STEM experiences that are relevant to teachers and students, important to stakeholders, and consistent with the philosophies, theories and models that underpin great STEM education experiences.

## 16.3 Case Cross-Reference

Table 16.2 is a cross-reference for those who would like to learn more about particular strategies. For a given strategy, the cases indicated offer the best examples, though not necessarily the only examples of the strategy. Two additional rows provide a cross-reference for the best examples of national context and cases that

**Table 16.2** Strategy to case cross-reference

| Strategy/Theme | Med | SA | Tai | Aus | Que | SC | Fund | Chn | Ind |
|---|---|---|---|---|---|---|---|---|---|
| Multi-sector network building | X | X | X | X | | X | | | |
| Buy-in from industry leaders | X | X | X | | | X | X | | |
| Intentionally include educators | X | X | X | | X | | | | |
| Include parents and students | X | X | X | | | | | X | |
| Leverage community needs | X | X | | | | X | | | |
| Adapt national programs | | X | | X | X | X | | X | X |
| Survive political change | X | X | | | | | X | | |
| Change school structure | X | | X | | | | | | |
| Teachers able to deliver | X | X | X | | | | | | |
| Opportunity in crisis | X | | | | | | X | | |
| *Other categories* | | | | | | | | | |
| National context | | | | X | X | X | | X | X |
| Specific policy connections | X | X | | | X | X | | X | X |

identify specific policy documents. The column headings are abbreviations as follows: Medellín, Colombia (Med); San Antonio, U.S. (SA); Taipei, Taiwan (Tai); Austin, U.S. (Aus); Querétaro, Mexico (Que); São Carlos, Brazil (SC); Fundão, Portugal (Fund); China (Chn); and India (Ind).

## 16.4 Making Policy

The title of this volume invokes the idea of regional policy (policy in cities and city-sized regions) as a powerful tool for creating STEM experiences. The volume's chapters and cases have covered numerous angles on how to identify, leverage, and build on the existing policies and plans of cities and regions. Leveraging local and regional policy is strongly advocated.

But what about *making* policy? Do the key policies and plans of your local government, chambers of commerce, schools, leading non-profits and leading industries intentionally incorporate the idea of STEM education, and especially K-12 STEM education, as part of development planning? Do those plans incorporate STEM education as a peer sector in the technopolis? Does your city or region have a STEM Technopolis?

Gibson and Butler (2015) shared a key insight about the people in a technopolis and how they facilitate cross-sector action. They identified two kinds of leaders, first-level influencers and second-level influencers. The first-level influencers are executives and recognized leaders with influence in the community. Their advocacy for plans, and the facilitation and protection they can provide for carrying out plans, is essential. The second-level influencers are those doing the ongoing work. They form connections, attend meetings, make proposals, make plans, and lead the execution of those plans. Both types of people are networkers and change-makers in their communities.

Working together, those influencers can affect policy. What steps might they take? Some possible steps include:

1. Identify existing city/regional government, chamber, school, non-profit, and/or industry policies and plans.
2. Identify existing STEM education content.
3. Identify possible hooks for new STEM content.
4. Attend the meetings of these organizations.
5. Meet offline with leaders from these organizations.
6. Propose additions that align with organizations' existing goals.
7. Propose metrics and make sure benefits are expressed in terms of those organizations.

## 16.5 Getting Started: Tools and Techniques From the Volume

If readers like the ideas in this chapter, other chapters, or throughout the volume, how might readers get started implementing in their communities? Table 16.3 contains a simple list of steps to begin the process of creating great STEM programs based on local priorities, appealing to stakeholders and volunteers, consistent with STEM pedagogy, and built with a foundation that can scale. Figures 16.1 and 16.2 and Tables 16.3 and 16.4 list the main models and ideas shared, making this location a quick reference for the models of this volume and their relationship. Consistent with the learn-by-doing approach advocated in this volume, do not treat the list

**Table 16.3** Getting started making a STEM technopolis program: Steps

| Step | Description |
|---|---|
| 1 | Identify an overall topic or goal |
| 2 | Using the STEM technopolis wheel (Fig. 16.1), identify people and organizations who may be stakeholders for your topic or goal. Keep in mind the idea of first-level influencers (executives with influence) and second-level influencers (leaders doing the work) |
| 3 | Meet with these people; attend meetings of their organizations. Watch for indicators of a virtuous cycle (Table 16.4) |
| 4 | Through meetings and your own research, identify policy planning documents that either do or should include policy that directly ties STEM education to local economic and development goals (Table 16.5). Strongly include K-12 STEM in this step |
| 5 | Review this volume's digital equity strategies (Table 16.5) |
| 6 | Re-evaluate topics and/or goals in light of the importance of the related industry cluster to the city or region |
| 7 | Brainstorm ideas of how the pedagogies (teaching methods) (Table 16.5) associated with STEM education can be brought to bear to create a great, community-based STEM experience for students |
| 8 | Create a white paper or vision document that (a) describes a specific STEM program for specific industry cluster/s; (b) uses STEM pedagogies; and (c) ties to the city's or region's economic and/or development goals |
| 9 | Develop a virtuous cycle model (Fig. 16.2) to include in a pitch |
| 10 | Using the STEM technopolis wheel again, identify organizations and influencers, and make proposals to those people and organizations |

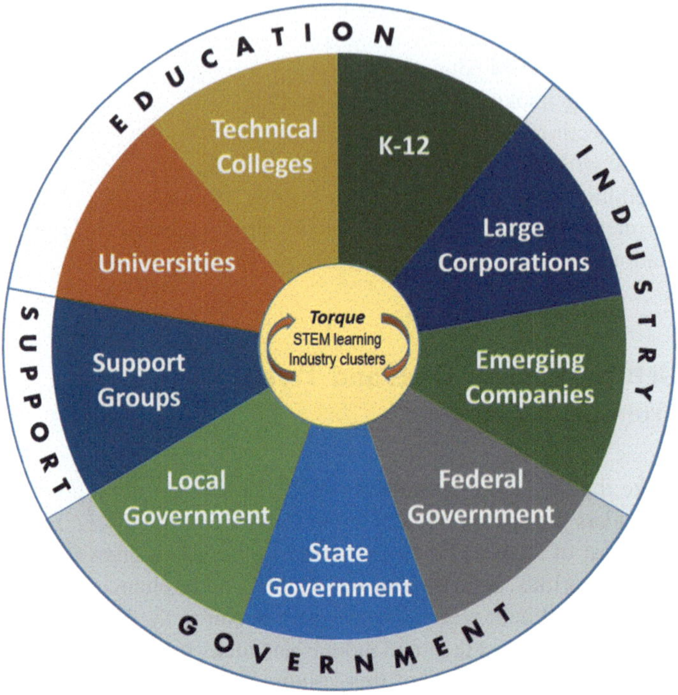


**Fig. 16.1** The STEM technopolis wheel Zintgraff, Kidwell, Pogue, Han and Butler, multiple chapters in this volume, adapted from Smilor et al. (1989)

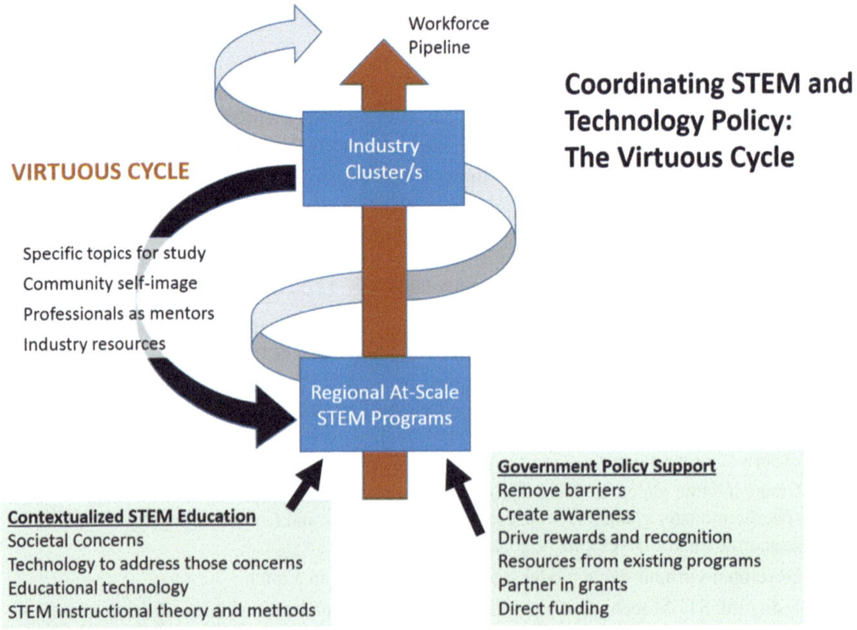


**Fig. 16.2** The STEM education, industry cluster, and society virtuous cycle

**Table 16.4** Indicators of a virtuous cycle (Zintgraff, this volume)

| Number | Description |
|---|---|
| 1 | Industry clusters inspire K-12 STEM education |
| 2 | Industry clusters support K-12 STEM education |
| 3 | Individual professionals support K-12 STEM education |
| 4 | Universities support K-12 STEM education |
| 5 | Government entities support K-12 STEM education |
| 6 | Non-profits support K-12 STEM education |
| 7 | Societal concerns inspire K-12 STEM education |
| 8 | STEM education feeds universities and industry |

**Table 16.5** Types of policy/planning documents, digital equity strategies, and STEM pedagogies

| Examples of policy/planning document types | Digital equity strategies (Adapted from Resta, this volume) | Pedagogies (Petrosino et al., this volume) |
|---|---|---|
| Medellín 2021, city-led development plan<br>SA2020, development plan from leading city non-profit in San Antonio<br>Educate Texas: State-level plan from non-profit<br>NIÑASTEM PUEDEN, part of Nuevo Modelo Educativo, national plan, Mexico | Industry cluster grants and equipment refurbish<br>City, state, telecom providers develop asset map and work to fill gaps<br>Access discounts for low-income families<br>Industry clusters consult to help identify open educational resources<br>Industry cluster volunteers, STEM professional development, internships<br>Industry clusters fund university research on digital technologies to enhance STEM learning | Constructivist philosophy: Learner's prior knowledge at center of experience<br>Social constructivism: Cultural knowledge and tools, social interaction in learning<br>Constructionism: Developing public artifacts and accepting critique<br>Problem-based instruction<br>Project-based instruction |

below as a magic formula or strictly ordered list. The order suggested is rational, but one should evaluate how it works in context and adapt accordingly.

## 16.6 Conclusion

This chapter and volume have been offered with a simple idea at the core, that the priorities and challenges of cities and regions are a great foundation for STEM education experiences for students. As shared by the volume's authors, there is evidence from around the world that the idea works when done right. The volume has focused in particular, though not exclusively, on STEM experiences for primary and secondary students (in the U.S., "K-12"). The volume has also shared how state and national initiatives are unfolding, and how they can be adapted as part of city and regional strategy.

This concluding chapter shared strategies for STEM Technopolis development: multi-sector network building, engagement of industry cluster stakeholders, building on the needs of cities and regions, including parents and students in development, adapting national programs for local use, designing for political change and resiliency, and preparing schools and teachers for rich STEM content. As seen in this chapter and the related cases, great STEM experiences can be part of a response to crisis, and part of long-term solutions.

Readers are encouraged to think about the place of STEM education in the context of their city or region, and about how each can help the other. Readers are encouraged to use policy and to make policy that will drive the creation of great STEM education experiences for students.

## References


Cuban, L. (1986). *Teachers and machines: The classroom use of technology since 1920*. New York: Teachers College Press.

Gibson, D. V., & Butler, J. S. (2015). Creating and sustaining high-technology development in Austin, Texas. In *The entrepreneurial university* (pp. 35–66). Routledge.

Gibson, D., & Oden, M. (2019). The launch and evolution of a technology-based economy: The case of Austin Texas. Growth and Change. Wiley.

Harris, J., Mishra, P., & Koehler, M. (2009). Teachers' technological pedagogical content knowledge and learning activity types: Curriculum-based technology integration reframed. *Journal of Research on Technology in Education, 41*(4), 393–416.

Mishra, P., & Koehler, M. (2006). Technological pedagogical content knowledge: A framework for teacher knowledge. *The Teachers College Record, 108*(6), 1017–1054.

Pajares, M. (1992). Teachers' beliefs and educational research: Cleaning up a messy construct. *Review of Educational Research, 62*, 307–332.

See, B. H., & Gorard, S. (2015). The role of parents in young people's education: A critical review of the causal evidence. *Oxford Review of Education, 41*(3), 346–366.

Smilor, R. W., Gibson, D. V., & Kozmetsky, G. (1989). Creating the technopolis: High-technology development in Austin, Texas. *Journal of Business Venturing, 4*(1), 49–67.

Spector, J. M. (2014). Program and project evaluation. In J. M. Spector (Ed.), *Handbook of research on educational communications and technology* (pp. 195–201). New York: Springer.

Storper, M. (2013). *Keys to the city—How economics, institutions, social interaction, and politics shape development*. Princeton: Princeton University Press.

Zintgraff, C., & Hirumi, A. (in press). Aligning learner-centered design philosophy, theory, research and practice. In J. M. Spector, B. B. Lockee, & M. Childress (Eds.), *Learning, Design, and Technology: An international compendium of Theory, Research, Practice, and Policy*. Springer.

Zintgraff, C. (2013). Philosophical implications of teacher skills demonstrated in the WhyPower project. In *Proceedings of the University of North Texas 2013 Research Exchange Conference* (pp. 46–50).

# Index




C. Zintgraff et al. (eds.), *STEM in the Technopolis: The Power of STEM Education in Regional Technology Policy*,
https://doi.org/10.1007/978-3-030-39851-4

**D**

**U**



**V**



**W**

The manufacturer's authorised representative in the EU is Springer Nature Customer Service Centre GmbH, Europaplatz 3, 69115 Heidelberg, Germany. If you have any concerns regarding our products, please contact ProductSafety@springernature.com

Printed and bound by CPI Group (UK) Ltd, Croydon, CR0 4YY
30/09/2026
02225648-0012